DATE DUE

2-13-14			

Mapping
Our World

GIS Lessons for Educators

ArcGIS® Desktop Edition

Lyn Malone • Anita M. Palmer • Christine L. Voigt • Eileen Napoleon • Laura Feaster

ESRI PRESS
REDLANDS, CALIFORNIA

ESRI Press, 380 New York Street, Redlands, California 92373-8100

Copyright © 2002, 2005 ESRI

All rights reserved
First edition published 2002. Second edition 2005.
10 09 08 07 06 05   1 2 3 4 5 6 7 8 9 10

Printed in the United States of America

Library of Congress Cataloging-in-Publication Data
Mapping our world : GIS lessons for educators / Lyn Malone ... [et al.].—
ArcGIS desktop ed., 2nd ed.
        p.   cm.
    Includes bibliographical references.
    ISBN 1-58948-121-6
    1. Geographic information systems—Study and teaching (Middle school).
    2. Geographic information systems—Study and teaching (Secondary).
    3. ArcGIS. I. Malone, Lyn.
    G70.212M288 2005
    910'.78'5—dc22   2005018078

Ask for ESRI Press titles at your local bookstore or order by calling 1-800-447-9778. You can also shop online at www.esri.com/esripress. Outside the United States, contact your local ESRI distributor.

ESRI Press titles are distributed to the trade by the following:

In North America, South America, Asia, and Australia:
Independent Publishers Group (IPG)
Telephone (United States): 1-800-888-4741
Telephone (international): 312-337-0747
E-mail: frontdesk@ipgbook.com

In the United Kingdom, Europe, and the Middle East:
Transatlantic Publishers Group Ltd.
Telephone: 44 20 8849 8013
Fax: 44 20 8849 5556
E-mail: transatlantic.publishers@regusnet.com

To our families, especially our husbands, Pat, Roger, and Brian—
Thank you for your support, guidance, patience, and love.

Lyn Malone, Anita M. Palmer, Christine L. Voigt

To Joe Ferguson and Ann Judge, devoted teachers and eager students
who lost their lives on September 11, 2001—
Your passion and dedication to geographic education lives on.

Contents

Foreword

I have always had a fascination with maps. I like maps that tell me where I am or where I can go and what I can do and see once I get there. I have always wanted to share that fascination with my colleagues and students.

As soon as I was introduced to GIS, I set upon finding out how to introduce and use it in the classroom. But I was lost and didn't have a clue where to begin. At a conference in Austin, Texas, I discovered a book called *Mapping Our World: GIS Lessons for Educators* and had the opportunity to talk with the authors. Little did I know that this would become my own map into the world of GIS in the classroom.

Like most teachers who see GIS being used in the classroom, I immediately had tons of questions, mostly about making different maps with the software. *Mapping Our World* gave me—and can give other teachers who are not experienced with GIS—the confidence to introduce it to their students and go forward with the lessons. With *Mapping Our World* we have the structure and support we need to integrate GIS into the curriculum.

What really clicked for me was seeing *Mapping Our World* move students and teachers into the realm of spatial inquiry. I finally understood that I had a tool with the flexibility to get students hooked into exploring their world beyond a paper map; they could learn to ask questions, to seek answers, and to develop solutions based upon their findings. As I used *Mapping Our World* more I could see students growing from making cool labels on maps and identifying countries and locations to thinking critically and solving problems. This was very exciting.

Mapping Our World can really begin to bring to students the ability to take charge of their own learning. Watching students help each other sort data, make decisions, come to a consensus, and finally make a finished map to explain how they would answer their questions is something that most teachers strive for in the classroom. This only becomes available during special, specific lessons, but with *Mapping Our World,* it can occur throughout the year within the curriculum.

During this past year, I have been taking notes on how teachers are using *Mapping Our World* in the classroom. Students have told me excitedly, "I like this better than using paper maps" and "I can explore on my own." One time I was working with a sixth-grade class, where part of the lesson was for students to study a thematic map to look at population patterns and the countries in the African continent. One student thought that two red polygons represented two population ranges within the same country, but when we used the Identify tool and discussed the importance of reading legends on a GIS map, she saw they were really two different countries. How often do we miss glimpses into students' misconceptions?

I am often asked about using GIS with students who have learning challenges. I have seen all students from the very young to gifted and talented to learning challenged who respond remarkably well to GIS. This is because GIS allows students to think visually or spatially (also because young people understand computers, often better than we do). I believe GIS is the tool to help all students develop their spatial thinking skills.

In its first incarnation, this book helped me to share my fascination with maps with other educators, teachers, and students. I am delighted that this new edition, updated to the latest software, is available because it can only widen the audience. *Mapping Our World* answers many questions and opens the door to more questions. I encourage all educators and teachers to use this resource; *Mapping Our World* has everything you need to be successful.

Eric Bowman
Instructional Technology Specialist
North East Independent School District
San Antonio, Texas

Acknowledgments

We would like to thank all those who helped make this book possible:

At ESRI: The K–12 team of George Dailey, Charlie Fitzpatrick, and Angela Lee, whose vision, leadership, and hard work made this book a reality; Kim Zanelli English, for viewing and testing selected exercises with her students; Brian Parr, who managed this project; Chris Zanger, who helped adapt some of the exercises for ESRI® ArcGIS® 9 and handled a variety of software details; Tim Ormsby, who tested the book's exercises; Amaree Israngkura, who designed the cover; Claudia Naber, who edited the content; Scott McNair, who put together the ArcGIS Software & Data CDs; Peter Schreiber, Lisa Horn, and Gail Hancock, who handled myriad legal issues; Deane Kensok, for timely help with Internet issues; Milton Ospina and Ann Johnson of the ESRI Higher Education team; Pam Spiva, who helped with a variety of administrative details; Christian Harder, ESRI Press publisher; Judy Hawkins, ESRI Press managing editor; Nick Frunzi, ESRI Educational Services director; and Jack Dangermond, ESRI founder and president, who made publication of this book a priority for his company.

All the lessons in this book were reviewed and tested by classroom teachers. We thank them for taking the time to work through these lessons with their students to ensure their accuracy and relevance: Brad Baker of Bishop Dunne Catholic School, Dallas, Texas; Gerry Bell of Port Colborne High School, Port Colborne, Ontario; Shanna Hurt of Arapahoe High School, Littleton, Colorado; Marsha MacLean of the Redlands Unified School District, Redlands, California; Bart Manson and Joe Myszkowski of Red River High School, Grand Forks, North Dakota; Cathy Pleau of V. J. Gallagher Middle School, Smithfield, Rhode Island; Cynthia J. Ryan of Barrington Middle School, Barrington, Rhode Island; Herb Thompson of Greenspun Junior High School, Henderson, Nevada; and Patricia Walls of Taylor County Middle School, Grafton, West Virginia.

We also thank Joseph Kerski, geographer with the United States Geological Survey, who reviewed the book for geographical accuracy.

Thanks also to Roger Palmer of Red River High School, Grand Forks, North Dakota, who spent countless hours to ensure that the book aligns with the national science and technology standards.

We also say thanks to Jim Trelstad-Porter, Director of International Student Advising at Augsberg College, who interpreted Spanish data and correspondence with the Instituto Nacional de Estadistica, Geografia e Informatica in Mexico.

Data acknowledgments

The authors and ESRI would like to thank the following data providers for contributing data and maps to this book:

Color shaded relief image of the world (modules 2 and 5) provided by WorldSat International, Inc. Copyright © 2004. All rights reserved (www.worldsat.com).

Agriculture (modules 3, 5 and 7), airports (module 7), energy (module 7), faults (module 2), lakes (module 5), oil and gas (module 5), population density (modules 2, 3, and 5), precipitation (modules 3 and 7), relief (module 3), rivers (module 5), tectonic plates (module 2), and volcanoes (module 2) are ArcAtlas™ layers based on data provided by Data+ and ESRI. Copyright © 1996 Data+ and ESRI. All rights reserved.

World economics and vital statistics data (modules 4 and 6) provided by CountryWatch.com, Inc. Copyright © CountryWatch.com, Inc. (www.countrywatch.com). All rights reserved.

2000 earthquake data (module 2) provided by Advanced National Seismic System (ANSS), formerly Council of the National Seismic System (CNSS). On the Web: quake.geo.berkeley.edu/cnss.

Supplemental earthquake data (module 2) provided by U.S. Geological Survey, National Earthquake Information Center. On the Web: gldss7.cr.usgs.gov/neis/epic.

Supplemental volcano data (module 2) provided by Smithsonian Institution, Global Volcanism Program. On the Web: www.volcano.si.edu/world.

Temperature and precipitation data for cities (module 3) provided by Worldclimate.com (www.worldclimate.com).

World physiographic features (module 2), world climate zones (module 3), and religion and language maps (module 5) provided courtesy of National Geographic Maps.

Historical city data from *Four Thousand Years of Urban Growth: An Historical Census* (module 4) provided by Edwin Mellen Press. Copyright © 1987.

Cities and towns, precipitation, roads, springs, streams, temperature, and water bodies for the Middle East (module 5) provided by U.S. Geological Survey (USGS), *Digital Atlas of the Middle East,* courtesy of the National Imagery and Mapping Agency.

Yemen border map data was constructed by the authors from information published by the British Yemeni Society.

California counties (module 6) provided by Tele Atlas and ESRI. Copyright © 2004. All rights reserved.

Economic and trade data for North America (module 6) provided by U.S. Census Bureau and Instituto Nacional de Estadistica, Geografia e Informatica, Mexico.

South Pole photograph (module 7) provided by National Oceanic and Atmospheric Administration (NOAA) Corps—South Pole Station.

AVHRR sensor images of Antarctica created from National Oceanic and Atmospheric Administration (NOAA) satellites and U.S. Geological Survey Digital Elevation Models (module 7) provided by USGS Terra Web. On the Web: terraweb.wr.usgs.gov.

Satellite image of Larsen Ice Shelf provided by the National Aeronautics and Space Administration (NASA).

Image files of flooded Earth (module 7) provided by National Oceanic and Atmospheric Administration (NOAA).

Precipitation data, hurricane tracks, and satellite images for Hurricane Mitch (module 7) provided by U.S. Geological Survey, *Digital Atlas of Central America,* courtesy of National Oceanic and Atmospheric Administration (NOAA) and National Aeronautics and Space Administration (NASA).

Airports, railroads, roads, and utilities for Central America (module 7) provided by U.S. Geological Survey, *Digital Atlas of Central America,* courtesy of National Imagery and Mapping Agency.

World Wildlife Fund ecoregions (modules 5 and 7) provided by the World Wildlife Fund.

Introduction

Dear Teacher,

Welcome to *Mapping Our World: GIS Lessons for Educators, ArcGIS 9 Desktop Edition.* This is a book of ideas, resources, exercises, and data that we believe will prove a valuable supplement to your World Geography course. It's not meant to take the place of your current textbook or curriculum, but to enhance it, to expand it into the world of high-speed computing, vast databases, the World Wide Web, and the supermaps of geographic information systems.

This book updates the materials from the original *Mapping Our World* book for use with the latest GIS software, ArcGIS 9 Desktop. If you have experience using "The Big Green Book" with your students, don't worry—all of your favorite lessons are still here. The intent and general design of the lessons, including the standards covered and the assessment rubrics, have not changed. In general, exercises were modified only where necessary to accommodate changes and new tools in the software. Exercise data has been reorganized to take advantage of the geodatabase format.

This book contains everything you need to create GIS projects: lessons with student handouts and data; answer keys; and assessments and rubrics. It also contains a one-year site license of ArcGIS 9 Desktop software. This means that your students will be using the same tools as the ones being used by professional planners, emergency- and disaster-response personnel, government agencies around the world, and businesses of every description. There is also a companion Web site *(www.esri.com/mappingourworld)* that has additional resources to help you integrate GIS technology and your curriculum as easily as possible.

We have designed the book with teachers in mind. For example, the physical shape of this book is one inch wider than the standard 8½-by-11 size, to make it easer for you to make copies of the student handouts. The unique binding allows the book to lie as flat as possible when making copies or on your desk. A digital copy of the book is provided on the data CD if you prefer to print the handouts from your computer. The lessons themselves have graphical cues to remind your students of questions they should answer and key items to notice.

Please take some time to read the How to Use this Book section. There you will find information on how the book is organized, what order to go through the lessons, and some important technical information related to the software and exercise data.

We hope you find *Mapping Our World* to be a useful tool for your classroom!

Sincerely,
The Mapping Our World Team

How to Use this Book

Using the lessons in this book, you and your class will investigate patterns of human life and the physical environment that span the globe, explore issues of concern to millions of people, and analyze data and information gathered from across the world to across the street. Your students will find, integrate, and use data culled from many sources, and build the core knowledge and skills essential for coping in a world increasingly characterized by vast quantities of raw information. Best of all, it will be fun.

Where to begin

Before using the *Mapping Our World* materials with your class, we recommend that you review and complete the following ordered list:

1 Finish reading this section and skim through the book to locate the various sections (modules, lessons, rubrics, answer keys, and so on).

2 Install software on your computer and the student computers. (Refer to "Setting up the software and data" later in this section and "Installing and Registering the ArcView 9 Demo Edition Software" at the back of the book.)

3 Install data on your computer and student computers. (Refer to "Setting up the software and data" later in this section and "Installing the Exercise Data" at the back of the book.)

4 Work through the module 1 activity by yourself. Once you've completed module 1, you will be ready to guide your students through it.

5 Introduce students to GIS using the suggestions in the module 1 lesson introduction.

6 Make copies and work through module 1 with your students. The one-lesson design of this module introduces students to the concept of GIS, basic ArcMap skills, and the steps of the geographic inquiry process.

The modules

Once you've finished module 1, you and your students are free to explore the content and lessons in modules 2–7 in any order you choose. You can teach each module or lesson independently of the others, and you can tailor the material to suit the specific needs of your class and curriculum.

Each module illustrates an important theme or concept of geographic knowledge. These concepts were derived from the *National Geography Standards—Geography for Life,* published in 1994. The following is a list of modules 2–7 and a brief description of the concepts they address:

Module 2: Physical Geography I—Landforms and Physical Processes
Powerful forces originating deep in the earth shape the landforms that characterize its surface.

Module 3: Physical Geography II—Ecosystems, Climate, and Vegetation
Four major determinants of climate are latitude, elevation, landforms, and proximity to the ocean.

Module 4: Human Geography I—Population Patterns and Processes
Many factors are involved in the distribution and migration of human populations.

Module 5: Human Geography II—Political Geography
Cultural concerns and conflicts continually reshape the political makeup of the world.

Module 6: Human Geography III—Economic Geography
Economic development, modernization, and trade illustrate the interrelatedness of the global community.

Module 7: Human/Environment Interaction
Physical processes influence patterns of human activity, just as human activities have an effect on the environment.

The lessons   Modules 2–7 approach a geographic theme from three perspectives: a global perspective, a regional case study, and an advanced investigation. These perspectives are presented as lessons. Like the modules, each lesson can be used in isolation; they do not need to be completed in any particular order.

 Global perspective explores a geographic concept from a worldwide point of view. The focus is on how a particular concept affects human society on a global level. These lessons are designed for the beginning GIS student and provide detailed step-by-step instructions in the GIS investigation.

 Regional case study targets the same geographic concept from a regional perspective. Students examine one world region and use examples and data specific to that area. These lessons also are designed for beginning GIS students and provide detailed step-by-step instructions.

 Advanced investigation addresses the same geographic concept, but provides the students an opportunity to research and input data from outside sources such as the Internet or other GIS datasets. These lessons are designed for experienced GIS students. They typically ask students to create and save their own ArcMap™ map documents and use the more advanced tools in the software. They require significant self-motivation and critical thinking.

The teacher notes   The first few pages of each lesson have a lesson overview and teacher notes that contain the following items.

Lesson overview: A short summary gives you a snapshot of the lesson.

Materials and time: What you'll need and how long it will take. This is included to help you plan.

Standards: A list of the National Geography Standards (as published in *Geography for Life: The National Geography Standards,* 1994) covered in this lesson for middle school and high school students is included here. (Note: A matrix that matches all lessons in the book to the National Geography Standards follows this section. It is followed by a matrix matching the lessons to the National Science and Technology Standards.)

Objectives: Specific learning objectives for each lesson give you more detailed information about the lesson.

GIS skills and tools: Important GIS skills and tools that are taught and used in this lesson are summarized here.

Geographic inquiry graphic: These are the five steps of geographic inquiry that are incorporated into each lesson. See the section "Geographic Inquiry and GIS" for more information about the geographic inquiry process.

Lesson introduction: This section incorporates practical information—when to pass out specific handouts, for example—along with lists of questions for a class discussion introducing the topic.

Student activity: The primary student activity in each lesson is the GIS Investigation performed at the computer. This section offers Teacher Tips and information specific to the GIS Investigation.

Assessment: A summary of what your students will do to demonstrate their understanding of concepts and the proficiency of their skills upon completion of each GIS Investigation. Complete student assessment handouts are located at the end of the lesson, along with their evaluation rubrics.

Extensions: Consult these lists of ideas if you wish to expand the lesson for your class or individual students.

Student answer sheets and handouts

All of the GIS investigations, with the exception of some advanced investigations, include questions for students to answer as they work. Separate answer sheets are provided for students to write their answers on. This design makes it easier for you to review student answers and allows you to reuse the investigation sheets if you wish. Answer keys for all investigations are located in the tabbed section at the back of the book.

Some lessons include other student handouts, such as a map or table to be completed in the lesson introduction or conclusion. A few lessons include transparency masters for use in a class discussion.

The "ArcMap Toolbar Quick Reference" and "ArcMap Zoom and Pan Tools" are two optional handouts that can be used with any of the lessons. Consider giving these to your students with module 1, and then have them save these sheets for use with later lessons. The "GIS Terms Quick Reference" can also be handed out to students if desired. All of these handouts are located at the back of the book.

Rubric-based assessment

The lessons in *Mapping Our World* allow and encourage your students to explore a variety of geographic concepts and topics. A single letter or number grade won't be an accurate representation of the depth or completeness of their understanding of all concepts they've dealt with. The rubrics included with each lesson will allow you to evaluate your student's performance in a number of different ways. A learner may show mastery of one particular concept, but perform another task at the introductory level. The rubrics will also help you provide specific feedback to your students, showing them exactly where they need additional assistance or practice.

Exemplary: The student has gone above and beyond a particular standard. He or she has a strong understanding of the concept and has the ability to mentor other students.

Mastery: This is the target level for all students. Performance at this level shows that they have a good understanding of the concept illustrated in the standard.

Introductory: The student has limited understanding of the standard. Or, the product he or she produced shows little evidence of meeting the standard.

Does not meet requirements: The student does not show any foundational knowledge of the standard and the products they produce show no evidence of their understanding.

Suggested ways to use the evaluation rubrics:

- Distribute a copy of the rubric to students when you return their evaluated work. Circle or highlight the student's level of achievement for each standard. This provides the greatest amount of feedback for the student on each particular standard. Use the back of the form to make additional comments.

- Use the rubric as a form of student self-evaluation. Give students an unscored copy of the rubric and ask them to evaluate their own work. In the case of group projects, you can also let each group member evaluate the performance of the other members of the group.

ArcGIS Desktop software

The title of this book includes the words "ArcGIS Desktop." The software that comes with the book is called ArcView®. The GIS investigations refer to ArcMap and ArcCatalog™. You may be wondering, why are three different terms used?

ESRI ArcGIS Desktop consists of three software products: ArcView, ArcEditor™, and ArcInfo®. These three products look and work the same—they differ only in how much they can do. This book uses ArcView, but the lessons apply to ArcEditor and ArcInfo as well.

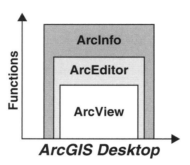

Each ArcGIS Desktop software product includes the same two applications—ArcMap and ArcCatalog. You use ArcMap to make maps, analyze them, and print them. You use ArcCatalog to browse and preview GIS data contained on your computer's hard disk or on a network. All of the lessons in this book use ArcMap exclusively, except for module 7's advanced investigation, which uses both ArcMap and ArcCatalog.

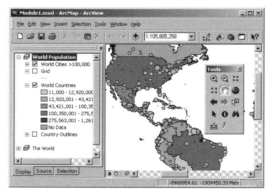
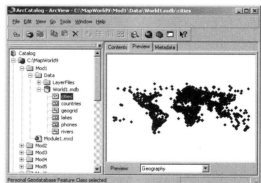

Setting up the software and data

Using the CDs

This book comes with two CDs. One CD contains a one-year licensed copy of ArcView 9 for Microsoft® Windows® operating system versions. The other CD contains the data and map documents used in the GIS investigations. Refer to the installation guides at the back of the book for detailed instructions on how to install the software and the data. If you do not feel comfortable installing programs on your computers, please be sure to ask your campus technology specialist for assistance. The software and data on these CDs need to be installed on your computer and all computers that the students will use to complete the GIS investigations.

Setting up the desktop and user interface

Once ArcView is installed, consider creating an ArcMap shortcut icon on the desktop of each student's computer so that students can quickly locate and start the ArcMap program. If students will be using ArcCatalog, create an ArcCatalog shortcut icon as well.

Instructions and graphics assume that students can see the file extensions, for example Global1.mxd or Cities.lyr. If your students cannot see file extensions (e.g., they see Global1 or Cities), you can either turn off this preference or have students ignore references to file extensions. Layer files, which sometimes have the same name as their data sources, are stored in separate folders to avoid confusion when file names are hidden. (To show file extensions, start ArcCatalog and go to Tools, Options. Click the General tab and uncheck the Hide file extensions box.)

Most exercises instruct students to add data to their map documents at some point. To navigate to the exercise data, a connection to the drive or folder where the data is stored is necessary. You may want to make sure this folder connection is set up in advance on each student's computer, or you may direct students to create the folder connection themselves during the exercise. If you choose to create the folder connection yourself, "Installing the Exercise Data" at the back of the book explains how to do this using ArcCatalog. Otherwise, follow the instructions in module 1, part 2, step 5 to create the folder connection when students add data for the first time.

Troubleshooting ArcGIS

Exercise instructions are written assuming the user interface and user preferences have the default settings. Unless students are working with a fresh installation of the software, however, chances are they will encounter some differences between the instructions and what they see on their screen. This is because ArcMap remembers settings from a previous session. For instance, if one student stretches the ArcMap window, the window will still be large when the next student starts ArcMap on that computer. Other potential differences include what toolbars are visible, where toolbars are located, the width of the table of contents, or whether or not the map scale changes when the window is resized.

Normally such differences will not be a problem, but you should be prepared to help individual students if they question an instruction or want to know why their ArcMap looks different than their neighbor's. A list of commonly encountered troubles and their solutions can be found on this book's Web site. You may want to print out this list for reference. If you have other software questions relating to this book, you can send e-mail to *workbooksupport@esri.com* with your questions.

Metadata

Metadata (information about the data) is included for all of the GIS data provided on the data CD. The metadata includes a description of the data, where it came from, a definition for each attribute field, and much other useful information.

To view metadata, open ArcCatalog and browse the Catalog Tree to the feature class of interest (e.g., C:\MapWorld9\Mod1\World1.mdb\phones). Click the Metadata tab on the right side of the window. You may find it easiest to view metadata using the FGDC FAQ stylesheet, which you can select from the Metadata toolbar.

If you want to know more about metadata and ArcGIS, we suggest that you work through the module 7 advanced investigation.

The companion Web site

This book's companion Web site is *www.esri.com/mappingourworld*. This Web site places a variety of GIS resources and other helpful information at your fingertips. For example:

- If you are new to GIS, you may want to visit the "Getting Started" section of the Web site before you begin the lessons.
- You'll want to check the Web site's "Resources by Module" section for specific resources, Web links, or changes when you get ready to use a particular lesson with your class.
- The Web site will have information on migrating from ArcView 3.x to ArcGIS Desktop 9 for those of you who also use the ArcView 3 *Mapping Our World* lessons.
- Any significant changes or corrections to the book will be posted here.

What's next?

After you and your students have used the lessons in this book for a while, the natural question is what comes next? Here are some suggestions:

- Incorporate one or more of the lesson extensions into your lesson plans.
- Challenge your students to apply the geographic inquiry process to current events or local issues. The book *Community Geography: GIS in Action* and the companion teacher's guide, also from ESRI Press, offer examples.
- Have your students put together a profile of your community and post it on the ESRI Community Atlas Web site *(www.esri.com/communityatlas)*. (Note: Your school may be able to earn software through this program.)
- Get connected. Find out who's doing what with GIS near you and contact them for ideas. The following resources can help you with this:
  - ESRI Education User Conference. For information, see *www.esri.com/educ*.
  - GIS Day™ Web site *(www.gisday.com)*
  - GIS.com Web site *(www.gis.com)*
  - KanGIS Web site *(kangis.org/learn)*
- Bring GIS professionals into the classroom. Most cities now use GIS; find out if the GIS coordinator would come to your classroom and present how they use GIS.
- Make GIS a permanent part of your classroom. Be sure to check with your district or state technology coordinator before you purchase an ArcView license for your school or classroom. A districtwide or statewide software license may already cover your school.

Geographic Inquiry and GIS

Geographic inquiry is at the core of *Mapping Our World: GIS Lessons for Educators*. In this book's lessons, you will use GIS as a tool kit to explore many issues and, as you use GIS, you will engage in the geographic inquiry process. This section introduces you to geographic inquiry and GIS.

Geography is the study of the world and all that is in it: its peoples, its places, and its environments, and all the connections among them. When you are investigating the physical world and its events, you are dealing with geography. Knowing where something is located, how its location influences its characteristics, and how its location influences relationships with other phenomena are the foundation of geographic thinking. To learn how to think geographically, you can use a process called geographic inquiry. Geographic inquiry asks you to see the world and all that is in it in spatial terms. Like other research methods, it also asks you to explore, analyze, and act upon the things you find.

The geographic inquiry process

STEPS	WHAT TO DO
1 Ask geographic questions	Ask questions to learn about the world around you
2 Acquire geographic resources	Identify data and information that you need to answer your questions
3 Explore geographic data	Turn the data into maps, tables, and graphs and look for patterns and relationships
4 Analyze geographic information	Go deeper into a geographic exploration and draw conclusions to answer your questions
5 Act upon geographic knowledge	Take your work to others to educate, make a decision, or solve a problem

The five steps of geographic inquiry are explicitly labeled in part 2 of the module 1 activity. The same steps form the foundation of the other investigations, but they are not labeled. You will naturally integrate geographic inquiry into the process of doing the exercises throughout this book.

What is GIS?
Chances are that GIS technology has already touched your life. If you flipped on a light switch today, chances are that GIS was used to help make sure the electricity was there to light up the room. When you drove down a highway, chances are that GIS managed the signs and streets along the way. If you received a delivery, chances are that GIS helped the driver find the way to your house. If you bought fresh vegetables, chances are that GIS helped manage the land and calculate the fertilizer needed for the crop. If you looked at a map on the Internet, chances are that GIS had a hand in that, too.

A *geographic information system* (GIS) uses computers and software to organize, develop, and communicate geographic knowledge. In simple terms, GIS takes the numbers and words from the rows and columns in databases and spreadsheets and puts them on a map.

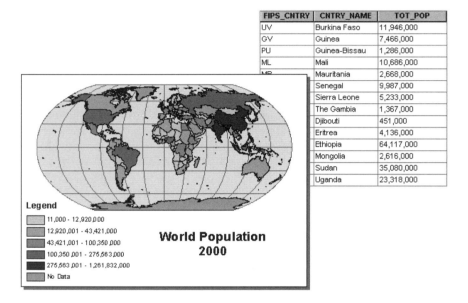

FIPS_CNTRY	CNTRY_NAME	TOT_POP
UV	Burkina Faso	11,946,000
GV	Guinea	7,466,000
PU	Guinea-Bissau	1,286,000
ML	Mali	10,686,000
MR	Mauritania	2,668,000
	Senegal	9,987,000
	Sierra Leone	5,233,000
	The Gambia	1,367,000
	Djibouti	451,000
	Eritrea	4,136,000
	Ethiopia	64,117,000
	Mongolia	2,616,000
	Sudan	35,080,000
	Uganda	23,318,000

Legend

	11,000 - 12,920,000
	12,920,001 - 43,421,000
	43,421,001 - 100,350,000
	100,350,001 - 275,563,000
	275,563,001 - 1,261,832,000
	No Data

World Population 2000

GIS is about visualizing information
The vast amounts of information available today require powerful tools like GIS to help people determine what it all means. GIS can make thematic maps (maps coded by value) to help illustrate patterns. To explore cities at risk of an earthquake, you might first make a map of where earthquakes have already occurred. You could explore further by coding earthquakes by magnitude. You might use one color to locate those that were strong and a second color for those that were weak, and then you might see a more complex pattern. Then you could analyze the patterns to answer questions about earthquakes. You will pursue this inquiry and more in a module 2 exercise.

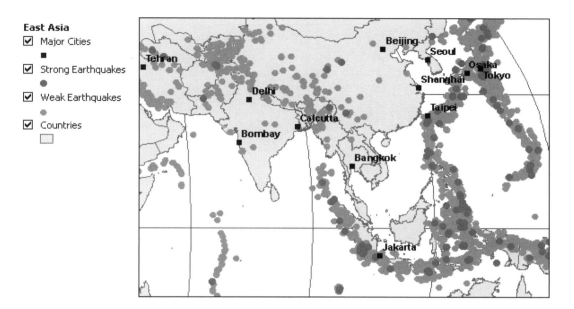

East Asia
- ☑ Major Cities
  ■
- ☑ Strong Earthquakes
  ●
- ☑ Weak Earthquakes
  ·
- ☑ Countries
  ☐

Putting it all together

GIS is a tool that can simplify and hasten geographic investigations. Like any tool, GIS has no answers packed inside it. Instead, for those who engage the tool and the process of geographic inquiry, it provides a means to discover pathways through our remarkable world of unending geographic questions.

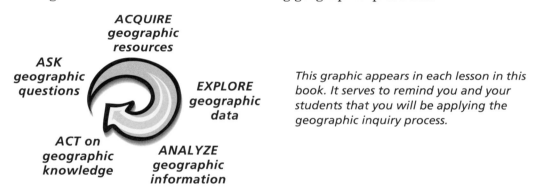

This graphic appears in each lesson in this book. It serves to remind you and your students that you will be applying the geographic inquiry process.

Now it's your turn to start investigating and mapping our world with GIS.

Correlation of National Geography Standards to *Mapping Our World* Lessons

STANDARD	MODULE 1 G	MODULE 2 G	R	A	MODULE 3 G	R	A	MODULE 4 G	R	A	MODULE 5 G	R	A	MODULE 6 G	R	A	MODULE 7 G	R	A
1 How to use maps and other geographic representations, tools, and technologies to acquire, process, and report information from a spatial perspective	✓	✓	✓	✓	✓	✓	✓	✓	✓	✓	✓	✓	✓	✓	✓	✓	✓	✓	✓
2 How to use mental maps to organize information about people, places, and environments in a spatial context																			
3 How to analyze the spatial organization of people, places, and environments on Earth's surface	✓							✓			✓								
4 The physical and human characteristics of places	✓	✓	✓			✓						✓					✓		
5 That people create regions to interpret Earth's complexity			✓																✓
6 How culture and experience influence people's perceptions of places and regions																			
7 The physical processes that shape the patterns of Earth's surface		✓	✓		✓	✓												✓	
8 The characteristics and spatial distribution of ecosystems on Earth's surface																			✓
9 The characteristics, distribution, and migration of human populations on Earth's surface									✓	✓									
10 The characteristics, distribution, and complexity of Earth's cultural mosaics													✓					✓	
11 The patterns and networks of economic interdependence on Earth's surface														✓	✓				
12 The processes, patterns, and functions of human settlement								✓								✓			
13 How the forces of cooperation and conflict among people influence the division and control of Earth's surface											✓	✓	✓		✓		✓		
14 How human actions modify the physical environment								✓											
15 How physical systems affect human systems		✓	✓	✓														✓	
16 The changes that occur in the meaning, use, distribution, and importance of resources																			
17 How to apply geography to interpret the past					✓														
18 How to apply geography to interpret the present and plan for the future		✓			✓			✓	✓		✓	✓		✓	✓		✓	✓	

G = Global investigation   R = Regional case study   A = Advanced investigation

National Science and Technology Standards

STANDARD	SCIENCE STANDARDS							TECHNOLOGY STANDARDS					
	A	B	C	D	E	F	G	1	2	3	4	5	6
Module 1 Global	MH					MH		✔	✔	✔		✔	✔
Module 2 Global	MH	MH		M		MH		✔	✔	✔		✔	✔
Regional	MH	MH		M		MH		✔	✔	✔		✔	✔
Advanced investigation	MH	MH		M		MH		✔	✔	✔	✔	✔	✔
Module 3 Global	MH		M		H	MH	H	✔	✔	✔		✔	✔
Regional						MH		✔	✔	✔		✔	✔
Advanced investigation	MH	M		MH	MH	MH	M	✔	✔	✔	✔	✔	✔
Module 4 Global						MH		✔	✔	✔		✔	✔
Regional	MH		M			MH		✔	✔	✔		✔	✔
Advanced investigation	MH							✔	✔	✔	✔	✔	✔
Module 5 Global	MH					MH		✔	✔	✔		✔	✔
Regional	MH					MH		✔	✔	✔		✔	✔
Advanced investigation								✔	✔	✔		✔	✔
Module 6 Global	M					M		✔	✔	✔		✔	✔
Regional	M					M		✔	✔	✔		✔	✔
Advanced investigation	M					M		✔	✔	✔	✔	✔	✔
Module 7 Global	MH			H	MH	MH		✔	✔	✔		✔	✔
Regional						MH		✔	✔	✔		✔	✔
Advanced investigation			M					✔	✔	✔		✔	✔

M = Middle school   H = High school

Mapping Our World

GIS Lessons for Educators

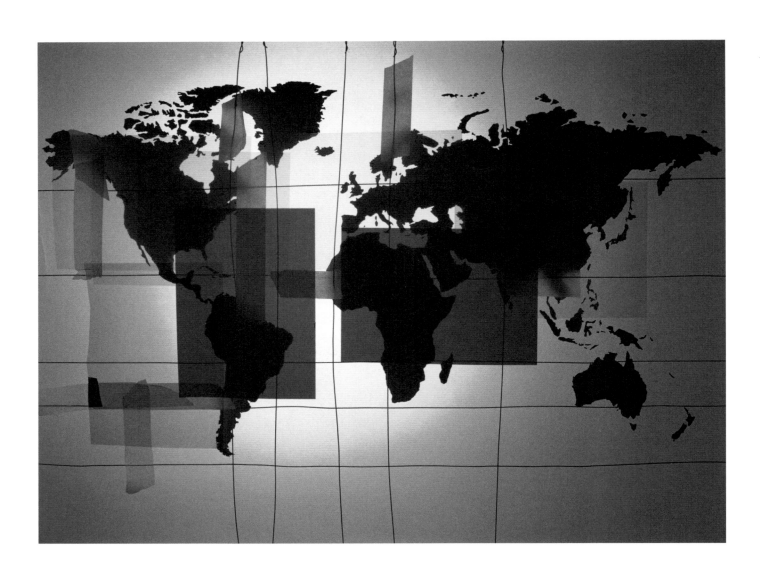

ArcMap: The Basics

This module covers the concepts and skills that will be discussed and used in the rest of the book. As your students step through the activities, they will be introduced to the geographic inquiry method and acquire a working understanding of GIS theory and practice—the "how to" of ArcMap as well as a sense of its conceptual framework, foundation, and purpose. By the end of the lesson, they will find themselves competent users of a basic GIS.

This module serves as a comprehensive introduction to the rest of the book. We suggest that you and your students complete it before moving on. Please note that the format of module 1 contains one lesson in two parts. Modules 2 through 7 each contain a global, a regional, and an advanced investigation lesson.

ArcMap: The Basics

Lesson overview

This module introduces the basic concepts and skills of GIS that will prepare your students for the lessons ahead. The activities laid out here will instruct them in starting ArcMap, navigating the computer to find the prepared map documents and data, and acquiring functional skills to help them use GIS. At the same time, students will also learn the steps involved in the process of geographic inquiry.

The format of this module differs from that of the rest of the book: it has one lesson in two parts, not the global, regional, and advanced investigation lessons found in modules 2 through 7. We suggest that you have the students spend two class periods on these activities to gain a functional knowledge of GIS, ArcMap skills, and the geographic inquiry method.

Estimated time   Two to three 45-minute class periods

Materials   ✔ Calculator

✔ Student handouts from this lesson to be copied:
  • GIS Investigation sheets (pages 11 to 37)
  • Geographic Inquiry: Thinking Geographically handout (page 9)
  • Student answer sheet (pages 38 to 45)
  • Assessment(s) (pages 46 to 49)

Standards and objectives

National geography standards

	GEOGRAPHY STANDARD	MIDDLE SCHOOL	HIGH SCHOOL
1	How to use maps and other geographic representations, tools, and technologies to acquire, process, and report information from a spatial perspective	The student knows how to make and use maps, globes, charts, models, and databases to analyze spatial distributions and patterns.	The student knows how to use technologies to represent and interpret Earth's physical and human systems.
3	How to analyze the spatial organization of people, places, and environments on Earth's surface	The student understands that places and features are distributed spatially across Earth's surface.	The student understands how spatial features influence human behavior.
4	The physical and human characteristics of places	The student understands how technology can shape the characteristics of places.	The student understands how the physical and human characteristics of place can change.

Standards and objectives (continued)

Objectives

The student is able to:

- Understand the basic concept of a geographic information system (GIS).
- Use a basic ArcMap skill set to build a map.
- Use the five-step geographic inquiry model.
- Print maps.

GIS skills and tools

 Identify a feature on the map

  Zoom in to a desired section of the map or to the center of the map

  Zoom out to a desired section of the map or to the center of the map

 Zoom to the full extent of all layers

 Pan to a different section of the map

 Find a feature in a layer and identify it

 Get help about a button

 Add a layer to the map

- Browse information about map features using MapTips
- Turn layers on and off
- Expand and collapse layers
- Activate a data frame
- Change the order of the table of contents to change the map display
- Create a bookmark for a map extent and return to it later
- Calculate the values for a field

For more on geographic inquiry and these steps, see Geographic Inquiry and GIS (pages xxiii to xxv).

Teacher notes

Lesson introduction

Begin this lesson by helping students understand that each map in a GIS has spreadsheet or database information attached to it. There are a number of Adobe® PDF and Microsoft PowerPoint® presentations on the *Mapping Our World* Web site dealing with this topic.

After completing part 1 of the activity, introduce part 2 by giving students the "Geographic Inquiry: Thinking Geographically" handout or show it to the class on an overhead projector. Have a brief discussion with your students about "thinking geographically." Refer to the "Geographic Inquiry and GIS" section in this book to familiarize yourself with the geographic inquiry model.

Student activity

 Before completing this lesson with students, we recommend that you complete it as well. Doing so will allow you to modify the activity to accommodate the specific needs of your students.

Ideally each student should be at an individual computer, but the lesson can be modified to accommodate a variety of instructional settings. On the first day, distribute the GIS Investigation sheets entitled "ArcMap: The Basics, Part 1: Introducing the software." Explain that in this activity the students will begin to learn the basic ArcMap skills they will need to create GIS maps. The worksheets will provide them with detailed instructions for their investigations. As they navigate through the lesson, they will be asked questions that will help keep them focused on key concepts. Some questions will have specific answers while others will require creative thought.

On the second day, introduce part 2 as described in the lesson introduction, and then distribute the GIS Investigation sheets entitled "ArcMap: The Basics, Part 2: The geographic inquiry model." Explain to your students that in this lesson they will practice the ArcMap skills they acquired in the first lesson, and be introduced to more advanced skills while exploring the geographic inquiry method.

 Teacher Tip: In part 1, steps 14–15 and part 2, step 7 of this activity, students are asked to use a calculator to divide two country statistics and come up with a third. Some of the countries have populations with nine or ten digits. If your students have hand-held calculators that only allow them to enter eight-digit numbers, you may want to have them use the calculator accessory on their computer.

 Teacher Tip: In part 2, step 8 of this activity, students are asked to calculate a field in an attribute table. This action results in permanent changes to the data. In order for this to work properly, each student must have their own copy of the Mod1 folder.

Steps 11 and 12 of part 2 do not involve the computer. You may wish to assign these steps as homework to be completed outside of class.

Things to look for while the students are working on this activity:

- Are students thinking spatially as they work through the procedure?
- Are students answering the questions as they work through the procedure?
- Are students using a variety of menus, buttons, and tools to answer the questions on the handout?
- Are students able to use the legends to interpret the data in the table of contents?
- Are students able to print out a map on the printer?

Conclusion   Before your students complete the assessment, conduct a brief discussion in which you ask them to brainstorm ideas about how GIS can be used in everyday life or how they could use GIS in their daily school assignments or classes. Ask them to describe the geographic inquiry process they learned as well as share how comfortable they are with using ArcMap. This discussion should also include an overview of which buttons they have used to build maps on screen, any ArcMap operations that were confusing, and printer operations.

Assessment   *Middle school: Highlights skills appropriate to grades 5 through 8*

The middle school "ArcGIS: The Basics" assessment asks students to create and print a map. They will be expected to turn on three layers and zoom in to a location of their choosing, use the Identify tool to obtain three pieces of data about that area, and write a brief paragraph explaining how they created their maps.

High school: Highlights skills appropriate to grades 9 through 12

The high school "ArcGIS: The Basics" assessment asks students to create and print a map. They will be expected to turn on four layers and zoom in to a location of their choosing, use the Identify tool to obtain three pieces of data about that area, write a brief paragraph explaining how they created their maps, and describe what they learned geographically about the area on their map.

Extensions   • Create several views with data from other folders on the data CD to answer a question. Make a connection between the data to make sure that the maps they create are meaningful.

• Have students calculate cell-phone density (the number of people per cell phone) in 1997 and in 2002. Use maps and tables to explore how the use of this technology has changed over this five-year period.

• Students can suggest other layers that might help explain the connection between the technological advances of telephone lines and cellular technology, and such political and social issues as GDP, education, health care, and so on.

• Students can develop a plan or outline for how to use GIS to fulfill a current class assignment.

• Ask students to choose a country in the news and use their GIS skills to find the country and to study the data associated with it.

• Check out the Resources by Module section of this book's Web site *(www.esri.com/mappingourworld)* for print, media, and Internet resources that educate the public on the uses of GIS.

Geographic Inquiry: Thinking Geographically

GEOGRAPHIC INQUIRY STEP	WHAT TO DO	EXAMPLES FROM THE GIS INVESTIGATION (MODULE 1, PART 2)
ASK A GEOGRAPHIC QUESTION	Think about a topic or place and identify something interesting or significant about it. Turn that observation into a geographic question or hypothesis that you can investigate. Types of geographic questions include: • Where are things located? • How do things change from one place to another? • Why do things change from one place to another?	• What countries have the most and the least phone lines? • Does the number of phone lines vary proportionately with the number of people among the world's most populous countries? • Why does Pakistan have fewer phone lines per person than India?
ACQUIRE GEOGRAPHIC RESOURCES	Identify the information and data needed to answer your geographic question: • What is the geographic focus of your research? • For what period of time do you need data? • For what subjects and specific topics do you need data?	Use data and maps from the Mod1 Data folder that contains: • World countries • Population data for world countries • Phone lines data for world countries
EXPLORE GEOGRAPHIC DATA	Turn the data into maps, tables, and graphs and look for patterns in the way things change from one place to another. Some ways to explore data in ArcMap include: • Create a map document and add data layers. • Turn layers on and off, and zoom and pan the map. Look at individual features and what surrounds them. • Change the symbols used to represent features. • Look for ways features in one layer relate to features in other layers.	Explore map layers and attribute tables containing population and phone line data for world countries.
ANALYZE GEOGRAPHIC INFORMATION	Focus on the information and maps that most seem to answer your questions. For example: • Find or identify particular features. • Select features with specific attributes to meet specific criteria. • Calculate new attributes from existing ones to get new information. Draw conclusions from what you have seen in the maps, tables, and graphs and answer your geographic question.	• Calculate phone line density for all countries. • Research and record population, phone line, and phone line density information for selected countries. • Rank selected countries by population and phone line density and compare the two lists. • Compare your answer with your initial hypothesis.
ACT ON GEOGRAPHIC KNOWLEDGE	Your conclusions are the result of turning pieces of data into geographic knowledge. Think about how you could share this knowledge, or how you or someone else could use it to make a decision, correct a problem, or help others.	Devise a plan of action for the phone system in one of the countries that you researched.

NAME _____   DATE _____

ArcMap: The Basics

Answer all questions on the student answer sheet handout

Part 1: Introducing the software

ArcGIS is made up of two programs: ArcMap and ArcCatalog. This computer activity will show you how to start the ArcMap program. You will be guided through the basics of using ArcMap to explore maps. After you do this activity, you will find it much easier to complete the other activities in this book.

Step 1   Start ArcMap

a   Double-click the ArcMap icon on your computer's desktop. If you do not have the icon on your desktop, click Start, Programs, ArcGIS, and ArcMap.

b   If the ArcMap start-up dialog box appears, make sure **A new empty map** is selected.

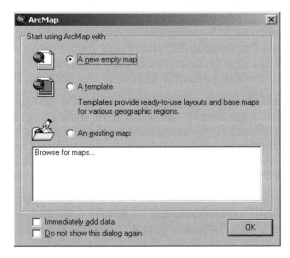

c **Click the OK button.**

The ArcMap window opens. Its title contains three pieces of information: the name of the map document (Untitled), the program (ArcMap), and the level of the program (ArcView).

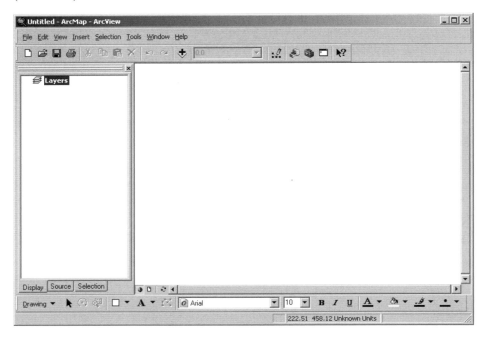

Step 2 **Open the Module1.mxd file**

a **A map document file has been created for you to use in this exercise. To open it, click the File menu and click Open.**

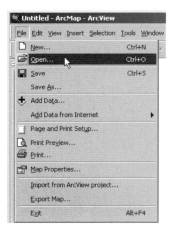

b **Navigate to the module 1 folder (C:\MapWorld9\Mod1).**

> *Hint: First make sure the correct drive is selected in the "Look in" drop-down list. If the disk drive you need is not listed, click the My Computer or My Network Places icon on the left side of the dialog to access the disk you want. Then double-click folders to open them.*

c   Choose **Module1.mxd** from the list.

> :💡: *Note: If file extensions are turned off (hidden) on your computer, you won't see .mxd, and you should choose Module1 from the list. Having file extensions turned on (visible) is not required to complete this investigation.*

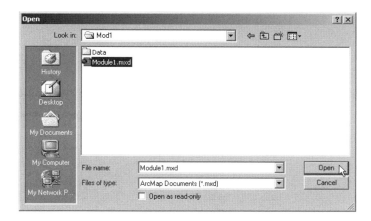

ArcGIS map document files end with the three-letter extension **.mxd**.

d   Click Open.

When the map document opens, you see a map of the world. The ArcMap window's title changed to show the name of the map document, Module1.mxd.

> :💡: *Remember: If file extensions are not visible on your computer, the name of the map document will be Module1.*

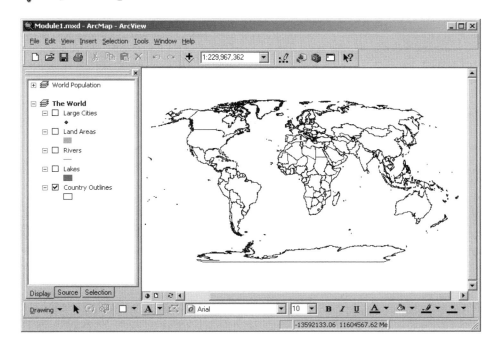

e   Take a closer look at the ArcMap window. On the right side you see a map. On the left side you see a column that contains a list. This column is the table of contents. The items in the list represent the two different maps contained in this map document (World Population and The World). The items listed under The World are the five different layers of information that can be shown on this map.

f   Look at the top of the ArcMap window and notice the different menus and buttons. The menus and buttons are grouped on different toolbars. Toolbars can be moved around, so your toolbars may not be arranged exactly as pictured below.

> :⚡: *Note: The Standard toolbar above shows the map scale (1:256,357,059). Your map scale may be different depending on the size and shape of your ArcMap window.*

You will use these menus and buttons to perform the various GIS functions. You might think this is a lot of buttons to learn, but you will not need to learn them all at once.

g   Locate the Tools toolbar. It may be floating (not attached to the window), or docked (attached to the window).

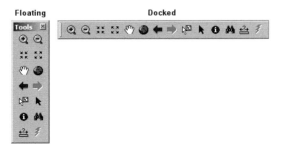

You can drag toolbars around and put them in a place that you like.

h   If your Tools toolbar is floating, click on its title and drag the toolbar to the gray area above the map near the other tools. The toolbar docks to the window. Otherwise, if your Tools toolbar is already docked, click on the small gray bar next to the magnifying glass, and drag the toolbar off of the ArcMap window. The toolbar floats.

i Now drag your Tools toolbar to the gray line between the map and the table of contents to dock it vertically next to the map. (Hint: If it docks below the table of contents instead, drag it to the bar again until it docks where you want it.)

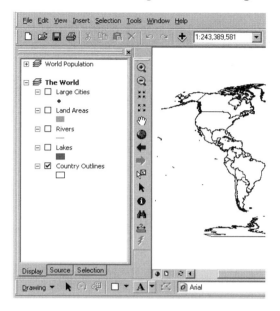

Step 3   Enlarge the ArcMap window

When the map document first opens, the ArcMap window may be small. If it is, you will want to enlarge it.

a In the upper right corner of your ArcMap window there are three buttons. Click once on the middle button that looks like a box.

The button you clicked is called the Maximize button. Now the ArcMap window fills your whole screen.

 b Look again at the three buttons in the upper right corner of your ArcMap window. Now the middle button looks like two boxes. (This is called the Restore Down button.) Click on it.

The window returns to the smaller size. You can also change the size of your ArcMap window by stretching it.

c Place the cursor on any corner of the ArcMap window that is not at the edge of your screen. The cursor changes to a diagonal double-headed arrow.

d Click and drag the window outward until the ArcMap window fills about two-thirds of the screen. Let go of the mouse button.

 Note: Enlarging the ArcMap window by stretching it allows you to choose a size that is large but does not cover the entire screen. As you work with ArcMap, other windows and dialog boxes will appear. You will find it useful to move them aside so you can also see the map on your screen.

Step 4   Work with layers

In ArcMap, a map is made up of layers that are grouped into a data frame. This map document has two data frames: World Population and The World.

a   Notice that The World data frame is listed in bold letters in the table of contents. The bold letters tell you which data frame is active. The active data frame is displayed in the map area.

The World data frame contains five layers. These are *Large Cities, Land Areas, Rivers, Lakes,* and *Country Outlines.* Next you will learn how to turn layers off and on and change their order.

b   Notice that each layer in the table of contents has a small box in front of it. Only the box for Country Outlines has a check mark in it.

c   Click the check mark next to Country Outlines. The check mark goes away. The display of Country Outlines in the map area also disappears.

d   Click the box next to the Land Areas layer. The green land areas layer is displayed. This is called turning on a layer.

e   Click the box next to the Large Cities layer to turn it on.

f   Turn on the Rivers and Lakes layers.

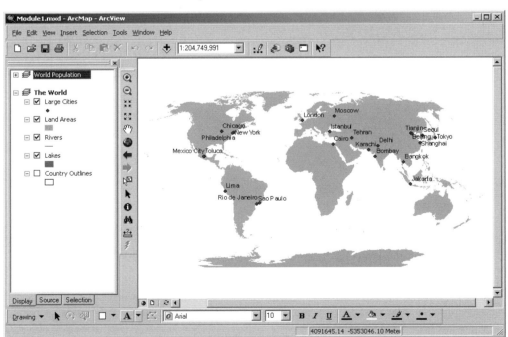

g   Answer the following question on your answer sheet:

?   *Which layers are not visible on the map but are turned on in the table of contents?*

Next you will change the order of the layers.

h   Notice that each layer in the table of contents has a name and a symbol. In the table of contents, place your cursor on the Rivers name.

i   Click and hold the mouse button. Drag the Rivers layer up above the Land Areas layer. Let go of the mouse button.

?   *What happened on your map?*

j   Drag the Lakes layer above the Land Areas layer.

?   *(1)   What happened on your map?*

?   *(2)   What would happen if you dragged the Lakes under Land Areas?*

> *Note: Whenever your maps don't appear as you think they should, check the following things:*
>
> 1. *Check to see if the layer you want is turned on.*
>
> 2. *Check the order of the layers in the table of contents. Layers that are represented by lines and points (streets, rivers, cities, etc.) will be covered up by layers that are represented by polygons (countries, states, etc.). You may need to drag line or point layers above the polygon layers in order to see them.*

Step 5   Change the active data frame

In this step you will make the World Population data frame active to display the other map.

> *Note: So far you have always clicked the mouse using the **left** mouse button. Some menus in ArcMap are accessed by clicking the **right** mouse button. The instructions use the words "right-click" when you need to use the right mouse button.*

a   Right-click the World Population data frame title in the table of contents. On the menu that appears, drag the mouse down and click **Activate** (with the left mouse button). The World Population data frame becomes bold and displays in the map area.

b Click the plus sign next to World Population in the table of contents. The table of contents expands to show the layers in the World Population data frame.

? *What is the name of the layer that is turned on in the World Population data frame?*

Step 6 **Widen the table of contents**

Now you will widen the table of contents so that the World Countries layer legend is not cut off.

a Move your cursor to the edge of the table of contents, in between the scroll bar and the Tools toolbar. When it is in the right place, it should look like this: ⟷

b Click and hold the mouse button. Drag the cursor to the right until you can see the full legend descriptions. Release the mouse button.

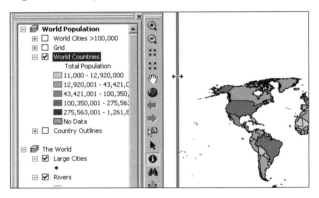

The table of contents becomes wider and the map becomes smaller.

Step 7 **Identify a country and record country data**

The value of GIS comes from the data (information) that is attached to each map. You will see how this works by using one of the tools to access data about countries.

a Move your cursor over the map and pause on any country (don't click). Notice that as your cursor pauses over different countries, the name of each country displays. This information display is called a MapTip.

b The Identify tool lets you see more data about your map by clicking on places you are interested in. Locate the Identify tool in the Tools toolbar.

Hint: To display a "tool tip" with a tool's name, pause your cursor over a tool without clicking. At the same time, a description of the tool appears at the bottom left corner of the ArcMap window.

c When a tool is selected, it looks like the button is pushed in. If your Identify tool already looks pushed in, proceed to step 6e. Otherwise, click to select it.

d The Identify Results window appears. Click the title bar of the Identify Results window and move the window so that it doesn't cover the map.

e Move your cursor over the map without clicking. Notice it changes to an arrow with an "i" next to it.

f Click on the United States on your map.

The Identify Results window displays information about the United States. The information you see is all the data that is available about the United States in the World Countries layer.

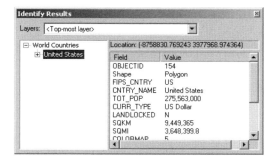

g In the Identify Results window, scan down the column labeled Field that begins with the word OBJECTID. Answer the following questions:

(1) What is the fourth listing in this column?

(2) What is the fifth listing in this column?

(3) What is the final listing in this column? (Hint: You will need to scroll down.)

h These words are names describing a characteristic, or attribute, of the United States. Another word for attributes is *fields.* Field names are often abbreviated.

(1) What do you guess the field entitled "SQMI" stands for?

(2) What is the number to the right of the field "SQMI"?

Step 8 **Compare the Identify Results data with the table data**

a The picture below shows part of the table that contains the data attached to the World Countries layer. Answer the following question using the picture.

Which row in this table has the attributes for the United States?

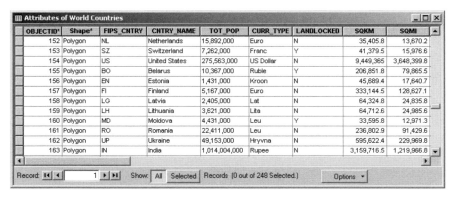

OBJECTID*	Shape*	FIPS_CNTRY	CNTRY_NAME	TOT_POP	CURR_TYPE	LANDLOCKED	SQKM	SQMI
152	Polygon	NL	Netherlands	15,892,000	Euro	N	35,405.8	13,670.2
153	Polygon	SZ	Switzerland	7,262,000	Franc	Y	41,379.5	15,976.6
154	Polygon	US	United States	275,563,000	US Dollar	N	9,449,365	3,648,399.8
155	Polygon	BO	Belarus	10,367,000	Ruble	Y	206,851.8	79,865.5
156	Polygon	EN	Estonia	1,431,000	Kroon	N	45,689.4	17,640.7
157	Polygon	FI	Finland	5,167,000	Euro	N	333,144.5	128,627.1
158	Polygon	LG	Latvia	2,405,000	Lat	N	64,324.8	24,835.8
159	Polygon	LH	Lithuania	3,621,000	Lita	N	64,712.6	24,985.6
160	Polygon	MD	Moldova	4,431,000	Leu	Y	33,595.8	12,971.3
161	Polygon	RO	Romania	22,411,000	Leu	N	236,802.9	91,429.6
162	Polygon	UP	Ukraine	49,153,000	Hryvna	N	595,622.4	229,969.8
163	Polygon	IN	India	1,014,004,000	Rupee	N	3,159,716.5	1,219,966.8

Record: 1 Show: All Selected Records (0 out of 248 Selected.) Options *b*   Compare the information in the Identify Results window with the information in the table shown above, and answer these questions:

?   *(1)   The field names in the Identify Results window display in the column starting with the word OBJECTID. Where are these field names displayed in the table?*

?   *(2)   Find the field in the table that represents square miles of land. How many square miles of land are in the United States?*

?   *(3)   Give a brief explanation of the relationship between the Identify Results window and the table.*

☒   *c*   Click the Close button that looks like an ✕ at the top right corner of the Identify Results window.

> *Caution: Be careful not to click the Close button on the ArcMap window by mistake. If you close ArcMap, the map document will close without saving your work.*

Step 9   Explore city data on the world map

a   Turn on the World Cities >100,000 layer.

b   Click the plus sign next to the World Cities >100,000 layer to expand its legend.

⊟ ⤢ **World Population**
   ⊟ ☑ World Cities >100,000
        Population
        ⬭ 5,000,001 and greater
        ◯ 1,000,001 to 5,000,000
        ◌ 500,001 to 1,000,000
        ▫ 250,001 to 500,000
        ▱ 100,000 to 250,000

Your map displays all the world cities with populations greater than 100,000. There are so many cities that they are all jumbled together on this small map. You need to zoom in to a smaller portion of the world to see distinctions between the cities.

🔍   *c*   Click the Zoom In tool to select it. Remember, the button looks pushed in when it is selected.

The Zoom In tool can be used two different ways. One way to zoom in is to drag a box around the area you want to display. You will zoom in on Europe and Africa.

d   Place your cursor on Greenland. (Hint: To find Greenland move your cursor over the map so the country names display.)

e   Click and hold down the mouse button. Drag the cursor down and to the right. When your cursor is near Australia, release the mouse button.

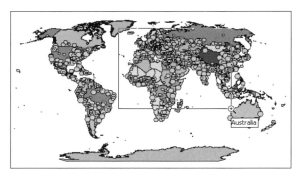

Another way to zoom in is to click on the place you want to be in the center of your map. Now you will zoom in closer to Europe.

f Click slowly three times on the yellow dots clustered around Europe.

⚡ Caution: Make sure to keep the mouse very still as you click. Otherwise, you may accidentally drag a tiny box and the map will zoom in too much. If this happens, go to the toolbar and click once on the Previous Extent button and do this step again.

You will use the Identify tool to get information about the cities.

g Click the Identify tool. Click the Layers box in the Identify Results window and select World Cities > 100,000.

h Click on a red or pink country area that is away from a yellow dot. Notice that the Identify Results window reports Nothing Found. This is because the Identify tool gets information only for the layer you selected in the Layers list.

i Using what you learned in steps g and h, use the Identify tool to find the name and country of any two cities you choose.

j Click the Close button on the Identify Results window.

Step 10   Explore Europe with an attribute table

When you used the Identify tool in step 7, you saw that a country on the map (the United States) was connected to attribute information about that country. The attributes were displayed in the Identify Results window.

In GIS, each object on your map is called a *feature.* For example, the United States is a feature in the World Countries Population layer, and Paris is a feature in the World Cities >100,000 layer. In this step, you will find out more about the connection between features and their attributes.

a Right-click the World Countries layer in the table of contents and choose Open Attribute Table. If your table is large, drag the right side of the table to the left to make it smaller, like the one pictured below. If your table is covering the map, click on its title bar and drag it out of the way until you can see all of Europe.

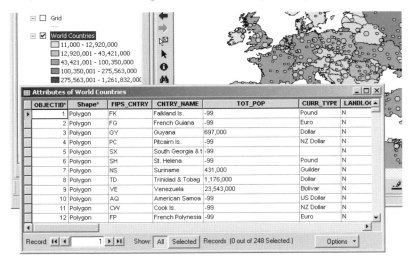

?   (1)   What is the name of the table you opened?

?   (2)   What country is listed in the first row of the table?

b   Scroll down to the end of the table.

?   What country is listed in the last row of the table?

c   Click the Options button at the bottom of the table and click Find & Replace.

d   Type **Poland** in the "Find what:" box in the Find and Replace dialog. Then click the Find Next button.

e   Click the Close button to close the Find and Replace dialog.

f   Click on the gray rectangle with the arrow at the beginning of the row for Poland.

OBJECTID*	Shape*	FIPS_CNTRY	CNTRY_NAME	TOT_POP
143	Polygon	PL	Poland	38,646,000
144	Polygon	LO	Slovakia	5,408,000
145	Polygon	SI	Slovenia	1,928,000
146	Polygon	SV	Svalbard	-99
147	Polygon	BE	Belgium	10,242,000

Attributes of World Countries

Notice that the row in the table turns blue to show that it is selected and so does the outline of Poland on the map.

g   Hold down the Ctrl key on your keyboard. In the table, click the gray boxes for these rows: Slovakia, Belgium, and Germany.

?   What happens to the map when you click on these rows in the table?

h   Click the gray box for the United States row. (You may need to scroll down a few rows.)

?   (1)   What happens to Poland and the other countries that were highlighted?

?   (2)   Did you see the United States become outlined in blue on the map? If not, why not?

i   Click the Close button on the table.

j   Click the Full Extent button. The map displays the whole world.

        Remember: Sometimes you may zoom in so close or zoom out so far that you lose your map entirely or can't tell where you are. You can always click the Full Extent button to get back to a map you can recognize.

?   Why can you see the United States now when you couldn't see it in the previous step?

k   Right-click World Countries in the table of contents, point to Selection, and click Clear Selected Features.

Whenever you select a feature on your map, it will be outlined in blue to indicate that feature has been selected. If you want to turn the blue outlines off, you need to clear the selected features.

Step 11 Practice identifying features

 a Click the Zoom In tool.

b Click and drag a box around the continent of South America.

     What do you see on your map?

 *c* Click the Identify tool. In the Identify Results window click the Layers drop-down list and choose <Top-most layer>. Now you can use the tool to identify either a country or a city.

> 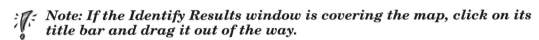 **Note: If the Identify Results window is covering the map, click on its title bar and drag it out of the way.**

d Identify the large South American country that is dark red.

    (1)   What country is it?

    (2)   What is this country's total population?

e Identify the large city in the northwest part of this country.

    (1)   What city is it?

    (2)   What population class is this city in?

f Look on the map for two cities in Brazil that are in a higher population range. (Hint: the yellow symbol for those two cities will be larger than the one for the city you just identified.)

 g Zoom in closer to the two cities to separate them from surrounding cities for identification purposes.

h Use the Identify tool to answer these questions:

    (1)   What are the names of these two large cities?

    (2)   What population class are these cities in?

i Close the Identify Results window.

Step 12 Practice zooming out

a Click the Zoom Out tool.

b Drag a two-inch box anywhere on your map.

c Drag another two-inch box anywhere on your map.

    (1)   What does your map look like?

    (2)   Which button could you use to return your map to full size?

d Zoom your map back to its original full extent.

Step 13   Practice finding a feature

In an earlier step you found a country by looking for it in the attribute table. In this step, you will find a country by looking for it on the map.

a   Turn off the World Cities >100,000 layer.

b   Click the Find button.

You will find the country of Sudan.

c   Click in the white Find box. Type **Sudan**.

d   Click the **In** drop-down list, scroll down, and click World Countries.

e   Under Search, click the white circle next to **In fields**.

f   Click the **In fields** drop-down list and select CNTRY_NAME. You want ArcMap to search for Sudan in the country name attribute field in the World Countries layer.

g   Click Find.

h   Notice that a results list appears at the bottom of the Find window and Sudan is listed. At the bottom of the window there is a message telling you "One object found."

> *Note: If you get a results list that is blank and the message at the bottom of the window says "No objects found," go back to steps 13c–13g. Make sure you spell the country name correctly and enter the correct information.*

i   Move the Find window off the map.

j   Right-click on Sudan in the results list and click Flash feature. The country of Sudan flashes on and off in the map. (If you didn't see it flash, repeat this step and watch Africa on the map.)

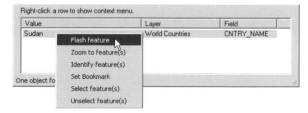

k   Right-click Sudan in the results list again and click Identify feature(s). Look at the Identify Results window and answer the questions below.

> *Note: All or part of the Identify Results window may be hidden behind the Find dialog. If so, move the Find dialog out of the way.*

(1)   How many tourists arrive in Sudan each year?

(2)   How many people live in Sudan?

(3)   Does this seem like a low or high number of tourists for this population?

l   Close the Identify Results window.

Step 14   Zoom to a feature and create a bookmark

a   Click the New Search button in the Find window.

b   Type **Qatar** in the Find box.

c   Click Find.

d   Right-click Qatar in the results list. Click Select feature(s).

Qatar is outlined in blue on the map. Is Qatar a large country or a small one?

e   Right-click Qatar in the results list and choose Zoom to feature(s).

f   Close the Find window.

g   Click three times on the Fixed Zoom Out button. (Its arrows point outward.)

The Fixed Zoom Out button is different from the Zoom Out tool. Clicking the Fixed Zoom Out button always zooms out from the center of the map.

> *Remember: Zooming with the Find tool placed Qatar in the middle of the map area. When you click the Fixed Zoom Out button, the map stays centered on Qatar.*

Next you will create a bookmark for Qatar. Later in the activity you will use the bookmark to quickly return to this map extent.

h   Click the View menu, point to Bookmarks, then click Create.

i   Press the Delete key to delete the bookmark name, Bookmark 1. Type **Qatar** for the new bookmark name.

j   Click OK.

k   Use the Identify tool to answer the following questions:

(1)   How many people live in Qatar?

(2)   How many cell phones do they have?

(3)   Divide the population of Qatar by the number of cell phones in Qatar. How many people are there for every cell phone in Qatar?

l   Close the Identify Results window.

m   Use MapTips to answer the following question:

?   *What large country is directly west of Qatar?*

n   Click the Fixed Zoom Out tool nine more times until you see the entire Arabian peninsula.

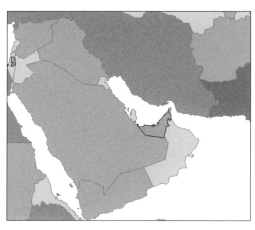

Step 15   Continue to explore the World Population map

Whenever you wish to center the map differently, you can use the Pan tool to move the map around.

  a   Click the Pan tool.

b   Use the MapTips to locate the country of Egypt at the left edge of the map.

c   Click and hold on Egypt. Drag the hand diagonally to the bottom right corner of the map area. Let the mouse button go. Answer the questions below:

?   *(1)   What boot-shaped country do you see on the map?*

?   *(2)   What is the population of that country?*

?   *(3)   How many cell phones does that country have?*

?   *(4)   Divide the answer to question 2 by the answer to question 3. How many people are there for every cell phone in this country?*

> *Note: If your calculator does not let you type in these long numbers, use the calculator on your computer. (Click Start, Programs, Accessories, Calculator.)*

d   Pan east to Japan and answer the following questions:

?   *(1)   What is the population of Japan?*

?   *(2)   How many cell phones does Japan have?*

?   *(3)   Divide the population of Japan by the number of cell phones in Japan. How many people are there for every cell phone in Japan?*

e   Close the Identify Results window.

f   Click the View menu. Point to Bookmarks and click Qatar. The map returns to Qatar using the bookmark you saved.

g   Right-click World Countries in the table of contents and point to Selection. Click Clear Selected Features.

❓   *What happened to Qatar?*

Step 16   Get help from the "What's This?" tool

If you forget what a particular button is for or how to use a tool, you can ask ArcMap to help you.

a   Click the What's This? tool on the Standard toolbar.

Notice your cursor turns into an arrow with a question mark.

b   Click the Find button.

A help message box is displayed that tells you about the Find button.

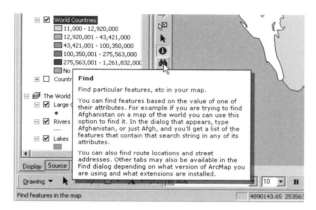

c   Click anywhere inside the pop-up message box to make it disappear.

You can use the What's This? button to help you use most ArcMap buttons.

d   Click the What's This? button.

e   Click any other button you want to know about.

f   When you are finished looking at the help, click the pop-up message box to close it.

Step 17   Label and print a map

In this step you will learn an easy way to label the countries and to print the map on your screen.

a   Use the Pan and Zoom tools to focus the map on a location of your choice.

b   Right-click World Countries and choose Label Features. Country names are added to the map.

❓   *Where do you think these labels come from?*

c   Click the File menu, then click Print.

d   In the Print dialog, click the Setup button.

e   Ask your teacher which printer you should use. Select its name from the drop-down list under Printer Setup in the Page and Print Setup dialog.

f   Under Map Page Size, check the box next to Use Printer Paper Settings.

g   Under Paper (the section above Map Page Size), make sure the paper size is set to "Letter." You may need to select it from the drop-down list.

h   Decide how you want the map to be placed on the paper. Click Portrait if you want the top to be a short side of the paper. Click Landscape if you want the top to be a long side.

i   At the bottom of the dialog, check the box next to "Scale Map Elements proportionally to changes in Page Size." Your dialog should look similar to the one pictured below.

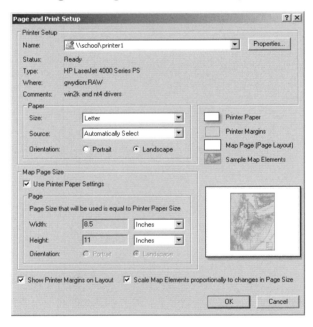

j   Click OK on both the Page and Print Setup window and the Print window.

k   Your map should print after a few moments.

Step 18   Close the map document and exit ArcMap

a   Click the File menu.

b   Click Exit.

c   Click No in the box that asks whether you want to save changes.

NAME _____   DATE _____

Part 2: The geographic inquiry model

Much like scientific analysis, geographic inquiry involves a process of asking questions and looking for answers. The geographic inquiry model is made up of the following five steps:

ACQUIRE

ASK

EXPLORE

ACT

ANALYZE

In this activity you will learn how maps and GIS can help you in the geographic inquiry process. At the same time, you will be practicing the ArcGIS skills you learned in part 1 and learning some new ones.

Step 1   Start ArcMap

a   Double-click the ArcMap icon on your computer's desktop.

ArcMap

b   If the ArcMap start-up dialog appears, click **An existing map** and click OK. Then go to step 2b.

Step 2 **Open the module1.mxd file**

   a You will use the same ArcMap map document that you used in part 1 for this exercise. To open it, go to the File menu and choose **Open**.

   b Navigate to the module 1 folder (**C:\MapWorld9\Mod1**) and choose **Module1.mxd** (or **Module1**) from the list.

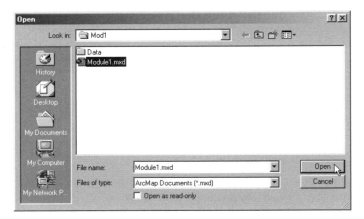

Step 3 **Adjust your window and activate the World Population data frame**

   a Enlarge the ArcMap window by stretching it.

   b In the table of contents, click the minus sign next to The World data frame to collapse it.

   c Click the plus sign next to World Population. The table of contents expands to show the layers in the World Population data frame.

   d Right-click World Population and left-click Activate. The World Population data frame becomes bold and displays in the map area. (Hint: You may need to widen the table of contents so that the World Countries legend is not cut off.)

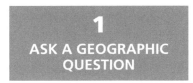

1

ASK A GEOGRAPHIC QUESTION

Step 4 **Ask a geographic question and develop a hypothesis**

It's important to think about the information on your maps both as you create them and when they've been completed. You might look at a map and then think of a question that it might help you answer. Or you might think of the question first, and then look for maps, or GIS layers, that might help you answer the question.

In this activity you will try to answer the following geographic question:

Geographic Question: Do the number of phone lines vary proportionately with the number of people among the world's most populous countries?

   a What makes this a geographic question?

   b Just as with scientific inquiry, you'll begin by constructing a hypothesis. Write a hypothesis that answers the geographic question stated above.

> *: Remember: A hypothesis is an educated guess. Your hypothesis may be*
> *right or wrong in the end. The goal is not to know the answer before*
> *you start your research, but to have a kind of "ballpark idea" or hunch*
> *about what the answer might be. Then you will set about supporting or*
> *rejecting it.*

2
ACQUIRE GEOGRAPHIC RESOURCES

Step 5   Add a layer to your map

Next, you need to identify the kind of information that will help you explore your question. You want to display this data as layers in your ArcGIS map.

? *a* Your map already has a layer with world countries and their population. What other attribute of countries do you need in order to investigate your hypothesis?

➕ *b* Click the Add Data button.

An Add Data dialog appears. It works similar to the Open dialog you used at the beginning of this activity.

c Click the Connect to Folder button. Navigate to the MapWorld9 folder (**C:\MapWorld9**). Click OK. The connection is added to the list of locations in the Add Data dialog.

> *: Note: If a folder connection to the MapWorld9 folder already exists on*
> *your computer, you may skip this step.*

d Navigate to the Module1 layer files folder (**C:\MapWorld9\Mod1\Data\LayerFiles**). Click **World Phone Lines.lyr**.

e Click Add.

? *What is the name of the layer that has been added to your table of contents?*

> *: Note: An ArcMap layer file contains the complete definition of a layer*
> *including its name, data source, symbology, and other properties. You*
> *can save a layer outside a map document as a layer file so it can be*
> *reused in other maps.*

3
EXPLORE GEOGRAPHIC DATA

Step 6   **Explore the World Phone Lines map**

Now it's time to look at the World Phone Lines layer and think about what the map tells you.

a   Turn off the World Countries layer.

b   Look at the legend for the World Phone Lines layer and answer the following questions:

World Phone Lines
    2002
    ☐ 10 - 2,175,290
    ☐ 2,175,291 - 8,317,008
    ☐ 8,317,009 - 23,213,319
    ☐ 23,213,320 - 40,560,159
    ☐ 40,560,160 - 70,614,903
    ☐ 70,614,904 - 212,929,439
    ☐ No Data

?   *(1)*   *What color in the legend indicates countries with the fewest phone lines?*

?   *(2)*   *What color indicates countries with the most phone lines?*

?   *(3)*   *What color indicates countries with no data available for this layer?*

c   Notice that the colors in the legend change gradually from the lightest color to the darkest. The name for this type of legend is a graduated color legend.

?   *What other layer in your map has a graduated color legend?*

d   Look at the map to answer the following questions. You may need to turn layers on or off. You may need to use MapTips or the Identify tool to find out a country's name. Answer the following questions:

?   *(1)*   *Which two countries had the most phone lines in 2002?*

?   *(2)*   *On which continent are most of the countries with the fewest phone lines?*

?   *(3)*   *Which two countries have the largest populations?*

?   *(4)*   *Name three countries that are in the same population class (color) as the United States.*

?   *(5)*   *Which of these three countries, if any, are in the same phone line class (color) as the United States?*

e   Turn off World Countries and turn on World Phone Lines.

f   With the Identify tool, click on the United States.

g   Read the geographic question again.

Geographic Question: Do the number of phone lines vary proportionately with the number of people among the world's most populous countries?

h   Scroll down slowly in the Identify Results window and look at the different fields.

?   *What two fields might help in answering the geographic question?*

i   Close the Identify Results window.

4
ANALYZE GEOGRAPHIC INFORMATION

Step 7   Research and record phone line and population data for China

Now it's time to locate China and record the population and phone lines in the appropriate columns on your answer sheet.

a   Click the Full Extent button on the Tools toolbar. This will return the map to its original size if you have zoomed or panned the map.

b   To unselect (turn the blue outline off) any countries that might be selected, click the Selection menu at the top of the ArcMap window, and click Clear Selected Features. If the menu choice is disabled (grayed out), it means that there are no features currently selected in the World Population data frame.

c   Click the Find button and type **China** in the Find box.

d   Click Find.

e   China appears in the results list at the bottom of the Find window.

f   Right-click on China in the results list and click Identify feature(s).

> *Note: If you get more than one result for China, right-click on the row for World Phone Lines.*

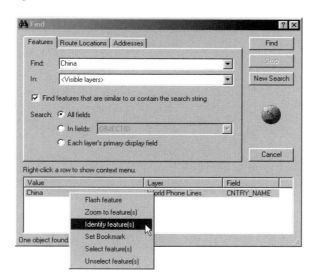

?   *g*   Find the POP_2000 attribute in the Identify Results window. Record the number in the Country Population column for China.

> *Note: All or part of the Identify Results window may be hidden behind the Find dialog. If so, move the Find dialog out of the way.*

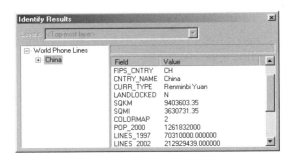

?   *h*   Scroll down in the Identify Results window and find the LINES_2002 attribute. Record the number in the Phone Lines 2002 column for China.

i   Close the Find and Identify Results windows.

Now you will use your calculator to figure out the number of people per phone line for China.

j   Enter into your calculator the population figure for China.

> *Note: If your calculator does not let you type in a number as large as the one for China's population, use the calculator on your computer. (Click Start, Programs, Accessories, Calculator.)*

k   Divide it by the number of phone lines for China.

?   *l*   Round off the result to the nearest two decimal places and record it in the last column on the answer sheet.

Step 8   Calculate the number of people per phone line for all countries

Now you will use ArcMap to calculate the number of people per phone line for all the countries in the World Phone Lines layer.

a   In the table of contents, right-click World Phone Lines and click Open Attribute Table.

b   Scroll all the way to the right until you see the last field in the table, PHONE_DENS (for phone density). Notice that the values in this field are Null, indicating that no value has been assigned.

c   Right-click on the PHONE_DENS field heading and click Calculate Values.

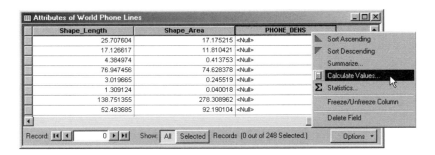

d   In the Field Calculator message box, click Yes to continue. The Field Calculator displays.

You want to divide the population values in the POP_2000 field by the number of phone lines in the LINES_2002 field.

e   In the Fields list, click POP_2000. It now appears in the white text box below.

   f   Click the division button.

g   In the Fields list, find LINES_2002 and click it.

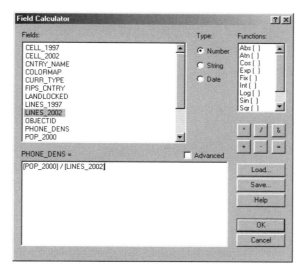

> *Note: If you make a mistake, highlight and delete any text in the white box. Then, repeat steps 8e–8g above.*

h   Click OK. The field calculator calculates the number of people per phone line and places the values in the PHONE_DENS field in the attribute table.

i   Scroll down in the table to see all the PHONE_DENS values.

j   Close the attribute table.

Step 9   **Research and record population, phone line, and phone line density data for all the countries**

a   Click the Find button. **China** appears in the Find box.

b   Click Find.

c   Right-click on China in the results list and click Identify feature(s).

(1)   What is the number of people per phone line (PHONE_DENS) for China?

(2)   Does this number agree with the value you calculated in step 7?

d   Use the Find and Identify tools to locate the remaining countries in the table on the answer sheet. Record the country population, phone lines, and the number of people per phone line for each country in the appropriate columns.

Step 10   Close the map document and exit ArcMap

You won't need to use ArcMap to complete the rest of this activity, so you'll close your map document at this time.

a   Click the File menu.

b   Click Exit.

c   Click No in the box that asks whether you want to save changes.

Ask your teacher whether you should continue doing steps 11–12 in class, or whether to do them as homework.

Step 11   Analyze results of your research

?   *a*   In the table on the answer sheet, the column on the left ranks the countries by population from highest to lowest. In the column on the right, rank the countries from the lowest number of people per phone line to the highest number of people per phone line, using the data you recorded in step 9. Then draw lines connecting the same country in each column.

b   Use the data in your table to answer the following questions:

?   (1)   Which country has the fewest people per phone line? How many people have to share a phone line in this country?

?   (2)   How does the country in question b-1 rank in population size with the other seven countries in your table?

?   (3)   Which country has the most people per phone line? How many people have to share a phone line in this country?

?   (4)   How does the country in question b-3 rank in population size with the other seven countries in your table?

?   (5)   What is the population of Japan? How many people have to share a phone line in Japan?

?   (6)   What country has the most phone lines? How does the number of people per phone line in this country compare with the seven other countries in your table?

?   (7)   Russia and Pakistan have about the same number of people. Why do you suppose these two countries have such a different number of people who have to share a phone line? What factors do you think contribute to this disparity?

?   (8)   Read the geographic question again. What do you think the answer to the geographic question is?

Geographic Question: Do the number of phone lines vary proportionately with the number of people among the world's most populous countries?

?   (9)   Compare your initial hypothesis (step 4b) with your answer in question 8. How does your hypothesis compare with your answer to the geographic question?

5
ACT UPON GEOGRAPHIC KNOWLEDGE

Step 12　**Develop a plan of action**

The last step of the geographic inquiry process is to act on what you have learned. Your action plan might be simply to repeat the process; thinking about what you've learned often leads to deeper, more complex, and interesting geographic questions.

For this step, choose one of the following four countries: China, Brazil, Indonesia, or the United States.

Imagine that you are an expert specializing in telecommunications. You need to devise a plan for your chosen country that deals effectively with the basic concern of your original geographic question.

To develop an effective plan, you may need to conduct further research on the phone system within your country. If you decided, for instance, that increasing the number of phone lines operating in your chosen country would improve the quality of life there, you could come up with a written plan of action for telecommunications officials, pointing out areas of strength and weakness, and explaining where and why expansion would be most beneficial.

?　　*a*　Use the information in your table to describe the current phone line situation in your chosen country. Record your chosen country on the answer sheet.

?　　*b*　Do you think that increasing the number of phone lines operating in your chosen country would improve the quality of life there? Why or why not?

?　　*c*　List three concerns you have about increasing the number of phone lines in your chosen country.

?　　*d*　List two new geographic questions that you would like to investigate to help you develop a sound plan.

NAME _____   DATE _____

Student answer sheet
Module 1
ArcMap: The Basics

Part 1: Introducing the software

Step 4   Work with layers

g   Which layers are not visible on the map but are turned on in the table of contents?

_____

_____

i   What happened on your map?

_____

j-1   What happened on your map?

_____

j-2   What would happen if you dragged the Lakes under Land Areas?

_____

_____

Step 5   Change the active data frame

b   What is the name of the layer that is turned on in the World Population data frame?

_____

Step 7   Identify a country and record country data

g-1   What is the fourth listing in this column? _____

g-2   What is the fifth listing in this column? _____

g-3   What is the final listing in this column? (Hint: You will need to scroll down.)

_____

h-1   What do you guess the field entitled "SQMI" stands for?

_____

h-2   What is the number to the right of the field "SQMI"?

_____

Step 8   Compare the Identify Results data with the table data

a   Which row in this table has the attributes for the United States? _____

b-1   Where are these field names displayed in the table?

_____

b-2   How many square miles of land are in the United States?

_____

b-3   Give a brief explanation of the relationship between the Identify Results window and the table.

_____

_____

_____

Step 9   Explore city data on the world map

i   Use the Identify tool to find the name and country of any two cities you choose.

CITY NAME	COUNTRY WHERE THE CITY IS LOCATED

Step 10   Explore Europe with an attribute table

a-1   What is the name of the table you opened?

_____

a-2   What country is listed in the first row of the table?

_____

b   What country is listed in the last row of the table?

_____

g   What happens to the map when you click on these rows in the table?

_____

_____

h-1   What happens to Poland and the other countries that were highlighted?

_____

_____

h-2   Did you see the United States become outlined in blue on the map? If not, why not?

_____

_____

_____

j   Why can you see the United States now when you couldn't see it in the previous step?

_____

_____

_____

Step 11   Practice identifying features

b   What do you see on your map?

_____

d-1   What country is it?   _____

d-2   What is this country's total population?   _____

e-1   What city is it?   _____

e-2   What population class is this city in?   _____

h-1   What are the names of these two large cities?

_____

_____

h-2   What population class are these cities in?   _____

Step 12   Practice zooming out

c-1   What does your map look like?

_____

c-2   Which button could you use to return your map to full size?

_____

Step 13   Practice finding a feature

k-1   How many tourists arrive in Sudan each year?   _____

k-2   How many people live in Sudan?   _____

k-3   Does this seem like a low or high number of tourists for this population?   _____

Step 14   Zoom to a feature and create a bookmark

d   Is Qatar a large country or a small one?   _____

k-1   How many people live in Qatar?   _____

k-2   How many cell phones do they have?   _____

k-3   How many people are there for every cell phone in Qatar?   _____

m   What large country is directly west of Qatar?   _____

Step 15   Continue to explore the World Population map

c-1   What boot-shaped country do you see on the map? _____

c-2   What is the population of that country? _____

c-3   How many cell phones does that country have? _____

c-4   How many people are there for every cell phone in this country? _____

d-1   What is the population of Japan? _____

d-2   How many cell phones does Japan have? _____

d-3   How many people are there for every cell phone in Japan? _____

g   What happened to Qatar?

_____

Step 17   Label and print a map

b   Where do you think these labels come from?

_____

_____

Part 2: The geographic inquiry model

Step 4 Ask a geographic question and develop a hypothesis

a What makes this a geographic question?

_____

_____

b Write a hypothesis that answers the geographic question.

_____

_____

_____

_____

_____

_____

Step 5 Add a layer to your map

a What other attribute of countries do you need in order to investigate your hypothesis?

_____

e What is the name of the layer that has been added to your table of contents?

_____

Step 6 Explore the World Phone Lines map

b-1 What color in the legend indicates countries with the fewest phone lines?

_____

b-2 What color indicates countries with the most phone lines?

_____

b-3 What color indicates countries with no data available for this layer?

_____

c What other layer in your map has a graduated color legend?

_____

d-1 Which two countries had the most phone lines in 2002?

_____

d-2 On which continent are most of the countries with the fewest phone lines?

_____

d-3 Which two countries have the largest populations?

_____

_____

d-4 Name three countries that are in the same population class (color) as the United States.

_____

_____

_____

d-5   Which of these three countries, if any, are in the same phone line class (color) as the United States?

_____

_____

_____

h   What two fields might help in answering the geographic question?

_____

_____

Step 7   Research and record phone line and population data for China

g, h   Use your Find and Identify tools to locate China. Record the population and number of phone lines in the appropriate columns.

COUNTRY NAME	COUNTRY POPULATION	PHONE LINES 2002	NUMBER OF PEOPLE PER PHONE LINE
China			

l   Record the number of people per phone line for China in the last column of the table above.

Step 9   Research and record population, phone line, and phone line density data for all the countries

c-1   What is the number of people per phone line (PHONE_DENS) for China?

_____

c-2   Does this number agree with the value you calculated in step 7?

_____

d   Use the Find and Identify tools to locate the countries in the table below. Record the population, phone lines, and number of people per phone line for each country. The first country, China, is already filled in for you.

COUNTRY NAME	COUNTRY POPULATION	PHONE LINES	NUMBER OF PEOPLE PER PHONE LINE
China	1,261,832,000	212,929,439	5.93
India			
United States			
Indonesia			
Brazil			
Russia			
Pakistan			
Japan			

Step 11 Analyze results of your research

a In the table below, the column on the left ranks the countries by population from highest to lowest. In the column on the right, rank the countries from the lowest number of people per phone line to the highest number of people per phone line, using the data you recorded in step 9. Then draw lines connecting the same country in each column.

RANKED BY POPULATION (HIGHEST TO LOWEST)	RANKED BY NUMBER OF PEOPLE PER PHONE LINE (LOWEST NUMBER OF PEOPLE PER PHONE LINE TO HIGHEST)
China	
India	
United States	
Indonesia	
Brazil	
Russia	
Pakistan	
Japan	

b-1 Which country has the fewest people per phone line? _____

How many people have to share a phone line in this country? _____

b-2 How does the country in question b-1 rank in population size with the other seven countries in your table?

_____

b-3 Which country has the most people per phone line? _____

How many people have to share a phone line in this country? _____

b-4 How does the country in question b-3 rank in population size with the other seven countries in your table?

_____

b-5 What is the population of Japan? _____

How many people have to share a phone line in Japan? _____

b-6 What country has the most phone lines? _____

How does the number of people per phone line in this country compare with the seven other countries in your table?

_____

b-7 Russia and Pakistan have about the same number of people. Why do you suppose these two countries have such a different number of people who have to share a phone line?

_____

What factors do you think contribute to this disparity?

_____
_____
_____

b-8   What do you think the answer to the geographic question is?

_____

_____

_____

_____

_____

b-9   How does your initial hypothesis (step 4b) compare with your answer to the geographic question?

_____

_____

_____

_____

_____

_____

Step 12   Develop a plan of action

a   Name your chosen country. _____

       Use the information in your table to describe the current phone line situation in your chosen country.

_____

_____

_____

_____

_____

b   Do you think that increasing the number of phone lines operating in your chosen country would improve
       the quality of life there? Why or why not?

_____

_____

_____

_____

_____

c   List three concerns you have about increasing the number of phone lines in your chosen country.

_____

_____

_____

_____

_____

d   List two new geographic questions that you would like to investigate to help you develop a sound plan.

_____

_____

_____

_____

_____

NAME _____   DATE _____

ArcMap: The Basics
Middle school assessment

Open the ArcMap document **Module1.mxd** (or Module1). Use the ArcMap skills you have learned in this module to do the following things:

1   Create a map with at least three different layers.

2   Zoom in on the map to an area of the world of your choosing.

3   Find out three pieces of specific information about the area you chose. (Hint: Use the Identify button.)

4   Write a geographic question that involves one of the pieces of information that you listed in question 3.

5   Print the map and attach it to this page.

ArcMap: The Basics

Assessment rubric

Middle school

STANDARD	EXEMPLARY	MASTERY	INTRODUCTORY	DOES NOT MEET REQUIREMENTS
The student knows how to make and use maps, globes, graphs, charts, models, and databases to analyze spatial distributions and patterns.	Creates and prints a map with more than three themes and focused on a portion of the world using a GIS.	Creates and prints a map with three different themes and focused on a portion of the world using a GIS.	Creates and prints a map with one or two different themes using a GIS.	Has difficulty creating the map without assistance and does not print it out.
The student knows and understands that places and features are distributed spatially across Earth's surface.	Identifies more than three pieces of information about the area of the world covered by his or her map and develops a geographic question based on that information.	Identifies three pieces of information about a particular area of the world and develops a geographic question based on that information.	Identifies one or two pieces of information about a particular area of the world and attempts to create a geographic question based on that information.	Identifies some information about a place, but does not create a geographic question based on the information gathered.
The student knows the role of technology in shaping the characteristics of places.	Successfully completes the assessment and develops a plan in step 11 of the lesson that illustrates an understanding of the importance of GIS technologies in analyzing the aspects of a region or place and solving geographic problems and questions.	Successfully completes the assessment and develops a clear and concise plan in step 11 of the lesson that illustrates an understanding of how GIS technologies contribute to geographic understanding of places and development of plans for changing them.	The student has a beginning understanding of the importance of GIS and related technologies in solving geographic questions.	The student does not see the relationship between GIS and related technologies in solving geographic questions.

This is a four-point rubric based on the National Standards for Geographic Education. The "Mastery" level meets the target objective for grades 5–8.

NAME _____   DATE _____

ArcMap: The Basics
High school assessment

Open the ArcMap document **Module1.mxd** (or Module1). Use the ArcMap skills you have learned in this module to do the following things:

1   Create a map with at least three different layers.

2   Zoom in on the map to an area of the world of your choosing.

3   Find out three pieces of specific information about the area you chose. (Hint: Use the Identify button.)

4   Write a geographic question that involves one of the pieces of information that you listed in question 3.

5   Write a brief paragraph explaining what you learned geographically about the area you zoomed to on your map.

6   Print the map and attach it to this page.

ArcMap: The Basics

Assessment rubric

High school

STANDARD	EXEMPLARY	MASTERY	INTRODUCTORY	DOES NOT MEET REQUIREMENTS
The student knows how to use technologies to represent and interpret Earth's physical and human systems.	Creates and prints a map with more than three themes and focused on a portion of the world using a GIS.	Creates and prints a map with three different themes and focused on a portion of the world using a GIS.	Creates and prints a map with one or two different themes using a GIS.	Has difficulty creating the map without assistance and does not print it out.
The student knows and understands the spatial behavior of people.	Identifies more than three pieces of information about the area of the world covered by his or her map and develops a geographic question about the human impact to that place, based on the information.	Identifies three pieces of information about a particular area of the world and develops a geographic question about the human impact to that place based on the information.	Identifies one or two pieces of information about a particular area of the world and attempts to create a geographic question based on that information.	Identifies some information about a place, but does not create a geographic question based on the information gathered.
The student knows and understands the changing physical and human characteristics of place.	In step 11 of the lesson, writes a clear and concise plan of action that takes into account the physical and human characteristics of a place or places.	In step 11 of the lesson, writes a clear and concise paragraph on the geographic characteristics of a particular place.	In step 11 of the lesson, identifies some physical and human characteristics of a place and attempts to formalize understanding in paragraph form.	In step 11 of the lesson, lists one or two characteristics of a place but does not show geographic understanding of the place.

This is a four-point rubric based on the National Standards for Geographic Education. The "Mastery" level meets the target objective for grades 9–12.

Physical Geography I
Landforms and Physical Processes

Explore the powerful forces that originate in the earth's interior and shape the landforms that characterize its surface.

The Earth Moves: A global perspective

Students will observe patterns of earthquake and volcanic activity on the earth's surface and the relationship of those patterns to the location of diverse landforms, plate boundaries, and the distribution of population. Based on their exploration of these relationships, students will form a hypothesis about the earth's distribution of earthquake and volcanic activity and identify world cities that face the greatest risk from those phenomena.

Life on the Edge: A regional case study of East Asia

Students will investigate the Pacific Ocean's "Ring of Fire," with particular focus on earthquake and volcanic activity in East Asia, where millions of people live with the daily threat of significant seismic or volcanic events. Through the analysis of volcanic location and earthquake depth, students will identify zones of subduction at tectonic plate boundaries and the location of populations in the greatest danger of experiencing a volcanic eruption or a major earthquake.

Mapping Tectonic Hot Spots: An advanced investigation

This lesson is designed to target the self-motivated, independent GIS students in your class. They will use the Internet to acquire the most recent data on earthquakes and volcanoes worldwide. By exploring this data through a GIS, students will construct a current World Tectonic Hot Spots map. This lesson could be an independent research project completed over several days.

The Earth Moves
A global perspective

Lesson overview

Students will observe patterns of seismic and volcanic activity around the world and the relationship of those patterns to the location of diverse landforms, plate boundaries, and the distribution of population. Based on their exploration of these relationships, students will form a hypothesis about the planet's distribution of earthquake and volcanic activity, and identify those cities that face the greatest risk from those phenomena.

Estimated time   Two to three 45-minute class periods

Materials

✔ Large map of the world

✔ Ten adhesive dots or map pins (two colors) for every four students

✔ Colored pencils for each student

✔ Student handouts from this lesson to be copied:
- World map (pages 58 to 60)
- GIS Investigation sheets (pages 61 to 71)
- Student answer sheets (pages 72 to 74)
- Assessment(s) (pages 75 to 79)

Standards and objectives

National geography standards

	GEOGRAPHY STANDARD	MIDDLE SCHOOL	HIGH SCHOOL
1	How to use maps and other geographic representations, tools, and technologies to acquire, process, and report information from a spatial perspective	The student knows and understands how to use maps to analyze spatial distributions and patterns.	The student knows and understands how to use geographic representations and tools to analyze and explain geographic problems.
4	The physical and human characteristics of places	The student knows and understands how to analyze the physical characteristics of places.	The student knows and understands the changing physical characteristics of places.
7	The physical processes that shape the patterns of Earth's surface	The student knows and understands how physical processes shape patterns in the physical environment.	The student knows and understands the spatial variation in the consequences of physical processes across Earth's surface.
15	How physical systems affect human systems	The student knows and understands how natural hazards affect human activities.	The student knows and understands how humans perceive and react to natural disasters.

Standards and objectives (continued)

Objectives

The student is able to:

- Locate zones of significant seismic and volcanic activity on the earth's surface.
- Describe the relationship between zones of high seismic activity, volcanic activity, and the location of tectonic plate boundaries.
- Identify world regions and cities where human populations face the greatest threat from earthquakes.

GIS skills and tools

    Add layers to the map

        Zoom in and out of the map

    Use the Identify tool to learn more about a selected record

    Pan the map to view different areas

    Find a specific feature

- Open the attribute table for a layer
- Sort data in ascending and descending order
- Select records in a table
- Select features based on an attribute
- Clear all selections for a layer

For more on geographic inquiry and these steps, see Geographic Inquiry and GIS (pages xxiii to xxv).

Teacher notes

Lesson introduction

Begin the lesson by distributing the provided outline maps of the world. Have the students individually write a "V" on locations on the map where they believe volcanoes are located, and an "E" on locations they think earthquakes typically occur. They should select at least eight volcanic areas and eight areas for earthquake activity. Once students have done this individually, divide them into groups of three or four and have them discuss their ideas within their small group.

Each group should take five minutes to consider the following questions and record their answers in the space provided on the worksheet:

- Are there any similarities between your group members' maps? What are they?
- How did the members of your group choose where to put a "V" and "E" on their maps?
- Are there locations where earthquakes and volcanoes occur close together? List them.
- Are there major cities close to these locations? What are they?

As a class, the students will incorporate each group's map into one large class map. Tell the students that one color of map pins will represent earthquakes and the other color will represent volcanoes. At the end of five minutes, each group should take 10 adhesive dots or map pins (five of each color) and place them on a large world map in the classroom. One color should be placed at locations where group members agree that there are volcanoes and the other color should be placed where group members agree that there are earthquakes. Before beginning the next part of the lesson, briefly discuss the patterns reflected on the map.

Student activity

 Before completing this lesson with students, we recommend that you complete it as well. Doing so will allow you to modify the activity to accommodate the specific needs of your students.

After the initial discussion, have the students work on the computer component of the lesson. Ideally, each student should be at an individual computer, but the lesson can be modified to accommodate a variety of instructional settings.

Distribute this lesson's GIS Investigation sheets to your students. Explain that in this activity, they will use GIS to investigate where earthquakes and volcanoes occur on the earth's surface. The worksheets will provide them with detailed instructions for their investigations. As they investigate, they will create a hypothesis about where these major physical events occur, identify zones of volcanic and seismic activity, and identify those densely populated places that face the greatest risk from these seismic activities.

In addition to the instructions, the investigation sheets include questions to help students focus on key concepts. Some questions will have specific answers while others will require creative thought.

Things to look for while the students are working on this activity:
- Are the students using a variety of tools?
- Are the students answering the questions as they work through the procedure?
- Students should be referring to their original maps and notes from class discussion.

Conclusion   Before beginning the assessment, engage students in a discussion of the observations and discoveries that they made during their exploration of the maps. Ask students to compare their initial ideas—as reflected by the pattern of colored dots on the classroom's world map—with the insights they acquired in the course of their GIS investigation.

- Has this investigation raised any questions that they would like to explore further?
- How can GIS help world cities better prepare for seismic events?
- How have their ideas about earthquakes and volcanoes changed since the start of the lesson?

Assessment   *Middle school: Highlights skills appropriate to grades 5 through 8*

Part 1 of The Earth Moves assessment for middle school students asks them to create an informative paper map based on their GIS investigation. Part 2 is a set of four questions that students will answer on a separate sheet of paper. An assessment rubric to evaluate student performance is provided at the end of the lesson.

High school: Highlights skills appropriate to grades 9 through 12

Part 1 of The Earth Moves assessment for high school students asks them to create an informative paper map based on their GIS investigation. Part 2 is a set of three questions that students will answer on a separate sheet of paper. An assessment rubric to evaluate student performance is provided at the end of the lesson.

Extensions
- By using ArcMap, students can create their own digital maps for the assessment. They can mark all of the required features and print the map using the layout view.
- Investigate your local area with ArcMap, and ask students to create a new layer that identifies natural hazards in your region.
- Look at historical data for notable earthquakes and volcanic eruptions and incorporate this data as new layers. From this information, ask students to make predictions on the next significant eruption or seismic event.
- Analyze fault line data included with this lesson to see if it provides additional insight to the location and movement of plate boundaries.
- Have students use supplemental earthquake data (see next page) to compare earth-quakes in 2000 with earthquakes in 2003 and 2004. Where did more recent earthquakes occur?
- Have students use supplemental volcano data (see below) to create maps for a region of the world by having small groups research specific details. For example, one group could focus on volcano type (morphology), another on volcano status, and another on volcano time frame.
- Check out the Resources by Module section of this book's companion Web site *(www.esri.com/mappingourworld)* for print, media, and Internet resources on the topics of earthquakes, volcanoes, and plate tectonics.

Supplemental data　　In addition to the earthquake and volcano data used in the investigation, supplemental earthquake and volcano data is provided. The supplemental layers, earthquakes_new and volcanoes2, contain additional features and attributes. They are stored in the World2 geodatabase in the module 2 data folder (C:\MapWorld9\Mod2\Data\World2.mdb).

- Earthquakes_new contains 52,284 earthquakes recorded in 2003 and 2004 that range in magnitude from 0 to 8.3.

- Volcanoes2 contains 1,509 volcanoes and numerous attributes, including the volcano type (morphology), evidence of eruption, and last known eruption.

Full metadata details for this and all data can be viewed in ArcCatalog. Instructions for viewing metadata in ArcCatalog are provided in the "How to Use This Book" section. See the "Extensions" section of these teacher notes for suggested activities using this data.

NAME _____   DATE _____

The Earth Moves
A map investigation

Part 1: Individual map

Volcanoes and earthquakes happen all over the planet. On the map provided, identify places where you think these physical events typically occur. Mark eight locations with a "V" for potential volcano sites, and mark eight locations with an "E" for places where you think earthquakes typically occur.

Part 2: Small group discussion

In your small group, compare your maps and answer the following questions:

Step 1   Are there any similarities between your group members' maps? What are they?

_____
_____
_____
_____
_____
_____
_____

Step 2   How did the members of your group choose where to put a "V" and "E" on their maps?

_____
_____
_____
_____
_____
_____

Step 3   Are there locations on your maps where earthquakes and volcanoes occur close together? List them.

_____
_____
_____
_____
_____
_____
_____
_____
_____

Step 4   Are there major cities close to these locations? What are they?

_____
_____
_____
_____
_____
_____
_____
_____
_____

NAME _____

DATE _____

The Earth Moves

Outline map of the world

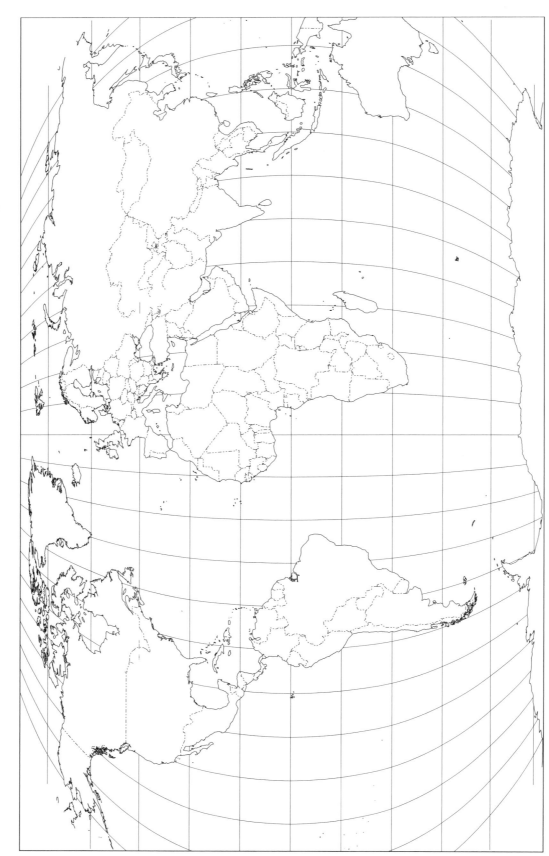

NAME _____   DATE _____

The Earth Moves
A GIS investigation

Answer all questions on the student answer sheet handout

Step 1   Start ArcMap

  a   Double-click the ArcMap icon on your computer's desktop.

  b   If the ArcMap start-up dialog appears, click **An existing map** and click OK. Then go
        to step 2b.

Step 2   Open the Global2.mxd file

a   In this exercise, a map document has been created for you. To open it, go to the File menu and choose **Open**.

b   Navigate to the module 2 folder (**C:\MapWorld9\Mod2**) and choose **Global2.mxd** (or **Global2**) from the list.

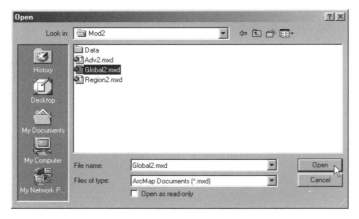

c   Click Open.

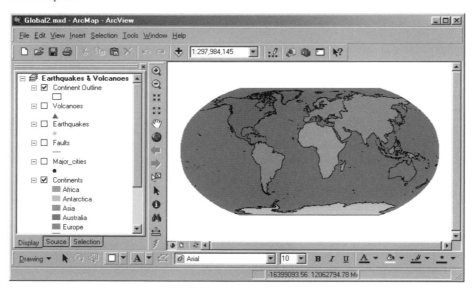

When the map document opens, you see a map with three layers turned on (Continent Outline, Continents, and Ocean). The check mark next to the layer name tells you the layer is turned on and visible in the map.

Step 3   Look at earthquake location data

In this step, you will compare your original theories about earthquake and volcano locations to actual data using GIS.

a   **Turn on the Earthquakes layer by clicking the box to the left of the name in the table of contents.**

This places a check mark in the box and adds a layer of points showing the locations of earthquakes on the map.

b   **Evaluate the map and write your answers to the following questions on the answer sheet:**

(1)   Do earthquakes occur in the places you predicted? List the regions you predicted correctly for earthquake locations.

(2)   What patterns do you see in the map?

Step 4   Sort and analyze earthquake magnitudes

You can take a closer look at the data behind the dots by looking at the attribute table of the Earthquakes layer. An attribute table contains specific information about the features in a layer. In the Earthquakes layer, each point represents an earthquake with a magnitude of greater than 5.0 on the Richter scale. In this step, you will use the attribute table to focus on the 15 strongest earthquakes.

a   **In the table of contents, right-click Earthquakes and click Open Attribute table.**

You see all the attribute data associated with the yellow earthquake points on the map.

Do not maximize this table. It will prevent you from viewing the map at the same time.

b   **Look at the table. Scroll down to see more records.**

Remember: Each record in this table represents one point on the map.

c   Click the field (column heading) labeled MAG to select it.

	OBJECTID*	Shape*	DATE_	LAT	LON	DEPTH	MAG
	67	Point	20000729	51.145	-179.328	49.1	5.6
	68	Point	20000730	33.933	139.35	10	5.5
	69	Point	20000730	-10.935	165.934	45.6	5.2
	70	Point	20000730	33.901	139.376	10	6.5
	71	Point	20000730	33.954	139.27	10	5.5
	72	Point	20000731	39.576	143.582	33	5.1
	73	Point	20000731	40.782	-29.517	10	5.2
	74	Point	20000731	-6.757	105.424	33	5.1
	75	Point	20000731	-16.697	174.542	10	6.1
	76	Point	20000731	-29.279	-176.35	10	6.1
	77	Point	20000801	-6.093	151.619	48.2	5.2
	78	Point	20000801	15.056	122.351	59.5	5.3

Attributes of Earthquakes

Record: |◄| ◄| 0 |►| ►| Show: All | Selected | Records (0 out of 1004 Selected.) | Options ▾

The heading appears depressed and the column is highlighted in blue when the field is selected. This field represents the magnitude of the earthquakes.

d   Scroll up to the top of the table. Now you will put the magnitudes in order from the largest magnitude to the smallest.

e   Right-click the MAG field heading and click Sort Descending.

The records have been rearranged from largest to smallest. Now you will select the 15 largest earthquakes.

f   Hold down the Ctrl key, click the small gray box to the left of the first record in the table, and drag your mouse until the first 15 records are highlighted in blue.

	OBJECTID*	Sh
	852	Point
	513	Point
	587	Point
	372	Point

Attributes of Earth

To make sure you have highlighted 15 earthquakes, look at the status bar at the bottom of the table. It should display:

Records (15 out of 1004 Selected.)

> *If you select too many records, click the Options button at the bottom of the table and click Clear Selection to clear the selections and try again.*

When you select a record in the attribute table, its point on the map will be highlighted also.

g   Move the attribute table out of the way so you can see where the 15 strongest earthquakes are located on the map. At this point, selected earthquakes may overlap, causing some selected earthquakes to "disappear." If this occurs, zoom in until you can see all the selected earthquakes.

> *Note: Refer to the ArcMap Toolbar Quick Reference for a brief explanation of the Zoom and Pan tools.*

?   *How do the 15 selected locations compare to your original paper map? List three ways.*

h   Click the Full Extent button to see the entire world on the map.

i   At the bottom of the attribute table, click the Options button and click Clear Selection.

> *Note: If you don't see the Options button, your table window may be too narrow. Widen your table window until you can see the Options button on the bottom right portion of the table.*

j   Close the attribute table.

Step 5   Look at volcano data

a   Turn off the Earthquakes layer and turn on the Volcanoes layer.

❓   *(1)   How do the volcano locations compare with your original predictions? List the regions of volcanic activity you predicted correctly.*

❓   *(2)   What patterns do you see in the volcano points and how do they compare with the earthquake patterns? (Hint: Turn the Earthquakes layer on and off.)*

The volcano data includes information on the status of each volcano (active, inactive, and potentially active). You will focus on the active volcanoes.

Step 6   Select all active volcanoes

a   In the table of contents, right-click Volcanoes and click Open Attribute Table.

The Type field of the table tells you if each volcano is Active, Potentially active, or Solfatara (emits gases, but is otherwise inactive).

b   Click on the Type field heading.

c   Right-click the Type field heading and click Sort Ascending. Scroll down and you will notice that there are many active volcanoes.

It would not be fun to highlight all of these as you did with the Earthquakes layer. This is a smart database—we can ask it to select all of the active volcanoes by using Select by Attributes.

d   At the bottom of the attribute table, click the Options button and click Select By Attributes.

e　In the Select by Attributes dialog, make sure Method is set to "Create a new selection." In the left-hand Fields list, double-click Type. Single-click the equals sign (=). On the right-hand side, click the Get Unique Values button. Three values appear in the Unique Values list. Double-click "Active" in this list.

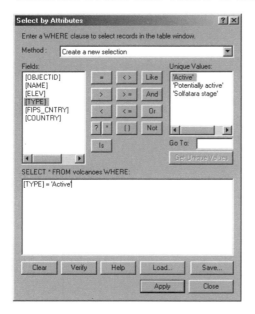

f　Near the bottom of the dialog, click Verify. If the expression is successfully verified, click OK.

　　Hint: If you receive a syntax error, check that your equation is exactly like the one in the graphic above. If it isn't, click Clear and try again.

g　At the bottom of the dialog, click Apply. All the active volcanoes are selected and highlighted blue.

h　Close the Select by Attributes dialog and close the attribute table to see the map. Use the Zoom and Pan tools to explore where the active volcanoes are located.

?　*(1)　Does this data provide any patterns that were not evident before? Identify those patterns.*

?　*(2)　Create a hypothesis as to why volcanoes and earthquakes happen where they do.*

Step 7　Identify the active volcanoes on different continents

In order to learn more about the active volcanoes, you can use the Identify tool.

a　Click the Identify tool. The Identify Results dialog displays. Click the Layers drop-down arrow and select Volcanoes in the list.

b　Move your cursor over the map display. Notice how the cursor has a small "i" next to it.

c Click on an active volcano on the map. The Identify Results dialog shows you the name of the volcano, its elevation, type, and country location. For example:

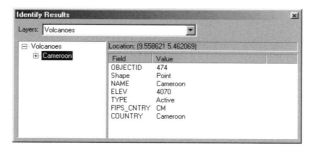

d Close the Identify Results window and zoom in to the continent of your choice.

e Use the Identify tool to find the name, elevation, activity level, and country location of three volcanoes.

Write that information on the answer sheet.

f Close the Identify Results dialog.

g In the table of contents, right-click Volcanoes, point to Selection, and click Clear Selected Features to unselect all of the active volcanoes.

h Click the Full Extent button to see the entire world on the map.

Step 8   **Add the plate boundaries layer**

The earth is always changing. The crust of the earth is composed of several tectonic plates that are always on the move. Movement occurs at the boundaries between the plates and on the surface of the plates themselves.

Based on the location of the earthquakes and volcanoes, where do you think the plate boundaries are? Draw them on your paper map.

There are four basic types of plate boundaries. In this part of the GIS activity, you will investigate where these boundaries exist and how they affect the landforms close to them. Here's a quick review of the different types of plate boundaries:

* *Divergent boundary.* One or two plates are splitting apart. New crust is being formed from the center of the earth, causing the plate to spread. Rift valleys are one example of this type of plate movement.
* *Convergent boundary.* Two plates are colliding, forcing one plate to dip down underneath another one. The plate that is folding under has old crust that is being destroyed, while the plate on top has mountains and volcanoes being formed. In the ocean, these appear as trenches.
* *Transform boundary.* Plates are sliding against each other, causing large faultlines and mountains to form. Here, the crust is neither created nor destroyed.
* *Plate boundary zones.* Plate boundaries appear erratic (zigzagged). Scientists believe there are actually microplates in these areas, but it is unclear what effect these zones have on the physical environment.

a Turn off the Volcanoes layer.

b Click the Add Data button.

c   Navigate to the module 2 data folder (**C:\MapWorld9\Mod2\Data**). Double-click **World2.mdb** to open it. Click **plates**.

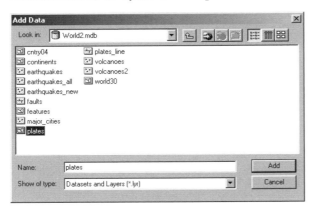

d   Click Add.

The plates layer is added to your table of contents.

e   In the table of contents, click the symbol beneath plates. The Symbol Selector opens.

f   On the right side, click the Fill Color down arrow and click No Color. This will create an outline of the plate boundaries.

g   Click the Outline Color down arrow. Pause your mouse over a color to see its name. Click the Electron Gold color.

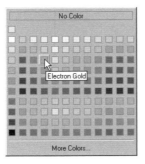

h   Increase the Outline Width to **2**.

i   Click OK. The gold outline symbol appears in the table of contents and on the map.

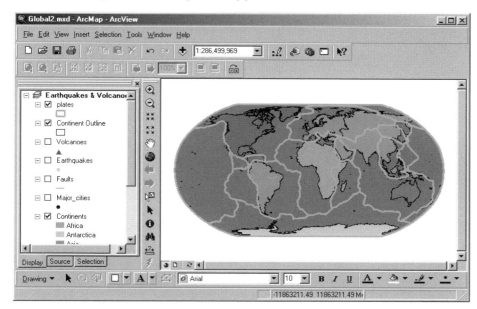

? Compare the actual plate boundaries to the ones you drew on your paper map. Record all similarities and differences.

Step 9   Add a layer file and an image

In order to get a closer look at landforms and boundaries, you will add two more layers:

- *Major Landforms.lyr:* a layer file containing major physical features of the planet.
- *earth_wsi.sid:* a color-shaded relief map of the earth, made from a satellite image.

a   Click the Add Data button, then click the Up One Level button to navigate to the Data folder. Double-click the LayerFiles folder to open it.

b   Double-click **Major Landforms.lyr** to add it to your map.

c   Click the Add Data button, then click the Up One Level button to navigate to the Data folder. Double-click the Images folder to open it.

d Click **earth_wsi.sid** and click Add.

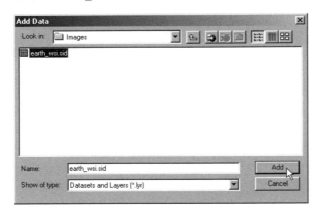

Any file with a .sid extension is a compressed image file. Please note that images do not have tables attached to them. Because earth_wsi.sid is an image file, the Identify tool will not call up data for the image.

e In the table of contents, click earth_wsi.sid and drag it above the Continents layer.

f Look at the map. Are there any areas where major landforms, plate boundaries, and seismic activity (earthquakes and volcanoes) overlap?

? *g* Use MapTips to find out the names of all major landforms formed at plate boundaries. Move the mouse pointer over a landform feature to see the MapTip. Write the names in the table on the answer sheet and label them on your paper map.

Predict how these features were formed by using your knowledge of plate boundaries (see the beginning of step 8 for a quick review).

First, you will identify the plates by labeling them in ArcMap.

? *h* In the table of contents, right-click on plates and click Label Features. Now, next to the name of each landform in the table on the answer sheet, write how you think the landform was created. The first one is completed for you.

Step 10   Identify major cities at high and low risk for seismic activity

Next, you will find cities with a high and low risk of earthquake and volcanic activity.

a   Turn off earth_wsi.sid. Move the Major_cities layer to the top of the table of contents and turn it on.

> ⊟ ⋐ **Earthquakes & Volcanoes**
>   ⊟ ☑ Major_cities
>   ●
>   ⊟ ☑ Major Landforms
>
>   ⊟ ☑ plates
>   ☐
>   ⊟ ☑ Continent Outline
>   ☐
>   ⊟ ☐ Volcanoes
>   ▲
>   ⊟ ☐ Earthquakes
>   ●
>   ⊟ ☐ Faults
>   —
>   ⊟ ☐ earth_wsi.sid

b   Use the Zoom, Pan and Identify tools to find the names of specific cities that are high-risk or low-risk for a seismic event. Write those names in the table on the answer sheet.

> *Hint: Remember to turn layers on and off as needed and to move the layers around in the table of contents.*

Step 11   Exit ArcMap

In this GIS investigation, you used different layers to determine where earthquake and volcanic activity is located around the world. From this information, you were able to determine cities at high or low risk for these natural disasters.

a   Ask your teacher for instructions on where to save this ArcMap document and on how to rename the map document.

b   If you are not going to save the map document, exit ArcMap by choosing Exit from the File menu. Click No when you are asked if you want to save changes to Global2.mxd (or Global2).

NAME _____   DATE _____

Student answer sheet
Module 2
Physical Geography I: Landforms and Physical Processes

Global perspective: The Earth Moves

Step 3   Look at earthquake location data

b-1   Do earthquakes occur in the places you predicted? List the regions you predicted correctly for earthquake locations.

_____

_____

_____

_____

_____

b-2   What patterns do you see in the map?

_____

_____

_____

_____

_____

Step 4   Sort and analyze earthquake magnitudes

g   How do the 15 selected locations compare to your original paper map? List three ways.

_____

_____

_____

_____

_____

Step 5   Look at volcano data

a-1   How do the volcano locations compare with your original predictions? List the regions of volcanic activity you predicted correctly.

_____

_____

_____

_____

_____

a-2   What patterns do you see in the volcano points and how do they compare with the earthquake patterns? (Hint: Turn the Earthquakes layer on and off).

_____
_____
_____
_____
_____
_____

Step 6   Select all active volcanoes

h-1   Does this data provide any patterns that were not evident before? Identify those patterns.

_____
_____
_____
_____
_____
_____

h-2   Create a hypothesis as to why volcanoes and earthquakes happen where they do.

_____
_____
_____
_____
_____
_____

Step 7   Identify the active volcanoes on different continents

e   Use the Identify tool to find the name, elevation, activity level, and country location of three volcanoes. Write that information in the space below.

_____
_____
_____
_____
_____
_____

Step 8   Add the plate boundaries layer

i   Compare the actual plate boundaries to the ones you drew on your paper map. Record all similarities and differences.

_____
_____
_____
_____
_____
_____

Step 9   Add a layer file and an image

g, h   Use MapTips to find out the names of all major landforms formed at plate boundaries. Write them below and label them on your paper map. Next to the name of each landform, write how you think the landform was created.

NAME OF LANDFORM	FORMED BY
Mid-Atlantic Ridge	The separation of South American and African plates

Step 10   Identify major cities at high and low risk for seismic activity

b   Find the names of specific cities that are high-risk or low-risk for a seismic event. Write those names in the table.

HIGH RISK	LOW RISK
1	1
2	2
3	3
4	4
5	5

NAME _____

The Earth Moves

DATE _____

Outline map of the world

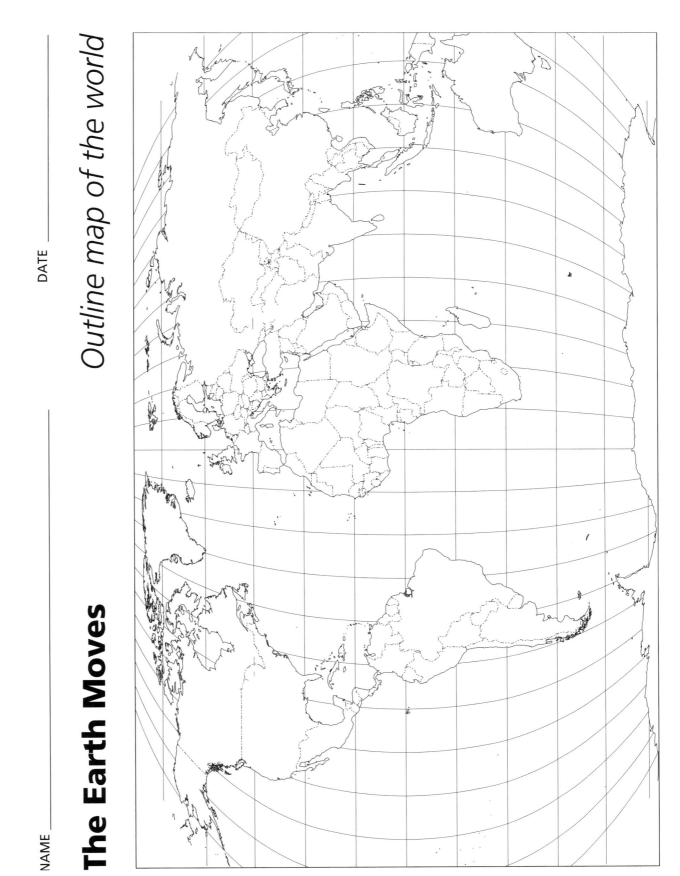

NAME _____   DATE _____

The Earth Moves
Middle school assessment

Part 1: Create an informative map

On the paper outline map provided, do the following three things:

1 Mark and identify all of the plate boundaries.

2 Identify three to five major landforms that are associated with the physical forces of volcanism and earthquakes by writing their locations on your map.

3 From the following list, label five world cities with high risk for a volcanic or seismic disaster.

San Francisco, USA	Bombay, India	Jakarta, Indonesia
Manila, Philippines	Bogotá, Columbia	Addis Ababa, Ethiopia
Mexico City, Mexico	Rome, Italy	Seattle, USA
Houston, Texas	Madrid, Spain	Tokyo, Japan
Hong Kong, China	Reykjavik, Iceland	
Sydney, Australia	Cairo, Egypt	

You may use the ArcMap map document and class notes to help you create this map. In the next section, you will be asked to explain why you chose the locations you did. Be sure to think carefully about your cartographic choices.

 Hint: You can use the Find tool in ArcMap to locate the world cities.

Part 2: Map analysis

On a separate piece of paper, answer each question below in paragraph form.

1 Describe the relationships you see between tectonic plate boundaries and areas at high risk for volcanic and seismic activity.

2 Explain why you selected *each* city on your map as a place at high risk for an earthquake or volcanic eruption.

3 Rank your five selected cities in the order of highest risk. The number one city will be the most at risk. Explain why you ranked them that way.

4 Compare and contrast your outline map from before the GIS investigation to the map you created in this assessment. How have your ideas changed?

The Earth Moves

Assessment rubric

Middle school

STANDARD	EXEMPLARY	MASTERY	INTRODUCTORY	DOES NOT MEET REQUIREMENTS
The student knows and understands how to use maps to analyze spatial distributions and patterns.	Uses GIS to analyze volcanic and earthquake data to identify five world cities most at risk for volcanic or seismic disasters, provides ample evidence for his or her choices, and draws conclusions on commonalities between the places.	Uses GIS to analyze volcanic and earthquake data to identify five world cities most susceptible to volcanic or earthquake disasters and provides ample evidence to support his or her decisions.	Uses GIS to identify four to five world cities most susceptible to volcanic or earthquake disasters and provides limited support for his or her decisions.	Uses GIS to identify some world cities susceptible to volcanic or earthquake disasters, but does not provide evidence to support his or her choices.
The student knows and understands how to analyze the physical characteristics of places.	Identifies at least five major landforms and the volcanic or seismic process that created them. Correctly identifies all plate boundaries.	Identifies three to five major landforms that were created because of volcanic or seismic activity.	Identifies two to three major landforms that were created by volcanic or seismic activity.	Identifies one major landform created by volcanic or earthquake activity.
The student knows and understands how physical processes shape patterns in the physical environment.	Clearly describes the relationship between zones of high earthquake and volcanic activity and the location of tectonic plate boundaries through the use of a variety of media.	Describes the relationship between zones of high earthquake and volcanic activity and the location of tectonic plate boundaries.	Provides limited evidence of the relationship between zones of high earthquake and volcanic activity and the location of tectonic plate boundaries.	Does not show evidence of understanding the relationship between zones of high earthquake and volcanic activity and the location of tectonic plates.
The student knows and understands how natural hazards affect human activities.	Ranks top five high-risk cities and provides ample evidence, supported with outside sources, for his or her decisions.	Ranks top five high-risk cities and provides a clear explanation for his or her choices.	Ranks the top five cities according to degree of risk, but does not provide an explanation for his or her decision.	Does not rank the cities in any particular order. Does not provide any explanation.

This is a four-point rubric based on the National Standards for Geographic Education. The "Mastery" level meets the target objective for grades 5–8.

NAME _____   DATE _____

The Earth Moves
High school assessment

Part 1: Create an informative map

On the paper outline map provided, do the following three things:

1   Mark and identify all of the plate boundaries.
2   Identify three to five major landforms that are associated with the physical forces of volcanism and earthquakes by writing their locations on your map.
3   From the following list, identify five world cities with high risk for a volcanic or seismic disaster:

San Francisco, USA	Bombay, India	Jakarta, Indonesia
Manila, Philippines	Bogotá, Columbia	Addis Ababa, Ethiopia
Mexico City, Mexico	Rome, Italy	Seattle, USA
Houston, USA	Madrid, Spain	Tokyo, Japan
Hong Kong, China	Reykjavik, Iceland	
Sydney, Australia	Cairo, Egypt	

You may use the ArcMap map document and class notes to help you create this map. In the next section, you will be asked to explain why you chose the locations you did. Be sure to think carefully about your cartographic choices.

   Hint: You can use the Find tool in ArcMap to locate the world cities.

Part 2: Map analysis

On a separate piece of paper, answer each question below in paragraph form.

1   Provide evidence for *each* of the cities you selected as high risk for a major earthquake or volcanic eruption.
2   Choose one of the cities on your map and research how this city is prepared (or not prepared) for a seismic disaster. Write a complete paragraph outlining the city's preparedness.
3   For each of the major landforms you selected, hypothesize how you think they formed. Use your knowledge of the four major types of plate boundaries to answer this question.

The Earth Moves

Assessment rubric

High school

STANDARD	EXEMPLARY	MASTERY	INTRODUCTORY	DOES NOT MEET REQUIREMENTS
The student knows and understands how to use geographic representations and tools to analyze and explain geographic problems.	Uses GIS to analyze volcanic and earthquake data to identify five or more world cities at high risk for volcanic or earthquake activity, and provides a detailed explanation of how he or she came to these conclusions.	Uses GIS to analyze volcanic and earthquake data to identify five world cities at high risk for volcanic or earthquake activity, and provides a brief narrative stating the logic for their conclusions.	Uses GIS to analyze volcanic and earthquake data to identify three to five world cities at high risk for volcanic or earthquake activity, and provides little explanation for their conclusions.	Uses GIS to analyze volcanic and earthquake data to identify one to five cities, but does not provide any explanation for their conclusions.
The student knows and understands the spatial variation in the consequences of physical processes across Earth's surface.	Clearly explains the correlation between the spatial distribution of volcanoes and earthquakes in relation to plate boundaries, and ties this in to their understanding of physical landforms.	Explains the correlation between the spatial distribution of volcanoes and earthquakes in relation to plate boundaries.	Identifies the correlation between volcanoes, earthquakes, and plate boundaries, but has difficulty explaining the relationship.	Identifies the relationship between the spatial distribution of volcanoes and earthquakes, but does not correlate them to the location of plate boundaries.
The student knows and understands the changing physical characteristics of places.	Clearly describes through various media how the physical process of plate tectonics shape five major landforms.	Describes with ample evidence how the physical process of plate tectonics shaped three to five major landforms.	Describes in some detail how the physical process of plate tectonics shaped two to three major landforms.	Identifies a few major landforms, but provides little description of how plate tectonics shaped them.
The student knows and understands how humans perceive and react to natural disasters.	Researches and explains how several of their high-risk cities have prepared for a major volcanic or seismic event.	Researches and explains how two of their high-risk cities have prepared for a major volcanic or seismic event.	Researches and provides limited explanation for how one of their high-risk cities has prepared for a major seismic or volcanic event.	Provides some explanation of how a high-risk city has prepared for a major seismic or volcanic event, but does not provide research to support it.

This is a four-point rubric based on the National Standards for Geographic Education. The "Mastery" level meets the target objective for grades 9–12.

Life on the Edge
A regional case study of East Asia

Lesson overview

Students will investigate the Pacific Ocean's "Ring of Fire," with particular focus on earthquake and volcanic activity in East Asia, where millions of people live with the daily threat of significant seismic or volcanic events. Through the analysis of volcanic location and earthquake depth, students will identify zones of subduction at tectonic plate boundaries and the location of populations in the greatest danger of experiencing a volcanic eruption or a major earthquake.

Estimated time   Two to three 45-minute class periods

Materials
✔ Colored pencils

✔ Student handouts from this lesson to be copied:
  • Map of East Asia (page 85)
  • GIS Investigation sheets (pages 87 to 93)
  • Student answer sheets (pages 95 to 96)
  • Assessment(s) (pages 97 to 101)

Standards and objectives

National geography standards

GEOGRAPHY STANDARD	MIDDLE SCHOOL	HIGH SCHOOL
1 How to use maps and other geographic representations, tools, and technologies to acquire, process, and report information from a spatial perspective	The student knows and understands how to use maps and databases to analyze spatial distributions and patterns.	The student knows and understands how to use geographic representations and tools to analyze and explain geographic problems.
4 The physical and human characteristics of places	The student knows and understands how to analyze the physical characteristics of places.	The student knows and understands the changing physical characteristics of places.
7 The physical processes that shape the patterns of Earth's surface	The student knows and understands how to predict the consequences of physical processes on Earth's surface.	The student knows and understands the spatial variation in the consequences of physical processes across Earth's surface.
15 How physical systems affect human systems	The student knows and understands how natural hazards affect human activities.	The student knows and understands how humans perceive and react to natural disasters.

Standards and objectives (continued)

Objectives

The student is able to:

- Locate zones of significant earthquake and volcanic activity in East Asia.
- Describe the relationship between zones of high earthquake activity, volcanic activity, and the location of tectonic plate boundaries.
- Identify subduction zones along plate boundaries.
- Identify densely populated areas that are most at risk for volcanic and/or seismic disasters.

GIS skills and tools

  Add layers to the map

  Measure distances between points on the map

  Identify a feature to learn more about it

      Zoom in and out of the map

  Pan the map to view different areas

  Find a specific feature

- Label features on the map
- Turn layers on and off
- Change the order of the table of contents to change the map display

For more on geographic inquiry and these steps, see Geographic Inquiry and GIS (pages xxiii to xxv).

Teacher notes

Lesson introduction
First, provide a brief overview of the region of East Asia and the Ring of Fire. Emphasize East Asia's dense population and its place in the Ring of Fire. Distribute the map handout to each student. The map contains outlines of countries, plate boundaries, and major cities in East Asia.

Ask your students to draw outlines of the areas where they believe there is the greatest risk for a major geophysical disaster. Give the students about three to five minutes to make their initial predictions. If time permits, ask some students to share their predictions with the class, or break out into small groups for discussion. Later in the lesson, students will be asked to make a hazards zone map of the Ring of Fire.

Student activity
 Before completing this lesson with students, we recommend that you work through it yourself. Doing so will allow you to modify the activity to accommodate the specific needs of your students.

After the initial discussion, students will work on the computer component of the lesson. Distribute the student GIS Investigation sheets, which provide step-by-step instructions for the activity. In this investigation, students will analyze several bits of data including information on population density, plate boundaries, volcanoes, and earthquake activity for the year 2000. They will identify areas of subduction along plate boundaries based on characteristics of these zones.

In addition to detailed instructions, the worksheets include questions to help students focus on key concepts. Some questions will have specific answers, while others will require creative thought. Students will also add information to their outline map of East Asia.

Things to look for when the students are working on this activity:
- Are the students answering the questions as they work through the procedure?
- Are the students referring to their original maps and notes from class discussion?
- Do some students complete the exercise quickly? If yes, refer to the Extensions section for ideas on how to make the GIS Investigation more challenging.

Conclusion
After the GIS Investigation, lead a class discussion that compares student predictions and student findings after the exercise.

How closely did their predictions match what they learned?
- Compare student observations and questions generated by their investigations of the maps.
- Have students identify what data was most important in determining hazard zones for this region. One way to do this is to list types of data on the board and have the students rank them in order of importance.
- Discuss why some data was more helpful than other data.
- Ask the students if there is other data that they think would be helpful to them and have them explain why.

Assessment

Middle School: Highlights skills appropriate to grades 5 through 8

Part 1 of the Life on the Edge middle school assessment asks students to create a Hazards Map of East Asia, identifying zones of high and low risk for volcanic and seismic events. In addition, students are required to create a legend for their unique map. Part 2 of the assessment requires students to analyze their maps when answering the four questions. An assessment rubric to evaluate student performance is provided at the end of the lesson.

High School: Highlights skills appropriate to grades 9 through 12

Part 1 of the Life on the Edge assessment for high school students asks them to create a Hazards Map of East Asia, identifying zones of low, medium, and high risk for volcanic and seismic events. In addition, students are required to create a detailed legend for their unique map. Part 2 of the assessment requires students to analyze their maps when answering a set of three questions. One of the questions asks students to develop an emergency action plan for a major city in a high-risk zone for a volcanic or seismic event. An assessment rubric to evaluate student performance is provided at the end of the lesson.

Extensions

- Students can select a city within East Asia and create a buffer zone around a volcano to determine the area of potential damage if there is a major eruption. They can then create a disaster plan for that city in the event that such an eruption occurs.
- Students can create a new layer that points out locations of natural hazards in their local area.
- Analyze historical data for the most notable earthquakes and volcanic eruptions and create new layers from this data. Students can then make predictions on the next significant eruption or seismic event for East Asia.
- Students can create their own digital maps for the assessment. By using ArcMap, they can mark all of the required features and print the map using the layout view.
- Students can present their maps to the class, comparing their initial hazards maps to their final theories.
- Check out the Resources by Module section of this book's companion Web site (*www.esri.com/mappingourworld*) for print, media, and Internet resources on the topics of earthquakes, volcanoes, plate tectonics, and East Asia.

NAME _____   DATE _____

Life on the Edge
A map investigation

East Asia sits on the western edge of the Ring of Fire. This region experiences an enormous amount of geophysical activity both seismic and volcanic in nature. On the outline map of East Asia, use your colored pencils to outline the areas where you think the greatest risk of volcanic or earthquake disasters is. The map includes country borders, some major cities, and thicker outlines that indicate plate boundaries.

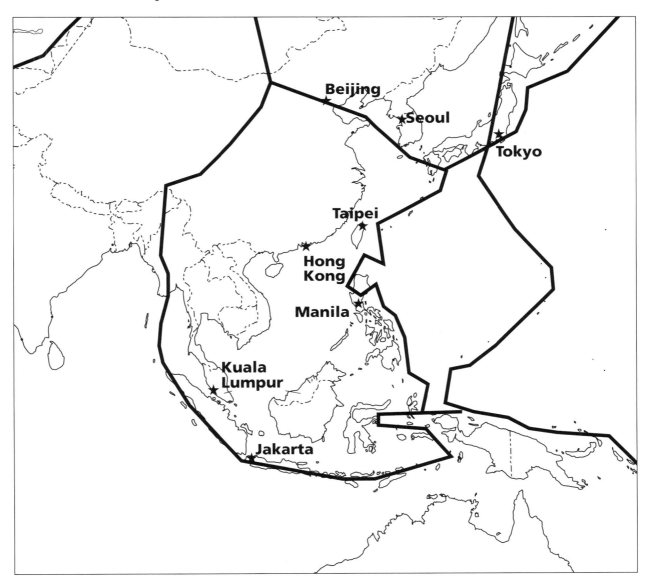

NAME _____   DATE _____

Life on the Edge
A GIS investigation

ACQUIRE

ASK

EXPLORE

ACT

ANALYZE

Answer all questions on the student answer sheet handout

In this investigation, you will investigate the Pacific Ocean's "Ring of Fire," with particular focus on earthquake and volcanic activity in East Asia, where millions of people live with the daily threat of significant seismic or volcanic events.

Step 1   Start ArcMap

   a   Double-click the ArcMap icon on your computer's desktop.

ArcMap

   b   If the ArcMap start-up dialog appears, click **An existing map** and click OK. Then go to step 2b.

Step 2   Open the Region2.mxd file and look at cities data

 a   In this exercise, a map document has been created for you. To open it, go to the File
        menu and choose **Open**.

 b   Navigate to the module 2 folder (**C:\MapWorld9\Mod2**) and choose **Region2.mxd**
        (or **Region2**) from the list.

 c   Click Open.

        When the map document opens, you see a map with two layers turned on (Major
        Cities and Countries). The check mark next to the layer name tells you the layer is
        turned on and visible in the map.

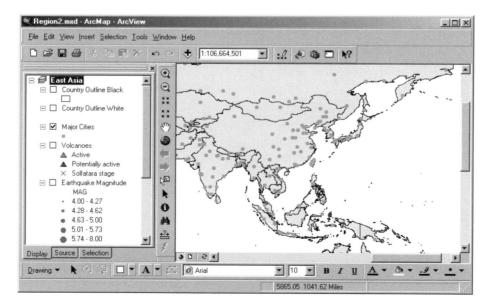

In order to identify the name of each city marked on the map, you can use the Identify tool.

d   Click the Identify tool. An Identify Results dialog appears.

e   In the Identify Results dialog, click the Layers drop-down arrow and click Major Cities in the list. Move the cursor over the map. Notice how the cursor has a small "i" next to it.

f   Click on any green dot on the map.

The Identify Results dialog appears similar to the graphic below:

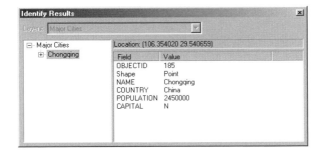

> *The right side of the Identify Results window tells you the name of the city, the country it's in, its population, and whether or not it's the capital city. In the example above, N = No (Chongqing is not the capital of China).*

g   Use the Identify tool to locate one city within India and Japan. Record each city's name and population in the table on your answer sheet.

h   When you are finished, close the Identify Results window by clicking the × in the upper right corner of the window.

Step 3   Look at population density

a   Scroll down in the table of contents until you see Population Density. (If necessary, widen the table of contents to see the Population Density legend.)

b   Turn on the Population Density layer by clicking the box next to its name.

A check mark appears and the layer is drawn on the map. The darker areas on the map represent areas of high population density. In this layer, density is measured by the number of people per square kilometer of space.

c   Use the Identify tool to locate two world cities in East Asia in areas where the population density is greater than 200 people per square kilometer.

Record the names of these two cities.

Step 4 Look at earthquake magnitudes

a Turn on the Earthquake Magnitude layer.

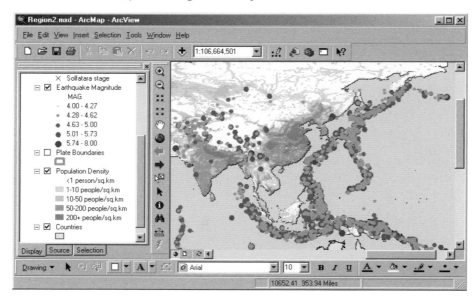

This adds a layer of graduated points to the map and displays earthquake locations according to magnitude. The larger dots represent greater earthquake magnitudes.

> *Note: Only earthquakes with a magnitude greater than 4.0 on the Richter scale are included in this layer.*

(1) *In general, where did the large earthquakes occur?*

(2) *Did large earthquakes occur near densely populated areas? Where?*

Step 5 Measure the distance between active volcanoes and nearby major cities

a Turn off the Earthquake Magnitude layer.

b Turn on the Volcanoes layer.

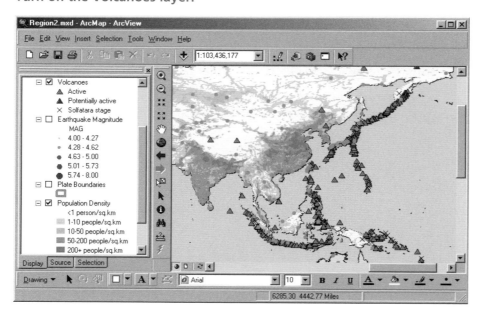

There are three types of volcanoes identified by this layer: active, potentially active, and solfatara state (venting primarily hot gases). Note the different symbols used for each type of volcano.

c   Choose a major city that's located near an active volcano. Click the Zoom In tool and then click your chosen city's dot on the map six times. Your map is now zoomed in and centered on that city.

You will use the Measure tool to measure the distance between the active volcano and the major city you selected.

d   Click the Measure tool.

e   Click the volcano that is close to your chosen major city. Now move your mouse pointer over to the city. A line is attached from the point where you first clicked to where you move your mouse pointer.

f   Once you have reached the city, **double-click** to end the line.

In the lower left corner of your ArcMap window, you see the distance you measured.

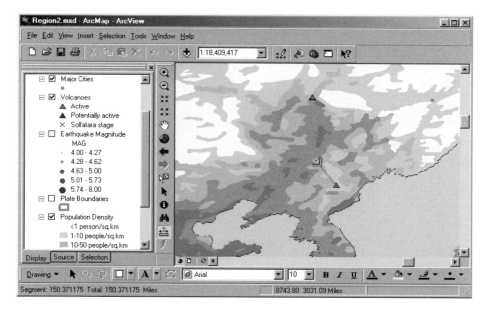

g   Use the Measure tool to measure the distance from other cities to nearby active volcanoes.

? *(1)*   *Are there many active volcanoes located close to highly populated areas? What is the closest distance you found? Record the name of the volcano and the city and their distance apart.*

? *(2)*   *What patterns do you see in the volcano points and how do they compare with the earthquake patterns? (Hint: You should turn the Earthquake Magnitude layer on and off.)*

You will use a bookmark to zoom back out to the East Asia region.

h   At the top of the ArcMap window, click View, point to Bookmarks, and click East Asia. This takes you back to the view of East Asia.

Step 6   Look at plate boundaries

a   Turn off all layers except Countries. Turn on the Plate Boundaries layer.

Now you will label the plates in the East Asia region.

b   In the table of contents, right-click Plate Boundaries and click Label Features. All the plates are labeled on the map.

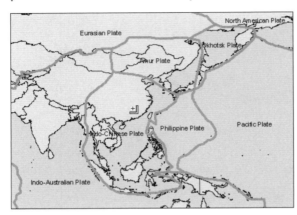

c   Record the labels on your paper outline map.

Many of the plate boundaries in the Pacific Rim have areas called subduction zones. These are places along plate boundaries where one plate is diving underneath another one. Subduction zones can be identified by underwater trenches and island arcs that are formed at these boundaries.

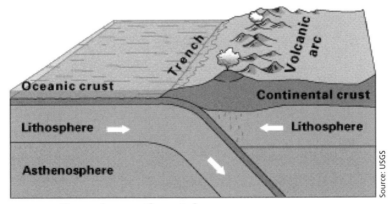

Oceanic-continental convergence

To get a closer look at landforms and boundaries, you will add the earth_wsi.sid image. This is a composite satellite image of the earth. It will allow you to see the detailed physical features of the earth.

Step 7   Add an image file

   a   Click the Add Data button and navigate to your Images folder (**C:\MapWorld9 \Mod2\Data\Images**).

b   Select earth_wsi.sid and click Add to add it to your map.

c   It displays beneath the other layers because it is at the bottom of the table of contents.

d   Look on the map to determine where subduction is occurring. Remember: Subduction zones are characterized by deep trenches and volcanic island arcs, and occur along plate boundaries.

> *Hint: You may need to turn the Plate Boundaries layer on and off.*

e   On your paper map, draw the zones of subduction.

Step 8   Investigate your map

a   Use the Zoom and Pan tools to focus on different areas of the map to get a closer look at all the physical features.

> *Note: Refer to the ArcMap Toolbar Quick Reference for a brief explanation of the Zoom and Pan tools.*

b   Use the Identify tool to find out the names of volcanoes or cities near them. To use the tool, click the Identify tool and click an object on the map.

Remember: The map is drawn beginning with the layer listed at the bottom of the table of contents first and ending with the layer listed at the top. Move the layers around in the table of contents by clicking and dragging.

c   Check with your teacher if there is an area of East Asia that you need to explore. When you're finished exploring the map, proceed to step 9.

Step 9   Exit ArcMap

In this GIS investigation, you used different layers to determine where earthquake and volcanic activity is located around the world. From this information, you were able to determine cities at high or low risk for these natural disasters.

a   Ask your teacher for instructions on where to save this ArcMap document and on how to rename the map document.

b   If you are not going to save the map document, exit ArcMap by choosing Exit from the File menu. Click No when you are asked if you want to save changes to Region2.mxd (or Region2).

PHOTOCOPY

NAME _____   DATE _____

Student answer sheet
Module 2
Physical Geography I: Landforms and Physical Processes

Regional case study: Life on the Edge

Step 2   Open the region2.mxd file and look at cities data

g   Use the Identify tool to locate one city within each country listed in the table below and record that city's population.

CITY NAME	COUNTRY NAME	CITY POPULATION
Kunming	China	1,280,000
	India	
	Japan	

Step 3   Look at population density

c   Use the Identify tool to locate two world cities in East Asia in areas where the population density is greater than 200 people per square kilometer. Record below.

WORLD CITIES THAT HAVE A POPULATION DENSITY GREATER THAN 200 PEOPLE PER SQUARE KILOMETER

Step 4   Look at earthquake magnitudes

a-1   Where did the largest earthquakes occur?

_____

_____

_____

_____

_____

a-2   Did large earthquakes occur near densely populated areas? Where?

_____

_____

_____

_____

_____

Step 5   Measure the distance between active volcanoes and nearby major cities

g-1   Are there many active volcanoes located close to highly populated areas? What is the closest distance you found? Record the name of the volcano and the city, and their distance apart.

_____

_____

_____

_____

_____

g-2   What patterns do you see in the volcano points and how do they compare with the earthquake patterns? (Hint: You should turn the Earthquake Magnitude layer on and off.)

_____

_____

_____

_____

_____

NAME _____   DATE _____

Life on the Edge
Outline map of East Asia

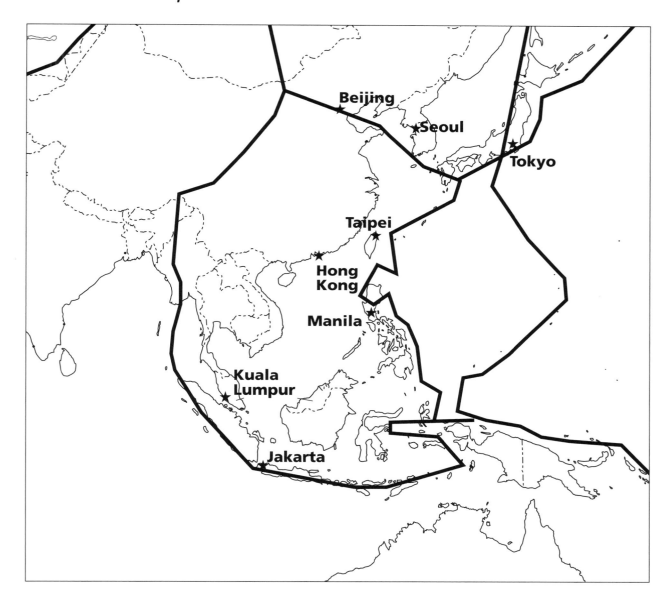

NAME _____   DATE _____

Life on the Edge
Middle school assessment

Part 1: Create a hazards map

On the paper outline map provided, do the following three things:
1 Mark and identify zones at high and low risk for volcanic activity.
2 Mark and identify zones at high and low risk for seismic activity.
3 Create a map legend that identifies the four zones described above.

You may use the ArcMap map document and class notes to help you create this map and determine zones of high and low risk. In the next section, you will be asked to explain why you chose the locations you did. Be sure to think carefully about your cartographic choices.

 Hint: You can use the Find tool in ArcMap to quickly locate different world cities.

Part 2: Map analysis

On a separate piece of paper, answer each question below in paragraph form:
1 What criteria did you use to define your hazard zones for volcanic and earthquake activities?
2 Describe the relationships you see between tectonic plate boundaries and areas at high risk for volcanic and seismic activity.
3 Describe the landforms that are located in and around the Pacific Rim. Identify their common characteristics.
4 Choose a city in one of the high-risk zones. Develop an emergency action plan for that city. Be sure to take into consideration the population of the city and possible evacuation plans.

Life on the Edge

Assessment rubric

Middle school

STANDARD	EXEMPLARY	MASTERY	INTRODUCTORY	DOES NOT MEET REQUIREMENTS
The student knows and understands how to use maps and databases to analyze spatial distributions and patterns.	Uses GIS to analyze volcanic and earthquake data to create an original hazards map, identifying zones of high and low volcanic and seismic activity.	Uses GIS to analyze volcanic and earthquake data to identify zones of significant seismic and volcanic activity and compares this with his or her original findings.	Uses GIS to identify the locations of major volcanoes and earthquakes.	Has difficulty correctly identifying major volcanoes and earthquakes.
The student knows and understands how to analyze the physical characteristics of places.	Describes in detail the characteristics of the Pacific Rim landforms and identifies the physical processes that formed them.	Clearly describes the characteristics of landforms of the Pacific Rim.	Describes the characteristics of the Pacific Rim landforms using little detail.	Identifies landforms of the Pacific Rim, but does not provide any specific characteristics.
The student knows and understands how physical processes shape patterns in the physical environment.	Clearly describes the relationship between zones of high earthquake and volcanic activity and the location of tectonic plate boundaries through the use of a variety of media.	Describes the relationship between zones of high earthquake and volcanic activity and the location of tectonic plates.	Provides limited evidence of the relationship between zones of high earthquake and volcanic activity and the location of tectonic plate boundaries.	Does not show evidence of understanding the relationship between zones of high earthquake and volcanic activity and the location of tectonic plates.
The student knows and understands how natural hazards affect human activities.	Using a variety of media, the student illustrates use of higher-order thinking to determine areas at high risk of human loss due to geographic hazards in East Asia.	Uses higher-order thinking to determine areas at high risk of human loss due to geographic hazards in East Asia, and provides a clear explanation.	Identifies areas of risk for possible human loss, but does not clearly explain the relationship between densely populated areas and geographic hazards of East Asia.	Identifies areas at risk for loss of human life in the region of East Asia, but does not provide explanation for his or her decision.

This is a four-point rubric based on the National Standards for Geographic Education. The "Mastery" level meets the target objective for grades 5–8.

NAME _____   DATE _____

Life on the Edge
High school assessment

Part 1: Create a hazards map

On the paper outline map provided, do the following four things:

1. Mark and identify zones of high, medium, and low risk for volcanic activity.
2. Mark and identify zones of high, medium, and low risk for seismic activity.
3. Create a complete map legend.
4. Include all major cities, plate boundaries, and subduction zones.

Before you draw your map, you need to determine the criteria for each level of risk (low, medium, and high). Use the following items to determine the different levels of risk:

1. Population density
2. Magnitude of earthquakes from year 2000 data
3. Location of active volcanoes and their proximity to major population centers
4. Location of plate boundaries and subduction zones

You may use the ArcMap map document and class notes to help you create this map and determine zones of high and low risk. In the next section, you will be asked to explain why you chose the locations you did. Be sure to think carefully about your cartographic choices.

    Hint: You can use the Find tool in ArcMap to quickly locate different world cities.

Part 2: Map analysis

On a separate piece of paper, answer each question below in paragraph form:

1. Define each level of hazard risk (high, medium, and low) you used in your map. Be sure to include each factor used to determine the different levels of risk.
2. Describe the relationships you observe between tectonic plate boundaries and areas at high risk for volcanic and seismic activity.
3. Explain how subduction zones affect the physical features of at least three places in the Pacific Rim.
4. Choose a major city in a high-risk zone. Develop an emergency action plan for the city. Be sure to take into consideration the population of the city and possible evacuation plans.

Life on the Edge

Assessment rubric

High school

STANDARD	EXEMPLARY	MASTERY	INTRODUCTORY	DOES NOT MEET REQUIREMENTS
The student knows and understands how to use geographic representations and tools to analyze and explain geographic problems.	Uses GIS to analyze volcanic and earthquake data to identify five or more world cities at high risk for volcanic or earthquake activity, and provides a detailed explanation of how he/she came to these conclusions.	Uses GIS to analyze volcanic and earthquake data to create an original hazards map of zones of high, medium, and low risk for volcanic and seismic activity and provides a brief narrative stating the logic of their choice.	Uses GIS to analyze volcanic and earthquake data to create an original hazards map of zones of high, medium, and low risk for volcanic and seismic activity and provides little explanation of how the map was created.	Uses GIS to analyze volcanic and earthquake data, but does not correctly identify zones of various levels of activity.
The student knows and understands the spatial variation in the consequences of physical processes across Earth's surface.	Clearly explains the correlation between the spatial distribution of volcanoes and earthquakes in relation to plate boundaries, and incorporates it into their understanding of physical landforms.	Explains the correlation between the spatial distribution of volcanoes and earthquakes in relation to plate boundaries.	Identifies the correlation between volcanoes, earthquakes, and plate boundaries, but has difficulty explaining the relationship.	Identifies the relationship between the spatial distribution of volcanoes and earthquakes, but does not correlate them to the location of plate boundaries.
The student knows and understands the changing physical and human characteristics of places.	Correctly identifies all areas of subduction in East Asia and explains in detail through words and images how this affects the physical aspects of at least three places in the region.	Correctly identifies all areas of subduction in East Asia, and explains how this affects the physical aspects of three places in the region.	Correctly identifies all areas of subduction in East Asia and explains how this affects the physical aspects of two to three places in the region.	Correctly identifies a few areas of subduction in East Asia and does not explain their effects on the physical environment.
The student knows and understands how humans perceive and react to natural disasters.	Using a variety of media, illustrates use of analytical thinking to formulate a thorough emergency action plan for a major city at high risk for a seismic event.	Uses analytical thinking to formulate an emergency action plan for a major city at high risk for a seismic event.	Identifies a major city at high risk for a seismic event and formulates an emergency action plan that lacks detail and evidence of higher-order thinking.	Identifies a major city at high risk for a seismic event, but does not include an emergency action plan.

This is a four-point rubric based on the National Standards for Geographic Education. The "Mastery" level meets the target objective for grades 9–12.

Mapping Tectonic Hot Spots
An advanced investigation

Lesson overview

This lesson is designed to target the self-motivated, independent GIS students in your class. They will use the Internet to acquire the most recent data on earthquakes and volcanoes worldwide. By exploring this data through a GIS, students will construct a current World Tectonic Hot Spots map. This lesson could be an independent research project completed over several days.

Estimated time   Three to four 45-minute class periods

Materials   ✔ Student handouts from this lesson to be copied:
- Volcano data sheet (page 106)
- GIS Investigation sheets (pages 107 to 112)

Standards and objectives   *National geography standards*

	GEOGRAPHY STANDARD	MIDDLE SCHOOL	HIGH SCHOOL
1	How to use maps and other geographic representations, tools, and technologies to acquire, process, and report information from a spatial perspective	The student knows and understands how to use maps and databases to analyze spatial distributions and patterns.	The student knows and understands how to use technologies to represent and interpret Earth's physical systems.
7	The physical processes that shape the patterns of Earth's surface	The student knows and understands how physical processes shape patterns in the physical environment.	The student knows and understands the spatial variation in the consequences of physical processes across Earth's surface.
15	How physical systems affect human systems	The student knows and understands how natural hazards affect human activities.	The student knows and understands how humans perceive and react to natural disasters.
18	How to apply geography to interpret the present and plan for the future	The student knows and understands how the interaction of physical and human systems may shape present and future conditions on Earth.	The student knows and understands how to use geographic perspectives to analyze problems and make decisions.

Objectives

The student is able to:

- Use the Internet to locate data for recent earthquake and volcanic activity.
- Use a GIS to map data acquired from the Internet.
- Compare the pattern of recent earthquake and volcanic events with the patterns of seismic events in the past.
- Identify world regions and cities where human populations face the greatest threat from current seismic events.

GIS skills and tools

- Locate data on the Internet and prepare it in a comma-delimited text file
- Add event layers
- Convert event layers to geodatabase feature classes
- Classify data according to different attributes
- Create a layout
- Print a presentation-quality map

For more on geographic inquiry and these steps, see Geographic Inquiry and GIS (pages xxiii to xxv).

Teacher notes

Lesson introduction

Lead a class discussion that addresses the following questions:

- Are you aware of any earthquake or volcanic activity on the earth within the last month?
- Write down the number of earthquakes you believe occurred throughout the world or in a specific area within the last month.

Explain to your students that they will use the Internet to find data on the most recent seismic events and they will use a GIS to map that data. This exercise will require the student to work independently, with little guidance.

Student activity

 Before completing this lesson with students, we recommend that you complete it as well. Doing so will allow you to modify the activity to accommodate the specific needs of your students.

Distribute this lesson's student GIS Investigation sheets, along with the volcano data sheet handout. Students should follow the guidelines to retrieve earthquake and volcano data from the Internet. They will create text files of that data and use them to create point layers of recent earthquake and volcano events in ArcMap. Once the data is in ArcMap, students will be able to perform different analyses on it.

Conclusion

Before beginning the assessment, engage students in a discussion of patterns of recent seismic activity revealed by their maps. Do the mapped events occur along known plate boundaries? Are some boundary zones more active at the present time than others? Are any patterns revealed by the current data? Where is future research and analysis needed?

Assessment

Students will create a layout titled "Tectonic Hot Spots, (Month, Year)." For example, Tectonic Hot Spots, August 2004 would be an appropriate title for data collected during August 2004. In addition to the map, you may wish to ask students to submit a written report that summarizes their observations and explains their analysis of the data. A written report could include any or all of the following points:

- Briefly summarize the earthquake and volcano activity for the thirty-day period.
- Were you surprised at the frequency or the location of tectonic activity in the past month? Why or why not?
- Compare the earthquake and volcano activity in the thirty-day period with the earthquakes 2000 data from the Earthquakes theme. What relationships can be observed between these two datasets?

 Note: Due to the independent nature of this lesson, there is no supplied assessment rubric. You are free to design assessment rubrics that meet the needs of your specific adaptation.

Extensions

- Have the students investigate their local area to find out if there were ever any volcanic or seismic activities in their hometown. Identify how these events (even if they were thousands of years ago) helped to shape the landscape of their area.
- Find the closest active volcano and/or closest fault line to your town. How does this affect your community and the communities that are closest to the volcano/fault line?
- Conduct research into current films and literature to find books and movies that deal with volcanoes and/or earthquakes. Do these films paint a realistic picture based on the research the students have conducted?
- Check out the Resources by Module section of this book's companion Web site *(www.esri.com / mappingourworld)* for print, media, and Internet resources on the topics of earthquakes, volcanoes, and plate tectonics.

NAME _____

DATE _____

Mapping Tectonic Hot Spots

Volcano data sheet

VOLCANO	COUNTRY	LATITUDE (DD)	LONGITUDE (DD)	ELEVATION (M)	DATE OF ACTIVITY

NAME _____   DATE _____

Mapping Tectonic Hot Spots
An advanced investigation

 Note: Due to the dynamic nature of the Internet, the URLs listed in this lesson may have changed, and the graphics shown below may be out of date. If the URLs do not work, refer to this book's Web site for an updated link: www.esri.com/mappingourworld.

In this investigation, you will use the Internet to acquire the most recent data on earthquakes and volcanoes worldwide. By exploring this data through a GIS, you will construct a current World Tectonic Hot Spots map.

Step 1   **Locate earthquake data on the World Wide Web**

The World Wide Web is a great source for geographic data. In this step, you will go to the U.S. Geological Survey (USGS) Web site to find earthquake data.

a   **Open your Web browser and go to the USGS National Earthquake Information Center at gldss7.cr.usgs.gov/neis/epic/epic.html.**

 ***Note: Be careful to copy the entire URL into your browser.***

b   For Search area, select **Global (Worldwide)**.

c   For Output File Type, select **Spreadsheet format (comma delimited)**.

d   For Search Parameters, select **USGS/NEIC (PDE) 1973 - Present**.

e   For Optional Search Parameters, type the range of dates for the preceding 30 days. (For example, if today is August 23, request data from 2004/7/23 to 2004/8/23). Set the Minimum Magnitude to 5 and the Maximum Magnitude to 9.

```
Optional Search Parameters:

Date

    2004    Starting Year 7    Starting Month 23    Starting Day
    2004    Ending Year  8     Ending Month  23     Ending Day

Magnitude

    5       Minimum Magnitude 9    Maximum Magnitude

Depth

            Minimum Depth          Maximum Depth

Intensity

            Minimum Intensity      Maximum Intensity

  [ Submit Search ]   [ Clear form ]
```

f   Do not type any values for Depth or Intensity. Click Submit Search.

Step 2   **Prepare data for ArcMap**

a   Use your mouse to highlight and select all the data shown. Be sure to include the field names, but do not include other text on the Web page.

b   From the Edit menu, select Copy.

c   Open Notepad. From the Edit menu, choose Paste.

For example:

```
Untitled - Notepad                              _ □ ×
File  Edit  Format  Help
Year,Month,Day,Time(hhmmss.mm)UTC,Latitude,Longitude,Magnitude,Depth
  2004,07,23,221706.78,-33.01,-178.44,5.1, 10
  2004,07,24,185459.19, 26.53, 128.70,5.5, 36
  2004,07,25,140109.84,-17.63, -69.09,5.3,156
  2004,07,25,143519.03, -2.43, 103.97,7.3,582
  2004,07,25,172837.32,  5.21, 125.30,5.0,174
  2004,07,25,194857.02, 12.40,  95.01,5.1, 23
  2004,07,26,044157.28, -8.51, 119.79,5.1,125
  2004,07,27,010652.82, -2.99, 147.54,5.4, 30
  2004,07,28,035628.70, -0.42, 133.11,6.5, 13
  2004,07,28,040159.74, -0.37, 133.02,5.4, 10
  2004,07,28,102328.73,-10.34, 161.09,5.7, 55
  2004,07,28,222213.77, 30.75,  83.57,5.1, 10
```

d   Delete the "(hhmmss.mm)UTC" section of the Time column heading. Shorten Latitude to **Lat**, Longitude to **Long**, and Magnitude to **Mag**.

e   Ask your teacher where to save this document. Save it as a text file and name it **Quake04.txt**. Exit the text editor program.

f   Minimize your Web browser.

You will use the Latitude and Longitude coordinates in this text file to create a new layer of point features that indicate the locations of the earthquakes.

Step 3   Start ArcMap and open a map document

a   Start ArcMap.

b   Open the **Adv2.mxd** (or **Adv2**) map document from your module 2 folder (**C:\MapWorld9\Mod2\Adv2.mxd**).

When the map document opens, you see a world map. Two layers are turned on (Continents and Ocean).

Step 4   Add x,y data to the map

a   From the Tools menu, choose Add XY Data. The Add XY Data dialog appears.

 b   Near the top of the dialog under "Choose a table from the map or browse for another table," click the Browse button. The Add dialog appears.

c   In the Add dialog, navigate to the location of Quake04.txt and click Add.

d   Confirm that the X Field is set to Long. Click the Y Field down arrow and choose Lat from the list.

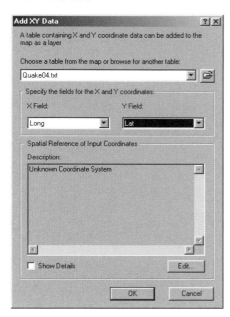

e   Click OK to create a new layer from the data in Quake04.txt. Quake04.txt Events appears in the table of contents.

> *Note: When you can add geographic locations in the form of x,y coordinates to your map, these geographic locations are referred to as "events."*

f   Examine the new layer to see where recent earthquakes have occurred. Because ArcMap randomly selects a symbol color, the layer may be difficult to see.

g   Change the symbol color to see the points better.

h   In the table of contents, right-click Quake04.txt Events, point to Data, and click Export Data. In the Export Data dialog, confirm that Export is set to All Features.

i   Choose "Use the same Coordinate System as the data frame."

j   Near the bottom of the dialog, under "Output shapefile or feature class," click the Browse button. The Saving Data dialog appears.

k   At the bottom of the Saving Data dialog, click the "Save as type" down arrow and choose Personal Geodatabase feature classes. Navigate to the World2 geodatabase (**C:\MapWorld9\Mod2\Data\World2.mdb**). Name the feature class according to your initials and the dates of the earthquake data. For example, if your initials are ABC and the data covers July 23–August 23, 2004, name the feature class **ABC_Quake072304**. Click Save.

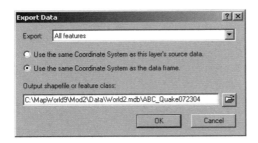

l   Click OK. Click Yes when you are asked if you want to add the exported data to the map as a layer.

m   Remove Quake04.txt Events from the table of contents.

Step 5   Save your work

a   Save your map document. Ask your teacher where you should save your work and how you should name the map document. It is recommended that you save your map document as **ABC_Adv2** where your initials are ABC.

Step 6   Locate and format volcano data

You will now follow a different procedure to collect and map volcano data.

a   Restore your Web browser and go to the Weekly Volcanic Activity Report produced by the Smithsonian and USGS at www.volcano.si.edu/gvp/reports/usgs.

b   Open each link for New Activity and Ongoing Activity (see example below). Record the data on your volcano data sheet.

> New Activity: | Rinjani, Indonesia | St. Helens, USA |
> Ongoing Activity: | Asama, Japan | Colima, México | Etna, Italy | Fuego,
> Guatemala | Galeras, Colombia | Kilauea, USA | Mauna Loa, USA | Santa Maria,
> Guatemala | Soufrière Hills, Montserrat | Tungurahua, Ecuador |

c   Open Notepad. Use the information on your volcano data sheet to create a comma-delimited file.

Note: When creating this file, be sure to eliminate all spaces. Convert all south latitudes and west longitudes to negative numbers. For date of activity, use the date the volcano first began erupting. See example for guidelines:

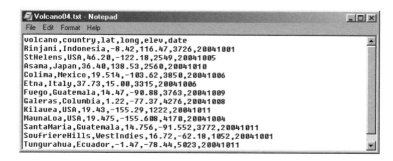

d   Ask your teacher where to save the file. Save it as a text file and name it **Volcano04.txt**.

e   Follow the procedures outlined in steps 4a–4g for the earthquake data to create a Volcano04.txt point layer on your map. You will need to select Volcano04.txt in the Add XY Data dialog.

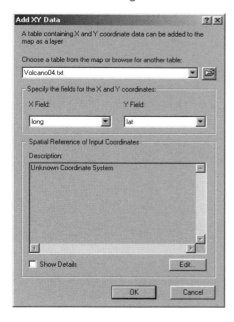

f   Follow the procedures outlined in steps 4h–4l for the earthquake data to convert the Volcano04.txt Events layer to a feature class in the World2 geodatabase (**C:\MapWorld9\Mod2\Data\World2.mdb**). Name the feature class according to your initials and the dates of the volcano data. For example, if your initials are ABC and the data covers October 1–October 11, 2004, name the feature class **ABC_Volcano100104**.

g   Remove the Volcano04.txt Events layer from the map document.

h   Save your map document according to your teacher's instructions. Be sure to keep the name you assigned earlier in the investigation.

Step 7   Explore the data

a   Explore the data you have mapped by classifying the data points in a variety of ways: earthquake magnitude, earthquake depth, date of most recent volcanic eruption, and so on. Your teacher may provide additional patterns to analyze.

As you explore the data, consider the following questions:
- Do the mapped events occur along known plate boundaries? Where?
- Are some boundary zones more active at the present time than others? Which ones?
- Where did the largest earthquakes occur?
- Are any of the volcanoes currently showing activity near cities or located in areas with high population density? Identify them.

Step 8   Map your observations and save the map document

a   Create a layout and title it **Tectonic Hot Spots, (Month, Year)**. For example, Tectonic Hot Spots, October 2004 would be an appropriate title for a map with data from October 2004.

Remember to include the following items in your layout:
- Title
- Legend
- Map scale
- North arrow
- Author of map
- Date of map creation
- Data sources

Note: If you experience difficulty creating the map layout, use ArcGIS Desktop Help.

b   Print your map.

c   Save your map document according to your teacher's instructions.

module 3

Physical Geography II

Ecosystems, Climate, and Vegetation

Physical Geography II explores a variety of characteristics that influence climate: latitude, elevation, landforms, proximity to ocean, and the El Niño and La Niña events.

Running Hot and Cold: A global perspective

Students will explore characteristics of the earth's tropical, temperate, and polar zones by analyzing monthly and annual temperature patterns in cities around the world. In the course of their investigation, students will observe temperature patterns associated with changes in latitude as well as differences caused by factors such as elevation and proximity to the ocean.

Seasonal Differences: A regional case study of South Asia

Students will observe patterns of monsoon rainfall in South Asia and analyze the relationship of those patterns to the region's physical features. The consequences of monsoon season on human life will be explored by studying South Asian agricultural practices and patterns of population distribution.

Sibling Rivalry: An advanced investigation

Students will study climatic phenomena El Niño and La Niña by downloading map images from the GLOBE (Global Learning and Observations to Benefit the Environment) Web site. They will incorporate these images into an ArcMap map document and identify patterns and characteristics of these phenomena, then assess the impact these anomalies have on the global and local environment.

Running Hot and Cold
A global perspective

Lesson overview

Students will explore characteristics of the earth's tropical, temperate, and polar zones by analyzing monthly and annual temperature patterns in cities around the world. In the course of their investigation, students will observe temperature patterns associated with changes in latitude as well as differences caused by factors such as elevation and proximity to the ocean.

Estimated time   Two to three 45-minute class periods

Materials   ✔ Colored pencils

✔ Student handouts from this lesson to be copied:
- Student handout (page 121)
- Running Hot and Cold outline map of the world (page 122)
- GIS Investigation sheets (pages 123 to 136)
- Answer sheet (pages 137 to 143)
- Assessment(s) (pages 144 to 147)

Standards and objectives   *National geography standards*

GEOGRAPHY STANDARD	MIDDLE SCHOOL	HIGH SCHOOL
1 How to use maps and other geographic representations, tools, and technologies to acquire, process, and report information from a spatial perspective	The student understands how to use maps, charts, and databases to analyze spatial distributions and patterns.	The student understands how to use technologies to represent and interpret Earth's physical and human systems.
5 That people create regions to interpret Earth's complexity	The student understands the elements and types of regions.	The student understands the structure of regional systems.
7 The physical processes that shape the patterns of Earth's surface	The student understands how Earth–Sun relationships affect physical processes and patterns on Earth.	The student understands spatial variation in the consequences of physical processes across Earth's surface.

Standards and objectives (continued)

Objectives

The student is able to:

- Locate tropical, temperate, and polar zones.
- Describe the characteristic yearly and monthly temperature patterns in those zones.
- Describe the influence of latitude, elevation, and proximity to the ocean on yearly temperature patterns.
- Compare and explain monthly temperature patterns in the Northern and Southern hemispheres.

GIS skills and tools

 Change the font color

 Zoom in on the map

 Add new text to the map

 Zoom to the full extent of the map

 Select and move text on the map

 Identify features to learn more about them

 Add layers to the map

 Select features on the map

 Pan the map

 Find features on the map

- Turn layers on and off
- Set the font, size, and style for feature labels
- Turn feature labels on and off
- Set the font, color, size, and style for text
- Add text to the map
- Activate a data frame
- Display a graph
- Set the selectable layers
- Determine latitude and longitude of map features
- Clear selected features on a map
- Open an attribute table
- Select records in a table
- Sort records in a table
- Freeze a field in a table
- Clear selected records in a table

ACQUIRE
geographic
resources

ASK
geographic
questions

EXPLORE
geographic
data

ACT on
geographic
knowledge

ANALYZE
geographic
information

For more on geographic inquiry and these steps, see Geographic Inquiry and GIS (pages xxiii to xxv).

Teacher notes

Lesson introduction

Begin the lesson by asking students to name places that they believe to be the coldest and hottest on the planet. Briefly compare their choices and the reasoning behind them. Distribute the Running Hot and Cold student handout. Working in pairs or small groups, students should identify the planet's three hottest cities in July and the three coldest cities in January. At the end of five minutes, each group should share their lists with the rest of the class. Use the blackboard or an overhead projector to tally the cities mentioned as each group reports. Based on the tally, circle the cities that were listed most often. Explain that they are going to do an activity that will explore temperature patterns in cities around the world. As they complete the GIS Investigation, they will have an opportunity to check their answers on this handout and reconsider them in view of what they learn.

Before beginning the computer activity, engage students in a discussion about the cities that are circled on the list.

• Why do you think this city is one of the coldest or hottest?

• What countries are these cities located in?

• Has anyone ever visited one of these cities?

Student activity

 Before completing this lesson with students, we recommend that you complete it as well. Doing so will allow you to modify the activity to accommodate the specific needs of your students.

After completing and discussing the Running Hot and Cold student handout, have your students work on the computer component of the lesson. Ideally, each student should be at an individual computer, but the lesson can be modified to accommodate a variety of instructional settings.

Distribute the GIS Investigation sheets to the students. Explain that in this activity they will use GIS to observe and analyze yearly and monthly temperature patterns in cities around the world. The worksheets will provide them with detailed instructions for their investigations. As they investigate, they will identify global and regional temperature variations and speculate on possible reasons for the patterns that they observe.

Teacher Tip: Step 9 asks students to rename their map document and save it. Make sure you inform your students on how to rename their map document and where to save it.

In addition to instructions, the worksheet includes questions to help students focus on key concepts.

Things to look for while the students are working on this activity:
- Are the students using a variety of tools?
- Are the students answering the questions as they work through the procedure?
- Are the students experiencing any difficulty navigating between windows in the map document?

Conclusion

When the class has finished the GIS Investigation, lead a discussion that summarizes the conclusions they reached. Be sure to address latitude in the Northern Hemisphere, latitude in the Southern Hemisphere, proximity to ocean, and elevation as factors that influence temperature. After students have had an opportunity to share their conclusions, discuss the similarities and differences among the ideas presented. Allow students to question each other and clarify confusing or contradictory statements. Develop a consensus about how each factor influences temperature.

Assessment

In the middle and high school assessments, students will use the GIS Investigation to draw conclusions about the factors that influence temperature patterns. They will write an essay offering data and examples from the GIS Investigation that support these statements.

Middle school students will be required to make a paper or ArcMap-generated map that illustrates the points of their essay. High school students will be required to use ArcMap only to make their maps.

Extensions

- Collect additional temperature data for cities in one specific world region. Use that data to create a regional version of this map document and analyze that data to create a regional temperature profile.
- Investigate the phenomenon of global warming. Use the Internet to collect monthly temperature data for recent years in one or more of the cities included in the map document. Compare actual recorded temperatures to average monthly temperatures to see if current temperatures are warmer than average. Compare changes in one region with global changes to see if there are differences.
- Collect rainfall data for the cities included in this map document. Use this data in combination with the temperature data to create an ArcMap layout that illustrates typical climate patterns for each of the climate types.
- Check out the Resources by Module section of this book's companion Web site *(www.esri.com/mappingourworld)* for print, media, and Internet resources on the topics of climate and global temperatures.

NAME _____   DATE _____

Running Hot and Cold
A map investigation

Directions: In the spaces below, list the three cities on the map on the next page that you believe are the hottest in July and the three cities that you believe are the coldest in January.

Hottest in July:

_____

_____

_____

Coldest in January:

_____

_____

_____

Running Hot and Cold

NAME

DATE

Outline map of the world

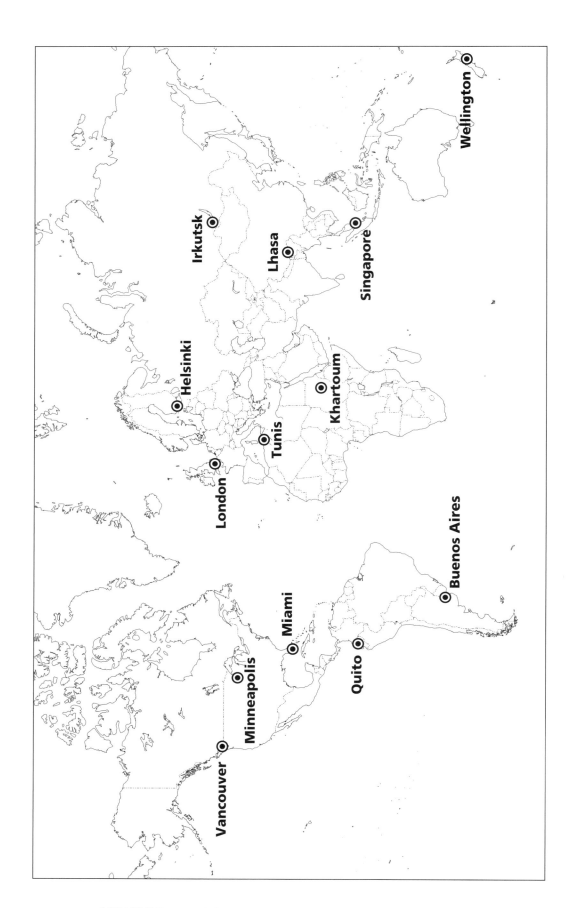

NAME _____   DATE _____

Running Hot and Cold
A GIS investigation

ACQUIRE
ASK
EXPLORE
ACT
ANALYZE

Answer all questions on the student answer sheet handout

In this activity, you will analyze monthly and annual temperature patterns in cities around the world. You will explore how latitude, elevation, and proximity to the ocean influence temperature patterns in the world's tropical, temperate, and polar zones.

Step 1   Start ArcMap

a   Double-click the ArcMap icon on your computer's desktop.

b   If the ArcMap start-up dialog appears, click **An existing map** and click OK. Then go to step 2b.

Global perspective: Running Hot and Cold   *GIS investigation*

Step 2   Open the Global3.mxd file

a   In this exercise, a map document has been created for you. To open it, go to the File menu and choose **Open**.

b   Navigate to the module 3 folder (**C:\MapWorld9\Mod3**) and choose **Global3.mxd** (or **Global3**) from the list.

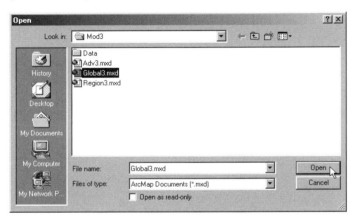

c   Click Open.

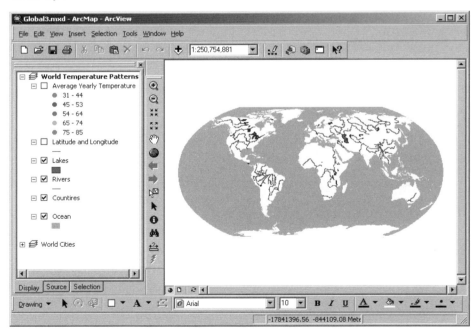

When the map document opens, you see a world map.

d   Turn on the Average Yearly Temperature layer.

Step 3   Observe annual world temperature patterns

The symbols on the map represent cities around the world. The color of each symbol reflects an average of temperatures recorded throughout the year in that city (in degrees Fahrenheit).

a   Look at the global temperature patterns displayed in the map.

b   Write three observations about the pattern of temperatures displayed on the map. Your observations should be global in scope, not focused on a specific country or city.

c   Click the check mark next to the Average Yearly Temperature layer to turn it off.

Step 4   Label the latitude zones

a   Turn on the Latitude and Longitude layer.

b   In the table of contents, right-click Latitude and Longitude and click Properties. Click the Labels tab.

c   At the top of the Labels tab, click the check box next to "Label features in this layer." Notice that NAME is already chosen as the field to use for labeling.

d   In the Text Symbol section, use the drop-down arrows to set the font to Arial and the size to 9. Set the style to Bold.

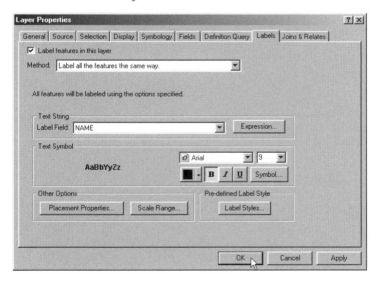

e   Click OK.

The major latitude and longitude lines are labeled on the map.

Note: In order to display the label for the prime meridian, you may need to zoom in or enlarge your map document window.

For the purposes of this exercise, the areas between the major latitude lines represent five zones of latitude. The table below names each latitude zone and the area it covers.

NAME OF LATITUDE ZONE	AREA IT COVERS IS BETWEEN THESE LATITUDES
North Polar Zone	Arctic Circle and North Pole
North Temperate Zone	Tropic of Cancer and Arctic Circle
Tropical Zone	Tropic of Cancer and Tropic of Capricorn
South Temperate Zone	Tropic of Capricorn and Antarctic Circle
South Polar Zone	Antarctic Circle and South Pole

Now you will label each of these zones on your map.

f   On the Draw toolbar at the bottom of the ArcMap window, use the drop-down arrows to set the font to Arial, the size to 9, and the style to Bold.

> *Note: If you don't see the Draw toolbar, right-click in the gray area near the top of the ArcMap window and click Draw to turn it on.*

g   Click the Font Color drop-down arrow and choose a deep violet color.

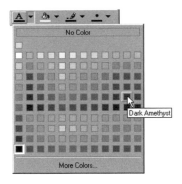

h   Zoom in to the area between the Arctic Circle and the North Pole.

i   On the Draw toolbar, click the New Text tool. The cursor turns into a plus sign with the letter A when you move it over the map.

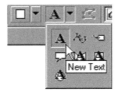

j   Click the map somewhere above the Arctic Circle. A Text box appears.

k   Type **North Polar Zone** in the box and press Enter. The North Polar Zone is now labeled in violet on the map.

l   Label each of the remaining zones using the same procedure you used to label the North Polar Zone.

> *Note: You can use the scroll bar to the right of the map display or the Pan tool to move up or down on your map.*

m   Click the Full Extent button to see the whole world.

n   In the table of contents, right-click Latitude and Longitude and click Label Features to turn off the latitude and longitude (black) labels. Now you are left with the new latitude zones (violet) labels.

> *Note: You can toggle labels on and off by right-clicking a layer and clicking Label Features. A check mark on the menu next to Label Features indicates that labels are turned on.*

o   Click the Select Elements tool. Click and drag any label on the map that you want to move. Move the labels so they cover ocean area and very little of the continents.

p   Turn on the Average Yearly Temperature layer. Observe the temperature patterns as they correspond to the latitude zones.

q   Click the Identify tool. The Identify Results window opens.

r   Click the Layers drop-down arrow and choose Average Yearly Temperature.

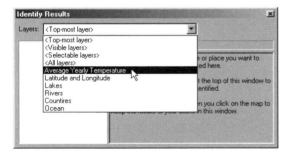

s   Use the Identify tool to click on cities and get the necessary information to complete the table on your answer sheet.

> *Remember: Each dot represents a city.*

    (1)   *Why do you think there aren't any cities in the North or South Polar Zones?*

    (2)   *How is the North Temperate Zone different from the South Temperate Zone?*

t   Turn off Average Yearly Temperature and close the Identify Results window.

Step 5   Observe climate distribution

 a   Click the Add Data button.

b   Navigate to the module 3 layer files folder (**C:\MapWorld9\Mod3\Data\LayerFiles**). Select **Climate.lyr** and click Add.

The Climate layer displays the regions of the world characterized by different types of climate.

(1)   Complete the table on your answer sheet.

(2)   Which zone has the greatest number of climates?

c   Turn on the Average Yearly Temperature layer.

Give an example of a city in each of the climate zones listed in the answer sheet.

Hint: Use the Identify tool to get the names of the different cities.

d   Ask your teacher for instructions on how to rename and save this map document. Record the new name of the map document and its new location on your answer sheet.

e   Ask your teacher if you should exit ArcMap now. Skip to step 6b if you are continuing to work.

f   From the File menu, click Exit.

Step 6   Observe monthly temperature patterns in the Northern Hemisphere

a   Start ArcMap and navigate to the folder where you saved your map document. Refer to your answer sheet for its name and location. Open the map document.

b   In the table of contents, click the minus sign in front of the World Temperature Patterns data frame to collapse it.

c   Click the plus sign in front of the World Cities data frame to expand it. Right-click World Cities and click Activate.

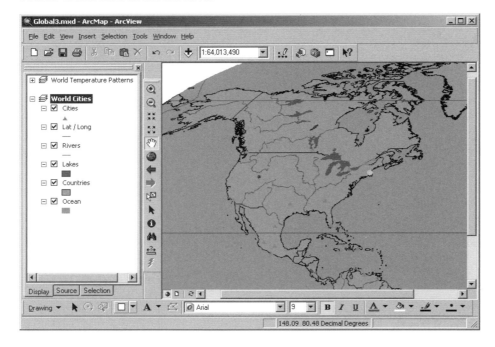

You see a map centered on North America showing cities, rivers, and lakes. The symbol for Boston is highlighted on the map.

> ***Note: If your map looks different than the one pictured above, use the Zoom In tool to drag a box around North America.***

d   Click the Tools menu, point to Graphs, and click Monthly Temperature.

The graph, Monthly Temperature, displays the monthly temperatures for Boston. You're going to select additional cities on the map, but first you'll reposition the graph and set the selectable layer to Cities.

e   Click the graph's title bar and drag it to a location where you can still see the map.

f   At the bottom of the table of contents, click the Selection tab. Notice that one city (Boston) is selected.

g   Click the check box for every layer except Cities to uncheck them.

```
☑ Cities (1 feature selected)
☐ Lat / Long
☐ Rivers
☐ Lakes
☐ Countries
☐ Ocean
```

h   Click the Display tab to show the table of contents again.

i   Click the Select Features tool.

j   Click on Miami at the southern tip of Florida.

> ***Note: Hold your mouse pointer over a city to display its name.***

Notice that both the map and the graph have changed.

   (1)   What does the graph show now?

   (2)   What city is highlighted on the map?

k   Hold down the Shift key and select Boston again on the map.

   (1)   What does the graph show now?

   (2)   What city or cities are highlighted on the map?

l   Use the Monthly Temperature graph to compare the pattern of monthly temperatures in Miami to the pattern of monthly temperatures in Boston.

   Complete the table on your answer sheet.

m   Hold down the Shift key and click on the city northeast of Boston.

   (1)   What is the name of the city?

   (2)   How does its monthly temperature pattern differ from Boston's?

n   Hold down the Shift key and click on the closest city south of Miami.

   (1)   What is the name of the city?

   (2)   How does its monthly temperature pattern differ from Miami's?

o   List the name of each of the cities displayed in the graph and complete the information in the table on the answer sheet.

Note: To find a city's latitude, pause your cursor over the city symbol on the map. When you do this, the coordinates of that city are displayed on the status bar at the bottom of the ArcMap window. The coordinates are displayed in decimal degrees, which means that the coordinates are expressed as a decimal rather than in degrees, minutes, and seconds. Latitudes north of the equator and longitudes east of the prime meridian are positive numbers, whereas latitudes south and longitudes west are negative numbers.

-71.33 46.91 Decimal Degrees

?   *p*   Based on the information displayed in the graph, the map, and the table on your answer sheet, state a hypothesis about how the monthly temperature patterns change as latitude increases.

Step 7   Test your hypothesis

a   At the top of the ArcMap window, click the Selection menu and click Clear Selected Features.

Note: The Monthly Temperature graph now shows many lines. ArcMap graphs all of the features in the Cities layer when none are selected.

b   Click the Pan tool. Click the map and pan over to Western Europe.

c   Click the Find tool. The Find dialog displays.

d   Type **Stockholm** in the Find text box and click Find. A record (row) displays in the white box at the bottom of the dialog.

e   Right-click the row and click Select Feature(s). Stockholm is highlighted in blue on the map.

f   Close the Find dialog.

g   Click the Select Features tool. Hold down the Shift key and select three more European cities that are increasingly south of Stockholm. The city names appear in the graph and the cities are highlighted on the map.

Note: To unselect a city that you selected by mistake, hold down the Shift key and click on it. To make the legend on the graph more readable, enlarge the graph window by dragging any of its borders.

?   (1)   *Complete the table on the answer sheet.*

?   (2)   *Do the cities you selected confirm or dispute your hypothesis? Explain.*

Step 8   Analyze temperature patterns in the Southern Hemisphere

You've already made a hypothesis about how latitude affects monthly temperature patterns in the Northern Hemisphere. Now you will explore the effect of latitude on the monthly temperature patterns within the Southern Hemisphere.

a   Click the Selection menu and click Clear Selected Features.

b   Click the Pan tool and reposition the map so it is centered on Australia.

c   Click the Find tool to locate and select the city of **Darwin** as you did in steps 7d–7e.

d   Close the Find dialog.

e   Click the Select Features tool. Hold down the Shift key and click the three cities on Australia's eastern and southern coasts.

Complete the table on your answer sheet.

f   Based on the information displayed on the graph, the map, and the table you just completed, compare the monthly temperature patterns in the Southern Hemisphere to those in the Northern Hemisphere.

Formulate a hypothesis about the relationship between monthly temperature patterns and increases in latitude.

Step 9   Test your hypothesis on how latitude affects monthly temperature patterns in the Southern Hemisphere

a   Click the Selection menu and click Clear Selected Features.

b   Click the Pan tool and reposition the map so it is centered on Africa.

Hint: Zoom in or out to include all of Africa if you need to.

c   Use the Find tool to locate **Cape Town**.

d   Close the Find dialog.

e   Select two or three more African cities located between Cape Town and the equator.

(1)   *Complete the table on your answer sheet.*

(2)   *Based on your observations, do the cities you selected confirm or dispute your hypothesis about how latitude affects monthly temperature patterns in the Southern Hemisphere? Explain.*

f   Click the Selection menu and click Clear Selected Features.

g   Reposition your map so it's centered on North America.

h   Ask your teacher if you should stop and exit ArcMap. If you should stop, be sure to save your changes. To exit ArcMap, click Exit from the File menu. If you do not need to exit ArcMap, proceed to the next step.

MODULE 3 • PHYSICAL GEOGRAPHY II: ECOSYSTEMS, CLIMATE, AND VEGETATION

Step 10   Investigate the ocean's influence on temperature

In addition to latitude and hemisphere, a city's proximity to the ocean also influences its temperature. Now you will investigate how the ocean influences the air temperature of coastal cities.

a   Make sure that World Cities is the active data frame.

b   One at a time, select all the cities in Canada.

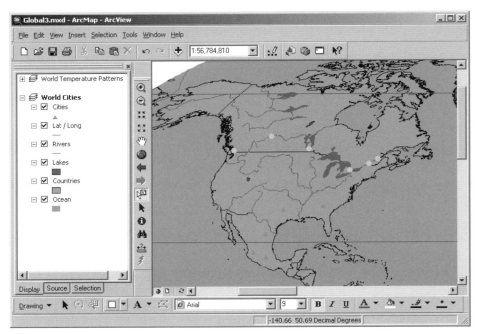

?   *(1)   In which Canadian city would you experience the coldest winter temperatures?*

?   *(2)   In which Canadian city would you experience the warmest winter temperatures?*

?   *(3)   Looking at the map, why do you think the warmest city has temperatures that are so much warmer than the others in the winter? (Hint: How is this city different from all the others in terms of its location?)*

c   Clear the selected features.

d   Reposition your map so that it is centered on Western Europe.

e   In the table of contents, right-click Cities and click Open Attribute Table.

f   Click the title bar for the Attributes of Cities table and move it so you can still see all of Western Europe on the map.

    Hint: Use your mouse to resize the table if it is too big.

g   Click the column heading called Name. The heading looks like a button that's been pushed in and the column highlights in blue.

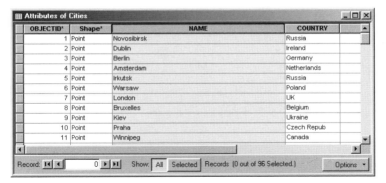

h   Right-click the column heading and click Sort Ascending to make the list alphabetical.

i   Select Amsterdam in the list by clicking the small gray box in the first column of that row. The row that Amsterdam is in highlights in blue.

j   Scroll down the list. Hold down the Ctrl key, and select the following cities, taking note of where each one is on the map as you select it: Berlin, Kiev, London, and Warsaw.

All of the cities are highlighted in the table and on the map, and they display on the graph.

k   Move the table out of the way or minimize it if necessary so you can see the map and graph. Analyze the map and graph.

(1)   Complete the table on your answer sheet.

(2)   What do these cities have in common as to their location on the earth?

(3)   Which cities have the mildest temperatures?

(4)   What happens to the winter temperatures as you move from London to Kiev?

(5)   Why do you think some cities have milder temperatures than the others?

l   Take a look at how you answered the questions about the Canadian city temperatures (step 10b).

Based on your observations of Canada and Western Europe, state a hypothesis about the influence of proximity to the ocean (or distance from it) on patterns of temperature.

MODULE 3 • PHYSICAL GEOGRAPHY II: ECOSYSTEMS, CLIMATE, AND VEGETATION

Step 11   Investigate the impact of elevation on temperature patterns

Elevation of a city significantly affects the temperature of that city. In this step, you will investigate the relationship between elevation and temperature.

 a   Click the Full extent button.

b   Restore the Attributes of Cities table if you minimized it.

c   Scroll up in the table until you find the city of Kisangani. Click on the small gray box to the left of this record to select it.

d   Hold down the Ctrl key, scroll down in the table, and select the following cities: Libreville, Quito, Singapore. Take note of where each one is on the map as you select it in the table.

e   Move the table out of the way or minimize it so you can see the map and the graph.

(1)   *Complete the table on your answer sheet.*

(2)   *What do these cities have in common as to their location on the earth?*

(3)   *What temperature pattern do these four cities have in common?*

(4)   *How is Quito different from the other three?*

(5)   *Since all these cities are located on or very near the equator, what other factor could explain the difference in their temperature patterns?*

f   Restore the Attributes of Cities table if necessary.

g   Click the Selected button at the bottom of the table. Only the selected records are displayed in the table.

h   Right-click on the NAME column heading and click Freeze/Unfreeze Column. Scroll to the right until the ELEV_FT field is adjacent to the NAME field.

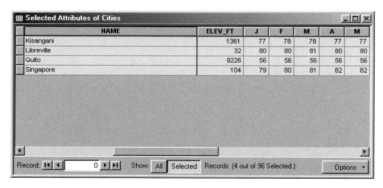

NAME	ELEV_FT	J	F	M	A	M
Kisangani	1361	77	78	78	77	77
Libreville	32	80	80	81	80	80
Quito	9226	56	56	56	56	56
Singapore	104	79	80	81	82	82

    Analyze the selected records and complete the table on your answer sheet.

i   Close the Attributes of Cities table.

j   Compare the elevation table on your answer sheet to the graph on your computer.

    Based on your observation of temperatures along the equator and the information in the table, state a hypothesis about the influence of elevation on patterns of temperature.

k   Click the Selection menu and click Clear Selected Features.

Step 12   Revisit your initial ideas

Before you began this GIS Investigation, you were asked to identify the three coldest cities in January and the three hottest cities in July from a map on which those cities were labeled.

a   Take out the paper map you used at the beginning of the GIS Investigation.

b   Select the 13 cities listed on your paper map in the Attributes of Cities table.

c   Click the Selected button.

d   Scroll to the right until you see the first column heading named **J**. (This is for January.) Note: The NAME field should still be frozen.

e   Click the first **J** column heading to select it. The column is highlighted yellow.

f   Right-click the **J** column heading and click Sort Ascending. The table is sorted from lowest (coldest) to highest (hottest) January temperatures.

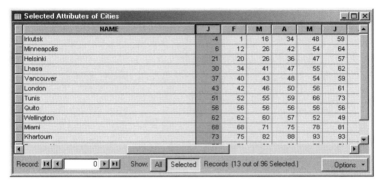

?   *g*   On your answer sheet, rank the 13 cities from coldest to hottest according to their average January temperatures.

h   Scroll to the right in the table and click on the third column heading named **J**. (This is for July.)

i   Right-click the **J** column heading and click Sort Descending. The table is sorted from highest to lowest July temperatures.

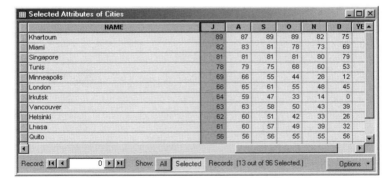

?   *j*   On your answer sheet, rank the 13 cities from hottest to coldest according to their average July temperatures.

k   Compare your original predictions from your paper map with the correct answers on your answer sheet.

?   *l*   On your answer sheet, put a check mark (✔) next to those answers that you predicted correctly.

     m   At the bottom of the attribute table, click the All button.

     n   Also at the bottom of the attribute table, click the Options button and click Clear Selection on the menu.

     o   Close the attribute table and the graph.

     There are no cities selected on the map at this time.

Step 13   **Exit ArcMap**

Through this GIS Investigation, you explored temperature data from 96 world cities. You explored the different latitude zones and have identified the variety of climates in each zone. You made different hypotheses to explain temperature patterns, and tested each hypothesis. Now you know how latitude, hemisphere, proximity to the ocean, and elevation affect temperature patterns around the world.

     a   Save your changes.

     b   Exit ArcMap by choosing Exit from the File menu.

NAME _____   DATE _____

Student answer sheet

Module 3
Physical Geography II:
Ecosystems, Climate, and Vegetation

Global perspective: Running Hot and Cold

Step 3   Observe annual world temperature patterns

b   Write three observations about the pattern of temperatures displayed on the map. Your observations
should be global in scope, not focused on a specific country or city.

_____

_____

_____

Step 4   Label the latitude zones

s   Use the Identify tool to get information on cities and complete the table below.

ZONE	TYPICAL TEMPERATURE RANGE	EXAMPLE CITY (IT REFLECTS TYPICAL TEMPERATURES OF THAT ZONE)	ANOMALIES (CITIES THAT DO NOT FIT THE PATTERN OF THEIR ZONE)
Tropical			
North Temperate Zone			
South Temperate Zone			

s-1   Why do you think there aren't any cities in the North or South Polar Zones?

_____

_____

_____

s-2   How is the North Temperate Zone different from the South Temperate Zone?

_____

_____

_____

Step 5   Observe climate distribution

b-1   Complete the table.

LATITUDE ZONES	CHARACTERISTIC CLIMATE(S)
Tropical zones	
Temperate zones	
Polar zones	

b-2   Which zone has the greatest number of climates?

_____

c   Give an example of a city in each of the following climate zones:

Arid   _____

Tropical Wet   _____

Tropical Wet and Dry   _____

Humid Subtropical   _____

Mediterranean   _____

Marine   _____

Humid Continental   _____

Subarctic   _____

Highland   _____

d   Write the new name you gave the map document and where you saved it.

_____   _____
(Name of map document.   **(Navigation path to where map document is saved.**
For example: ABC_Global3.mxd)   **For example: C:\Student\ABC)**

Step 6   Observe monthly temperature patterns in the Northern Hemisphere

j-1   What does the graph show now?

_____

j-2   What city is highlighted on the map?   _____

k-1   What does the graph show now?

_____

k-2   What city or cities are highlighted on the map?   _____

l   Use the Monthly Temperature graph to complete the table below.

CITIES	COLDEST MONTH	LOWEST TEMPERATURE	HOTTEST MONTH	HIGHEST TEMPERATURE	TEMPERATURE RANGE OVER 12 MONTHS
Boston					
Miami					

m-1 What is the name of the city? _____

m-2 How does its monthly temperature pattern differ from Boston's?

_____

_____

n-1 What is the name of the city? _____

n-2 How does its monthly temperature pattern differ from Miami's?

_____

_____

o   List the name of each of the cities displayed in the graph and complete the information in the table below.

CITY	LATITUDE	COLDEST MONTH	LOWEST TEMPERATURE (°F)	HOTTEST MONTH	HIGHEST TEMPERATURE (°F)	TEMPERATURE RANGE OVER 12 MONTHS
Boston						
Miami						

p   Based on the information displayed in the graph, the map, and the table on your answer sheet, state a hypothesis about how the monthly temperature patterns change as latitude increases.

_____

_____

Step 7   Test your hypothesis

g-1 Complete the table below.

CITY	LATITUDE
Stockholm	

g-2   Do the cities you selected confirm or dispute your hypothesis? Explain.

_____

_____

_____

Step 8   Analyze temperature patterns in the Southern Hemisphere

e   Complete the table below.

CITY	LATITUDE	COLDEST MONTH	LOWEST TEMPERATURE (°F)	HOTTEST MONTH	HIGHEST TEMPERATURE (°F)	TEMPERATURE RANGE OVER 12 MONTHS
Darwin						

f   Compare the monthly temperature patterns in the Southern Hemisphere to those in the Northern Hemisphere.

_____

_____

_____

Formulate a hypothesis about the relationship between monthly temperature patterns and increases in latitude.

_____

_____

_____

Step 9   Test your hypothesis on how latitude affects monthly temperature patterns in the Southern Hemisphere

e-1   Complete the table below.

CITY	LATITUDE
Cape Town	

e-2   Based on your observations, do the cities you selected confirm or dispute your hypothesis about how latitude affects monthly temperature patterns in the Southern Hemisphere? Explain.

_____

_____

_____

_____

Step 10   Investigate the ocean's influence on temperature

b-1   In which Canadian city would you experience the coldest winter temperatures?

_____

b-2   In which Canadian city would you experience the warmest winter temperatures?

_____

b-3   Looking at the map, why do you think the warmest city has temperatures that are so much warmer than the others in the winter?

_____

_____

_____

_____

k-1   Complete the table below.

CITY	LATITUDE
London	
Amsterdam	
Berlin	
Warsaw	
Kiev	

k-2   What do these cities have in common as to their location on the earth?

_____

k-3   Which cities have the mildest temperatures?

_____

k-4   What happens to the winter temperatures as you move from London to Kiev?

_____

k-5   Why do you think some cities have milder temperatures than the others?

_____

l   Based on your observations of Canada and Western Europe, state a hypothesis about the influence of proximity to the ocean (or distance from it) on patterns of temperature.

_____

_____

_____

_____

Step 11   Investigate the impact of elevation on temperature patterns

e-1   Complete the table below.

CITY	LATITUDE
Kisangani	
Libreville	
Quito	
Singapore	

e-2   What do these cities have in common as to their location on the earth?

_____

e-3   What temperature pattern do these four cities have in common?

_____

e-4   How is Quito different from the other three?

_____

e-5   Since all these cities are located on or very near the equator, what other factor could explain the difference in their temperature patterns?

_____

h   Analyze the selected records and complete the table below.

CITY	ELEVATION (FEET)
Kisangani	
Libreville	
Quito	
Singapore	

j   Based on your observation of temperatures along the equator and the information in the table above, state a hypothesis about the influence of elevation on patterns of temperature.

_____

_____

_____

_____

Step 12   Revisit your initial ideas

g   Rank the 13 cities from coldest to hottest according to their average January temperatures.

1. _____   8. _____
2. _____   9. _____
3. _____   10. _____
4. _____   11. _____
5. _____   12. _____
6. _____   13. _____
7. _____

j   Rank the 13 cities from hottest to coldest according to their average July temperatures.

1. _____   8. _____
2. _____   9. _____
3. _____   10. _____
4. _____   11. _____
5. _____   12. _____
6. _____   13. _____
7. _____

l   Put a check mark (✔) next to those answers that you predicted correctly.

NAME _____   DATE _____

Running Hot and Cold
Middle school assessment

Part 1

Use the ArcMap map document and your answer sheet to complete the tables below. For each city, circle each factor that influences its temperature pattern.

THREE HOTTEST CITIES IN JULY	CIRCLE THE FACTORS THAT INFLUENCE TEMPERATURE PATTERNS		
	Latitude	Proximity to the ocean	Elevation
	Latitude	Proximity to the ocean	Elevation
	Latitude	Proximity to the ocean	Elevation

THREE COLDEST CITIES IN JANUARY	CIRCLE THE FACTORS THAT INFLUENCE TEMPERATURE PATTERNS		
	Latitude	Proximity to the ocean	Elevation
	Latitude	Proximity to the ocean	Elevation
	Latitude	Proximity to the ocean	Elevation

Part 2

Use your GIS Investigation, Global3 map document, and other resources such as an atlas to write an essay that compares monthly and annual temperature patterns typical of the Tropical Zone and the North and South temperate zones. Your essay should provide example cities and data to support your conclusions.

Create a paper map or an ArcMap-generated map that illustrates the conclusions you make in your essay.

Running Hot and Cold

Assessment rubric

Middle school

STANDARD	EXEMPLARY	MASTERY	INTRODUCTORY	DOES NOT MEET REQUIREMENTS
The student understands how to use maps, graphs, and databases to analyze spatial distributions and patterns.	Uses GIS to analyze various aspects of climate such as monthly and annual temperature and precipitation and identifies cities with specific climate characteristics. Creates a detailed map that illustrates the points highlighted in the essay.	Uses GIS to analyze various aspects of climate such as monthly and annual temperature and precipitation and identifies cities with specific climate characteristics. Creates a map that illustrates points highlighted in the essay.	With some assistance, can use GIS to analyze various aspects of climate such as monthly and annual temperature and precipitation. Correctly identifies some cities with specific climate characteristics. Creates a map that illustrates some of the points highlighted in the essay.	Has difficulty using GIS to analyze various aspects of climate and identifying cities with specific climate characteristics. Creates a map, but has difficulty illustrating the points highlighted in the essay.
The student understands the elements and types of regions.	Writes an essay and creates a map that shows clear understanding of the climate patterns for various zones of Earth, including the Tropics and the North and South temperate zones.	Writes an essay and creates a map that shows an understanding of the climate patterns for various zones of Earth, including the Tropics and the North and South temperate zones.	Writes an essay and creates a map that shows some understanding of climate patterns, but does not clearly define differences between different zones, or only identifies characteristics of some zones.	Writes an essay and creates a map that shows limited understanding of climate patterns and cannot identify differences between various zones.
The student understands how Earth–Sun relationships affect physical processes and patterns on Earth.	Identifies key reasons and provides clear examples of why cities experience variations in climate patterns at different latitudes, in different hemispheres, at different elevations, and at different distances from the ocean.	Identifies key reasons why cities experience variations in climate patterns at different latitudes, in different hemispheres, at different elevations, and at different distances from the ocean.	Identifies some key reasons why cities experience variations in climate patterns due to some of the following: different latitudes, in different hemispheres, at different elevations, or at different distances from the ocean.	Identifies one or two reasons why cities experience variations in climate patterns at different latitudes, in different hemispheres, at different elevations, or at different distances to the ocean.

This is a four-point rubric based on the National Standards for Geographic Education. The "Mastery" level meets the target objective for grades 5–8.

NAME _____ DATE _____

Running Hot and Cold
High school assessment

Part 1

Use the ArcMap map document and your answer sheet to complete the tables below. For each city, circle each factor that influences its temperature pattern.

THREE HOTTEST CITIES IN JULY	CIRCLE THE FACTORS THAT INFLUENCE TEMPERATURE PATTERNS		
	Latitude	Proximity to the ocean	Elevation
	Latitude	Proximity to the ocean	Elevation
	Latitude	Proximity to the ocean	Elevation

THREE COLDEST CITIES IN JANUARY	CIRCLE THE FACTORS THAT INFLUENCE TEMPERATURE PATTERNS		
	Latitude	Proximity to the ocean	Elevation
	Latitude	Proximity to the ocean	Elevation
	Latitude	Proximity to the ocean	Elevation

Part 2

Use your GIS Investigation, Global3 map document, and other resources such as an atlas to write an essay that compares monthly and annual temperature patterns typical of the Tropical Zone and the North and South temperate zones. Your essay should provide example cities and data to support your conclusions.

Create an ArcMap-generated map that illustrates the conclusions you make in your essay.

Running Hot and Cold

Assessment rubric

High school

STANDARD	EXEMPLARY	MASTERY	INTRODUCTORY	DOES NOT MEET REQUIREMENTS
The student understands how to use technologies to represent and interpret Earth's physical and human systems.	Uses GIS to analyze various aspects of climate such as monthly and annual temperature, precipitation, elevation, and proximity to the ocean and identifies cities with specific climate characteristics. Uses GIS to create a detailed map that illustrates the points highlighted in the essay.	Uses GIS to analyze various aspects of climate such as monthly and annual temperature, precipitation, elevation, and proximity to the ocean and identifies cities with specific climate characteristics. Uses GIS to create a map that illustrates points highlighted in the essay.	With some assistance, can use GIS to analyze various aspects of climate such as monthly and annual temperature and precipitation. Correctly identifies some cities with specific climate characteristics. Uses GIS to create a map that illustrates some of the points highlighted in the essay.	Has difficulty using GIS to analyze various aspects of climate and identifying cities with specific climate characteristics. Uses GIS to create a map, but has difficulty illustrating the points highlighted in the essay.
The student understands the structure of regional systems.	Writes an essay and creates a GIS-generated map that shows clear understanding of the climate patterns for various zones of Earth, including the Tropics and the North and South temperate zones.	Writes an essay and creates a GIS-generated map that shows an understanding of the climate patterns for various zones of Earth, including the Tropics and the North and South temperate zones.	Writes an essay and creates a GIS-generated map that shows some understanding of climate patterns, but does not clearly define differences between different zones, or only identifies characteristics of some zones.	Writes an essay and creates a GIS-generated map that shows limited understanding of climate patterns and cannot identify differences between various zones.
The student understands spatial variation in the consequences of physical processes across Earth's surface.	Identifies key reasons and provides clear examples of why cities experience variations in climate patterns at different latitudes, in different hemispheres, at different elevations, and at different distances from the ocean.	Identifies key reasons why cities experience variations in climate patterns at different latitudes, in different hemispheres, at different elevations, and at different distances from the ocean.	Identifies some key reasons why cities experience variations in climate patterns due to some of the following: different latitudes, in different hemispheres, at different elevations, or at different distances from the ocean.	Identifies one or two reasons why cities experience variations in climate patterns at different latitudes, in different hemispheres, at different elevations, or at different distances from the ocean.

This is a four-point rubric based on the National Standards for Geographic Education. The "Mastery" level meets the target objective for grades 9–12.

Seasonal Differences
A regional case study of South Asia

Lesson overview

Students will observe patterns of monsoon rainfall in South Asia and analyze the relationship of those patterns to the region's physical features. The consequences of monsoon season on human life will be explored by studying South Asian agricultural practices and patterns of population distribution.

Estimated time

Two to three 45-minute class periods

Materials

✔ Four large pieces of butcher paper

✔ Four or more markers

✔ Student handouts from this lesson to be copied:

- GIS Investigation sheets (pages 153 to 159)
- Student answer sheets (pages 161 to 165)
- Assessment(s) (pages 166 to 169)

Standards and objectives

National geography standards

GEOGRAPHY STANDARD	MIDDLE SCHOOL	HIGH SCHOOL
1 How to use maps and other geographic representations, tools, and technologies to acquire, process, and report information from a spatial perspective	The student understands how to use maps, graphs, and databases to analyze spatial distributions and patterns.	The student understands how to use technologies to represent and interpret Earth's physical and human systems.
4 The physical and human characteristics of places	The student understands how physical processes shape places.	The student understands the changing human and physical characteristics of places.
15 How physical systems affect human systems	The student understands how variations within the physical environment produce spatial patterns that affect human adaptation.	The student understands how the characteristics of different physical environments provide opportunities for or place constraints on human activities.

Objectives

The student is able to:

- Describe the patterns of monsoon rainfall in South Asia.
- Explain the influence of landforms on patterns of precipitation in South Asia.
- Describe the impact of South Asia's climate and physical features on agriculture and population density in the region.

GIS skills and tools

 Select features on a map

 Measure distance on a map

 Add layers to a map

- Display graphs stored with the map document
- Analyze graphs in relation to a map
- Understand the relationship between a graph and a map
- Set selectable layers
- Rearrange layers in the table of contents
- Turn layers on and off

For more on geographic inquiry and these steps, see Geographic Inquiry and GIS (pages xxiii to xxv).

Teacher notes

Lesson introduction

Tell your students that they are going to explore seasonal differences in South Asia. They may be surprised to learn that students in South Asia would probably describe their year in terms of three seasons rather than four. Engage them in a discussion of local and personal perceptions and assumptions about seasons. Tack up four large pieces of butcher paper and have them list images, descriptions, and memories relating to each season.

- How does the physical environment change from season to season?
- How are those changes reflected in the activities, foods, and clothing they may have listed on the sheets?
- To what extent do seasonal changes in their environment affect their day-to-day lives?

Student activity

 Before completing this lesson with students, we recommend that you complete it yourself. Doing so will allow you to modify the activity to suit the specific needs of your students.

After the initial discussion, have the students work on the computer component of the lesson. Ideally, each student should be at an individual computer, but the lesson can be modified to accommodate a variety of instructional settings.

Distribute the GIS Investigation sheets to your students. Explain that in this activity, they will use GIS to observe and analyze the variable patterns of rainfall in South Asia that result from the region's seasonal monsoon winds. In South Asia it is rainfall, rather than temperature, that defines the seasons. The activity sheets will provide them with detailed instructions for their investigations. As they investigate, they will explore the relationship between South Asia's monsoon rains and its physical features and analyze the climate's impact on agriculture and population.

In addition to instructions, the handout includes questions to help students focus on key concepts. Some questions will have specific answers while others require creative thought.

Things to look for while the students are working on this activity:

- Are the students using a variety of GIS tools?
- Are the students answering the questions as they work through the procedure?
- Are the students experiencing any difficulty working with multiple windows and toggling between windows in the map document?

Conclusion

Use a projection device to display the Region3.mxd in the classroom. As a group, compare student observations and conclusions from the lesson. Students can take turns being the "driver" on the computer to highlight patterns and relationships that are identified by members of the class. Focus on the following concepts about South Asia's monsoon climate in your discussion.

- Rainfall is limited to one season of the year in South Asia except in the desert west, where little rain falls at all. (This would be an excellent point at which to elaborate on the seasonal shift in monsoon winds that produces the patterns of rainfall students observed in the map document.)

Conclusion (continued)

- Typically, the rainy season lasts from June through September, although the actual length of the season and amounts of rainfall vary across the subcontinent. (Be sure to note the orographic patterns of precipitation along India's southwest coast and in northeast India on the southern slopes of the Himalayas.)
- Agricultural activities are directly related to patterns of rainfall.
- In general, population density varies with patterns of rainfall. However, the importance of South Asia's rivers as an additional source of water for agriculture is apparent from the high density of population along their paths.

Close the lesson by challenging the students to identify the three seasons in South Asia. In general these seasons are the following:

- The rainy season (approximately June–September)
- The dry lush season after the rains when everything is growing and green (approximately October–January)
- The dry dusty season before the rains come (approximately February–May)

Assessment

Middle school: Highlights skills appropriate to grades 5 through 8

Students will assume the role of an American student living for a year in South Asia as an exchange student. They can choose to live in or around Bombay, New Delhi, or Calcutta. They will write a letter to friends back home on October 1, January 1, April 1, and July 1. Their letter will describe seasonal changes in their location and ways that their daily lives and the lives of people around them reflect those changes.

High school: Highlights skills appropriate to grades 9 through 12

Students will assume the role of an American student traveling for a year in South Asia. They will write a letter to friends back home on October 1, January 1, April 1, and July 1. Each letter will be from a different South Asian city. Their letter will describe seasonal characteristics in each city and ways that their daily lives and the lives of people around them reflect those characteristics.

Extensions

- Use the Internet to find rainfall data for South Asian cities in specific years such as 1990, 1995, and 2000 to see if the average patterns observed in this project are relatively consistent or if they vary significantly from year to year.
- Import downloaded data into ArcMap.
- Research South Asian farming methods to find out how activities such as planting and harvesting are coordinated with patterns of rainfall.
- Research the monthly and yearly rainfall patterns in your own location and compare these to the patterns observed in South Asia.
- Check out the Resources by Module section of this book's companion Web site *(www.esri.com/mappingourworld)* for print, media, and Internet resources on the topics of South Asia and monsoons.

NAME _____   DATE _____

Seasonal Differences
A GIS investigation

ACQUIRE

ASK

EXPLORE

ACT

ANALYZE

Answer all questions on the student answer sheet handout

In this activity, you will analyze the variable patterns of precipitation in South Asia that result from the region's seasonal monsoon winds. As you investigate those patterns, you will explore relationships between rainfall and physical features and analyze the climate's impact on agriculture and population.

Step 1   Start ArcMap

a   Double-click the ArcMap icon on your computer's desktop.

b   If the ArcMap start-up dialog appears, click **An existing map** and click OK. Then go to step 2b.

PHOTOCOPY

Step 2   Open the Region3.mxd file

a   In this exercise, a map document has been created for you. To open it, go to the File menu and choose **Open**.

b   Navigate to the module 3 folder (**C:\MapWorld9\Mod3**) and choose **Region3.mxd** (or **Region3**) from the list.

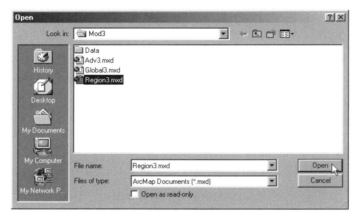

c   Click Open.

When the map document opens, you see a map of South Asia.

d   Click the Tools menu, point to Graphs, and click Monthly Rainfall. A graph of monthly rainfall for the city of Bombay opens.

e   Click the graph's title bar and position it anywhere on your desktop that does not cover your map. Stretch or shrink the ArcMap window if you need to.

f   Repeat step d to open the Annual Rainfall graph. Position it next to the Monthly Rainfall graph.

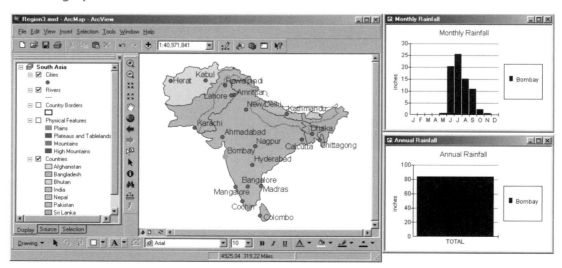

Step 3   Observe patterns of rainfall

The map allows you to explore and compare variations in the patterns of rainfall throughout the South Asian region. Look at the map and notice that the city of Bombay is selected—it is highlighted blue. The graphs to the right display rainfall information for the selected city—in this case, Bombay.

a • Analyze the graphs and answer the following questions on your answer sheet.

?   *(1)   Which month gets the most rainfall in Bombay?*

?   *(2)   Which months appear to get little or no rainfall in Bombay?*

?   *(3)   Approximately how much rainfall does Bombay get each year (in inches)?*

?   *(4)   Write a sentence summarizing the overall pattern of rainfall in Bombay in an average year.*

b • At the bottom of the table of contents, click the Selection tab. Click the check boxes to uncheck all the layers except Cities.

☑ **Cities (1 feature selected)**
☐ Rivers
☐ Country Borders
☐ Physical Features
☐ Countries

c • Click the Display tab to return to the table of contents.

 d • Click the Select Features tool. Click a dot for another city.

?   *(1)   How did this change the map?*

?   *(2)   How did this change the graphs?*

e • Click the city of Mangalore to select it in the map.

?   *Analyze the graphs and fill in the Mangalore section of the table on your answer sheet.*

f • Hold down the Shift key and click the cities of Bombay and Ahmadabad.

 Note: To enlarge the graphs and make them easier to read, drag any border with your mouse.

?   *(1)   Complete the table on the answer sheet.*

?   *(2)   As you move northward along the subcontinent's west coast, how does the pattern of rainfall change?*

?   *(3)   Although the monthly rainfall amounts differ, what similarities do you see among the overall rainfall patterns of these three cities?*

Step 4   Compare coastal and inland cities

a • Make sure the Select Feature tool is still active and select Bangalore.

?   *Use the Monthly Rainfall and Yearly Rainfall graphs to complete the table on your answer sheet.*

b • Hold down the Shift key and select Mangalore.

?   *How does the rainfall pattern of Bangalore compare with that of Mangalore?*

c • Click the Measure tool. Your cursor turns into a right-angle ruler with crosshairs ⌐.

d Click the dot that represents Bangalore once, then move it to the dot that represents Mangalore and double-click.

Note: If you accidentally clicked the wrong spot, you can double-click to end the line and start over.

A segment and total length appear on the status bar at the bottom left of the ArcMap window.

? *What is the distance between the two cities?*

Although Bangalore is located only a short distance inland from Mangalore, it receives far less rainfall than the coastal city.

e Turn on the Physical Features layer.

? *How can this data help you explain the differences between patterns of rainfall in inland Bangalore and coastal Mangalore?*

f Turn off Physical Features.

Step 5 Compare eastern and western South Asian cities

 a Click the Select Features tool. One at a time, select the Afghan cities of Kabul and Herat.

? (1) *Analyze the graphs and complete the table on your answer sheet.*

Note: Your first impression may be that the Afghan cities get a fair amount of rainfall. But, notice that the inches scale along the left side of each graph (y-axis) changed to reflect the rainfall range of the selected cities.

? (2) *Describe the pattern of rainfall in these two cities.*

? (3) *How do you think Afghanistan's rainfall pattern will affect the way of life in that country?*

b Select the eastern cities of Calcutta and Dhaka.

? (1) *Analyze the graphs and complete the table on your answer sheet.*

? (2) *Describe the pattern of rainfall in these two cities.*

c Hold down the Shift key and select four cities: Calcutta, Dhaka, Herat, and New Delhi.

? *What is happening to the patterns of rainfall as you move from west to east across South Asia?*

d Click in the white space surrounding South Asia to unselect the four cities.

e Close the Monthly Rainfall and Annual Rainfall graphs.

Step 6   Observe yearly precipitation

You've already looked at the monthly precipitation patterns for individual cities across South Asia. In this step, you will add data and look at the total yearly rainfall for regions of South Asia.

a   **Click the Add Data button.**

b   **Navigate to the module 3 layer files folder (C:\MapWorld9\Mod3\Data\LayerFiles).**

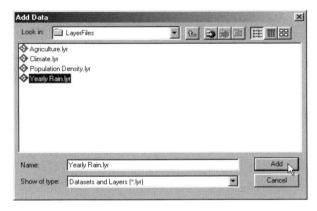

c   **Select Yearly Rain.lyr. Click Add.**

d   **Drag the layer down in the table of contents so that it is just below the Country Borders layer.**

e   **Turn off the Cities layer and turn on Yearly Rain.**

Amounts of rainfall are given in millimeters rather than inches. Here is a conversion table that compares millimeters to inches (25.4 mm. = 1 in.).

MM	100	200	600	1,600	2,800	5,600	12,000
IN	3.9	7.9	23.6	62.9	110.2	220.5	472.4

(1)   *Which regions within South Asia get the least rainfall?*

(2)   *Which regions within South Asia get the most rainfall?*

(3)   *In step 5c you were comparing Calcutta, Herat, New Delhi, and Dhaka. Does the map of yearly rainfall that is on your screen now reflect the observation you made at that time? Explain.*

f   **Turn off Yearly Rain and turn on Physical Features.**

What relationships do you see between South Asia's patterns of yearly rainfall and its physical features?

Step 7   **Explore the monsoon's impact on agriculture and population density**

a   Turn on the Country Borders layer.

The rain patterns and physical features of an area have a significant impact on the way of life of the people who live there. Now you will look at those layers and determine the kind of impact they have on individual countries.

b   Turn the Physical Features and Yearly Rain layers on and off to make your observations and to answer the questions below.

?   *(1)   Which regions or countries of South Asia are suitable for agriculture and which are not? Explain.*

?   *(2)   In which regions of South Asia do you expect to see the lowest population density? Explain.*

?   *(3)   In which regions of South Asia do you expect to see the highest population density? Explain.*

c   Turn off Physical Features and Yearly Rain layers.

Now you will add agricultural data for the region and will see if your predictions are correct.

d   Click the Add Data button. Navigate to the module 3 layer files folder (**C:\MapWorld9\Mod3\Data\LayerFiles**). Select **Agriculture.lyr** and click Add.

e   Drag Agriculture down in the table of contents so that it is just below the Country Borders layer. Turn on the Agriculture layer.

?   *(1)   Does the agriculture layer reflect the predictions you made in step 7b? Explain.*

?   *(2)   Why are grazing, herding, and oasis agriculture the major activities in Afghanistan?*

?   *(3)   What do you know about rice cultivation that would help explain its distribution on the agriculture map?*

?   *(4)   Is there any aspect of the agriculture map that surprised you? Explain.*

f   Turn off the Agriculture layer.

You will now examine population density in relation to precipitation and land use.

g   Click the Add Data button. Navigate to the module 3 layer files folder (**C:\MapWorld9\Mod3\Data\LayerFiles**). Add **Population Density.lyr**.

h   Drag Population Density below Country Borders in the table of contents.

i   Turn on Population Density.

?   *(1)   Does the Population Density layer reflect the population predictions you made in step 7b? Explain.*

?   *(2)   Why is Afghanistan's population density so low?*

?   *(3)   Since most of Pakistan gets little to no rainfall, how do you explain the areas of high population density in that country?*

?   *(4)   What is the relationship between population density and patterns of precipitation in South Asia?*

?   *(5)   What is the relationship between population density and physical features in South Asia?*

Step 8   Exit ArcMap

In this GIS Investigation, you explored the patterns of monsoon rainfall in South Asia. You used ArcMap to compare monthly and annual patterns of precipitation in cities throughout the region and explore the relationship between those patterns and the region's physical features. After analyzing this data, you added layers reflecting patterns of agriculture and population density and analyzed the relationship between those human characteristics and the region's climate and landforms.

a Ask your teacher for instructions on where to save this ArcMap map document and on how to rename the map document.

b If you are not going to save the map document, exit ArcMap by choosing Exit from the File menu. When asked if you want to save changes to Region3.mxd (or Region3), click No.

NAME _____   DATE _____

Student answer sheet
Module 3
Physical Geography II:
Ecosystems, Climate, and Vegetation

Regional case study: Seasonal Differences

Step 3   Observe patterns of rainfall

a-1   Which month gets the most rainfall in Bombay?   _____

a-2   Which months appear to get little or no rainfall in Bombay?   _____

a-3   Approximately how much rainfall does Bombay get each year (in inches)?   _____

a-4   Write a sentence summarizing the overall pattern of rainfall in Bombay in an average year.

_____

_____

_____

d-1   How did this change the map?

_____

_____

d-2   How did this change the graphs?

_____

_____

e   Analyze the graphs and fill in the Mangalore section of the table below.

CITY	MONTHS WITH RAINFALL	HIGHEST MONTHLY RAINFALL (INCHES)	TOTAL ANNUAL RAINFALL (INCHES)
Mangalore			
Bombay			
Ahmadabad			

f-1   Complete the rest of the table in step e, above.

f-2   As you move northward along the subcontinent's west coast, how does the pattern of rainfall change?

_____

_____

f-3   Although the monthly rainfall amounts differ, what similarities do you see among the overall rainfall patterns of these three cities?

_____

_____

Step 4   Compare coastal and inland cities

a   Complete the table below.

CITY	MONTHS WITH RAINFALL	HIGHEST MONTHLY RAINFALL (INCHES)	TOTAL ANNUAL RAINFALL (INCHES)
Bangalore			

b   How does the rainfall pattern of Bangalore compare with that of Mangalore?
   Similarities: _____
   Differences: _____

d   What is the distance between the two cities?   _____

e   How can this data help you explain the differences between patterns of rainfall in inland Bangalore and coastal Mangalore?

   _____
   _____
   _____
   _____
   _____
   _____
   _____
   _____
   _____

Step 5   Compare eastern and western South Asian cities

a-1   Analyze the graphs and complete the table below.

CITY	MONTHS WITH RAINFALL	HIGHEST MONTHLY RAINFALL (INCHES)	TOTAL ANNUAL RAINFALL (INCHES)
Kabul			
Herat			

a-2   Describe the pattern of rainfall in these two cities.

   _____
   _____

a-3   How do you think Afghanistan's rainfall pattern will affect the way of life in that country?

   _____
   _____

b-1   Analyze the graphs and complete the table below.

CITY	MONTHS WITH RAINFALL	HIGHEST MONTHLY RAINFALL (INCHES)	TOTAL ANNUAL RAINFALL (INCHES)
Calcutta			
Dhaka			

b-2   Describe the pattern of rainfall in these two cities.

_____

_____

c   What is happening to the patterns of rainfall as you move from west to east across South Asia?

_____

_____

Step 6   Observe yearly precipitation

e-1   Which regions within South Asia get the least rainfall?

_____

_____

e-2   Which regions within South Asia get the most rainfall?

_____

_____

e-3   In step 5c you were comparing Calcutta, Herat, New Delhi, and Dhaka. Does the map of yearly rainfall
      that is on your screen now reflect the observation you made at that time? Explain.

_____

_____

_____

_____

f   What relationships do you see between South Asia's patterns of yearly rainfall and its physical features?

_____

_____

_____

_____

Step 7   Explore the monsoon's impact on agriculture and population density

b-1   Which regions or countries of South Asia are suitable for agriculture and which are not? Explain.

_____

_____

_____

_____

b-2   In which regions of South Asia do you expect to see the lowest population density? Explain.

_____

_____

_____

_____

b-3   In which regions of South Asia do you expect to see the highest population density? Explain.

_____

_____

_____

_____

e-1   Does the Agriculture layer reflect the predictions you made in step 7b? Explain.

_____

_____

_____

_____

e-2   Why are grazing, herding, and oasis agriculture the major activities in Afghanistan?

_____

_____

_____

_____

e-3   What do you know about rice cultivation that would help explain its distribution on the agriculture map?

_____

_____

_____

_____

e-4   Is there any aspect of the agriculture map that surprised you? Explain.

_____

_____

_____

_____

i-1   Does the Population Density layer reflect the population predictions you made in step 7b? Explain.

_____

_____

_____

_____

i-2   Why is Afghanistan's population density so low?

_____

_____

_____

_____

i-3   Since most of Pakistan gets little to no rainfall, how do you explain the areas of high population density in that country?

_____

_____

_____

_____

i-4   What is the relationship between population density and patterns of precipitation in South Asia?

_____

_____

_____

_____

i-5   What is the relationship between population density and physical features in South Asia?

_____

_____

_____

_____

NAME _____   DATE _____

Seasonal Differences
Middle school assessment

For this activity, you are to assume the role of an American student who is spending a year living in South Asia as an exchange student. Your task is to write four letters to friends or family back home about your experiences and observations during your year in South Asia. Your four letters should be dated January 1, April 1, July 1, and October 1. Using the ArcMap map document as a guide, describe seasonal changes in your city and ways that your daily life and the lives of people around you reflect those changes. You may choose to spend your year in or near any one of the following locations: Bombay, Calcutta, or Dhaka. You may use additional sources such as your geography book, encyclopedias, and the Internet to help you develop your letters.

Use the space below to brainstorm for your essay.

January 1

April 1

July 1

October 1

Seasonal Differences

Assessment rubric

Middle school

STANDARD	EXEMPLARY	MASTERY	INTRODUCTORY	DOES NOT MEET REQUIREMENTS
The student understands how to use maps, graphs, and databases to analyze spatial distributions and patterns.	Uses GIS as a tool to analyze the patterns of monsoon rains in South Asia through mapping and creating original charts based on the data provided. Accurately depicts the seasonal weather conditions in a given city.	Uses GIS as a tool to analyze the patterns of monsoon rains in South Asia through mapping and viewing charts. Accurately depicts most of the seasonal weather conditions in a given city.	Identifies patterns of precipitation on maps and charts using GIS. Accurately depicts some of the seasonal weather conditions for a given city.	Has difficulty identifying patterns of precipitation using maps and charts in GIS. Has difficulty depicting any seasonal weather conditions for the given city.
The student understands how physical processes shape places.	Shows understanding of the effects of seasonal rains and other climate factors on the characteristics of many South Asian cities throughout an entire year. Creatively incorporates this into each letter.	Shows understanding of the effects of seasonal rains and other climate factors on the characteristics of one particular South Asian city throughout an entire year. Incorporates this into each letter.	Shows limited understanding of the effects of seasonal rains on the characteristics of one particular South Asian city throughout an entire year. Incorporates some of this information into each letter.	Attempts to describe seasonal climate changes in the region of South Asia. Has difficulty incorporating this information into each letter.
The student understands how variations within the physical environment produce spatial patterns that affect human adaptation.	Uses great detail and specific examples to illustrate the impact of seasonal climate changes on the daily life of people in a variety of South Asian cities throughout an entire year. Creatively incorporates this into each letter.	Clearly illustrates the impact of seasonal climate changes on the daily life of people in a particular South Asian city throughout an entire year. Incorporates this into each letter.	Shows limited understanding of the impact of seasonal climate changes on the daily life of people in a particular South Asian city throughout an entire year. Incorporates some of this information into each letter.	Attempts to describe the impact of seasonal climate changes on the daily life of people in South Asia. Has difficulty incorporating this information into each letter.

This is a four-point rubric based on the National Standards for Geographic Education. The "Mastery" level meets the target objective for grades 5–8.

NAME _____ DATE _____

Seasonal Differences
High school assessment

For this activity you are to assume the role of an American student who is spending a year traveling in South Asia. Your task is to write four letters to friends or family back home about your experiences and observations during your year abroad. Each letter should be written from a different South Asian city. Your four letters should be dated January 1, April 1, July 1, and October 1. Using the ArcMap map document as a guide, describe seasonal characteristics of each city on the date you are writing and ways that your daily life and the lives of people around you reflect those characteristics. You may use additional sources such as your geography book, encyclopedias, and the Internet to help you develop your letters.

Use the space below to brainstorm for your essay.

January 1

April 1

July 1

October 1

Seasonal Differences

Assessment rubric

High school

STANDARD	EXEMPLARY	MASTERY	INTRODUCTORY	DOES NOT MEET REQUIREMENTS
The student understands how to use technologies to represent and interpret the earth's physical and human systems.	Uses GIS as a tool to analyze the patterns of monsoon rains in South Asia through mapping and creating original charts based on the data provided. Accurately depicts the seasonal weather conditions in the four cities.	Uses GIS as a tool to analyze the patterns of monsoon rains in South Asia through mapping and viewing charts. Accurately depicts most of the seasonal weather conditions in the four cities.	Identifies patterns of precipitation on maps and charts using GIS. Accurately depicts some of the seasonal weather conditions for some of the cities.	Has difficulty identifying patterns of precipitation using maps and charts in GIS. Has difficulty depicting the seasonal weather conditions for the cities.
The student understands the changing human and physical characteristics of places.	Shows understanding of the effects of seasonal rains and other climate factors on the characteristics of four different South Asian cities. Describes the seasonal changes for an entire year for each city. Creatively incorporates this into each letter.	Shows understanding of the effects of seasonal rains and other climate factors on the characteristics of four different South Asian cities throughout an entire year (one city for each season). Incorporates this into each letter.	Shows limited understanding of the effects of seasonal rains on the characteristics of one or two South Asian cities throughout an entire year. Incorporates some of this information into each letter.	Attempts to describe seasonal climate changes in the region of South Asia. Has difficulty incorporating this information into each letter.
The student understands how the characteristics of different physical environments provide opportunities for or place constraints on human activities.	Uses great detail and specific examples to illustrate the impact of seasonal climate changes on the daily life of people in four South Asian cities. Describes the seasonal changes for an entire year for each city. Creatively incorporates this into each letter.	Clearly illustrates the impact of seasonal climate changes on the daily life of people in four different South Asian cities throughout an entire year (one city for each season). Incorporates this into each letter.	Shows limited understanding of the impact of seasonal climate changes on the daily life of people in one or two South Asian cities throughout an entire year. Incorporates some of this information into each letter.	Attempts to describe the impact of seasonal climate changes on the daily life of people in South Asia. Has difficulty incorporating this information into each letter.

This is a four-point rubric based on the National Standards for Geographic Education. The "Mastery" level meets the target objective for grades 9–12.

Sibling Rivalry
An advanced investigation

Lesson overview

Students will study climatic phenomena El Niño and La Niña by downloading map images from the GLOBE (Global Learning and Observations to Benefit the Environment) Web site. They will incorporate these images into an ArcMap map document and identify patterns and characteristics of these phenomena, then assess the impact these anomalies have on the global and local environment.

Estimated time   One to two 45-minute class periods

Materials   ✔ Internet access
✔ Student handouts from this lesson to be copied:
  • GIS Investigation sheets (pages 175 to 180)
  • Student answer sheets (pages 181 to 184)

Standards and objectives   *National geography standards*

	GEOGRAPHY STANDARD	MIDDLE SCHOOL	HIGH SCHOOL
7	The physical processes that shape the patterns of Earth's surface	The student understands how to predict the consequences of physical processes on Earth's surface.	The student understands the dynamics of the four basic components of Earth's physical systems: the atmosphere, biosphere, lithosphere, and hydrosphere.
15	How physical systems affect human systems	The student understands human responses to variations in physical systems.	The student understands strategies to respond to constraints placed on human systems by the physical environment.
18	How to apply geography to interpret the present and plan for the future	The student understands how to apply the geographic point of view to solve social and environmental problems by making geographically informed decisions.	The student understands how to use geographic knowledge, skills, and perspectives to analyze problems and make decisions.

Objectives

The student is able to:
• Create online maps using GLOBE datasets and download map images into ArcView for analysis.
• Compare and contrast various controls of climate to determine characteristics of El Niño and La Niña weather patterns.
• Describe the potential impact of El Niño at a global, regional, and local level.

GIS skills and tools

 Add layers to a map

 Zoom in to specific areas of a map

- Create maps using the GLOBE Web site and download them
- Georeference GLOBE map images using a world file
- Add georeferenced images to ArcMap
- Analyze images to determine temperature and precipitation patterns
- Compare images and feature classes

For more on geographic inquiry and these steps, see Geographic Inquiry and GIS (pages xxiii to xxv).

Teacher notes

Lesson introduction

Begin the lesson with a brainstorming session on what El Niño is. Record your students' responses on the board. Discuss their ideas: identify correct impressions and important points. Be sure to record this information in a way and place that will allow your students to compare investigative findings with preconceptions. After the initial discussion, provide a brief overview of what El Niño and La Niña are. Use the section of the companion Web site (*www.esri.com/mappingourworld*) to locate additional information if you need it. Explain to the students that in the following GIS Investigation, they will be making observations and comparing El Niño to its countervariation, La Niña.

Student activity

 Before completing this lesson with students, we recommend that you complete it yourself. Doing so will allow you to modify the activity to suit the specific needs of your students. The lesson is designed for students working individually at the computer, but it can be modified to accommodate a variety of instructional settings.

Distribute the GIS Investigation sheets to the class. Explain that this investigation will have them create maps on the Internet using data from the GLOBE Web site. The GLOBE program is a joint effort between several U.S. agencies: NASA (National Aeronautics and Space Administration), NSF (National Science Foundation), and the U.S. State Department, along with more than 100 other countries, to bring real-world environmental science into the classroom. The data displayed at the GLOBE site is a combination of the work of students from around the world and scientists using techniques ranging from visual observation with the naked eye to analysis of data from satellites. GLOBE has a special series of maps dedicated to observation of El Niño and La Niña global events. Your students will create maps online, then download the images into ArcMap with the GIS Investigation as their guide. They will then be free to download other maps and data that might help them to predict the overall impact of El Niño.

 The GIS Investigation has the students saving images from the Internet into a folder named Images in the module 3 data folder. If you would like students to save to another location, be sure to provide them instruction on where and how to save.

Step 5 can be done as part of the assessment. It includes instructions for downloading additional images to analyze El Niño and La Niña.

Things to look for while the students are working on this activity:

- Are students selecting appropriate datasets for gaining an understanding of El Niño and La Niña?
- Are they saving their images into an appropriate folder?

Conclusion   After students have identified a variety of characteristics of El Niño and La Niña, have them share their observations with the class. Compare those observations with notes from the original discussion.

Assessment   *Middle school: Highlights skills appropriate to grades 5 through 8*

Students will write an essay that describes the effects of El Niño and La Niña years on their local area or an area you designate for research. Their observations should be focused on weather patterns (temperature and precipitation). They should provide data and printed maps to support their findings.

High school: Highlights skills appropriate to grades 9 through 12

Students will write an essay that describes the effects of El Niño and La Niña years on their local area or an area you designate for research. Their observations should include weather patterns (temperature and precipitation) and other factors that affect the local economy. For example, if the research area is highly agricultural, then greater than normal rainfall could cause damage to crops. They should provide data and printed maps to support their findings.

Extensions   - Download local precipitation and temperature data for the past 10 years. Identify patterns and attempt to predict future weather events based on historic patterns.
- Have students collect and enter their local climate data into ArcMap for analysis.
- Select a region as unlike your own as possible, and have students study the consequences of El Niño in an unfamiliar setting.
- Check out the Resources by Module section of this book's companion Web site *(www.esri.com / mappingourworld)* for print, media, and Internet resources on the topics of El Niño, La Niña, and the Southern Oscillation.

NAME _____   DATE _____

Sibling Rivalry
An advanced investigation

 Note: Due to the dynamic nature of the Internet, the URLs listed in this lesson may have changed, and the graphics shown below may be out of date. If the URLs do not work, refer to this book's Web site for an updated link: www.esri.com/mappingourworld.

Answer all questions on the student answer sheet handout

Every two to seven years, the climatic phenomenon El Niño occurs over the tropics just south of the equator and off the western coast of Central and South America. In this GIS Investigation, you will record your observations of this event and identify characteristics of both El Niño and its counterpart, La Niña.

Step 1   Visualize and acquire GLOBE data

The GLOBE program (Global Learning and Observations to Benefit the Environment) is a joint effort between several U.S. agencies: NASA (National Aeronautics and Space Administration), NSF (National Science Foundation), and the U.S. State Department, along with more than 100 other countries, to bring real-world environmental science into the classroom. The data that is displayed at the GLOBE site is a combination of the work of students and scientists from around the world using everything from the naked human eye to satellites to make their observations. GLOBE has a special series of maps dedicated to El Niño and La Niña. You will create your maps online and then download the images into ArcMap for further analysis.

a   Go to the GLOBE home page on the Internet at www.globe.gov and click **Enter the GLOBE Site**.

b   Under **GLOBE DATA**, click **Maps and Graphs**.

c   click the **GLOBE Maps** link.

GLOBE Maps   Student and reference data
GLOBE Graphs   Time plots of student data
Diurnal Graphs   For measurements taken more than once per day
GLOBE Sites   Special visualizations of individual GLOBE experiment sites.
Search   Search for schools to map or graph
What's New?   Map and graph enhancements, shapefiles!
Features   Highlighted features of the visualization website
Image Gallery   Additional visualizations related to GLOBE
GLOBE PVA   Personal Visualization Archive - create collections of GLOBE images

The Globe Maps page opens and you see options to click and create your own map.

d   Change the following parameters to create your map:
  - Change the Map size to large.
  - Change the Data Category to La Niña/El Niño/SO

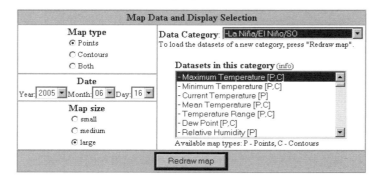

e   Click the Redraw Map button   Redraw map .

Now you see the default map for this particular data source. It is the El Niño Predicted Temperature Anomaly for the month ending March 31, 1999. This is the default map because the winter of 1998–1999 was determined to be an El Niño year. You will change the map parameters so you get a map for sea surface temperatures.

f   Scroll down and change the map parameters to match the settings in the graphic below.

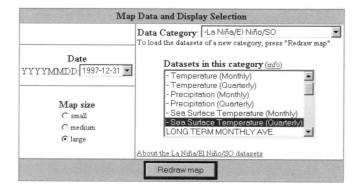

g Click the Redraw map button [Redraw map].

An anomaly means any significant variation from the norm. In this case, the map shows any deviation from the normal average quarterly sea surface temperatures. Any temperature that is within the normal range is identified by the color gray on the map. Temperatures that are above normal are identified in the color range of yellow-orange-red. Temperatures that are below normal are in the color range of aqua-blue-purple. A color legend is located at the top left-hand corner of the map image.

In order to se this map image in ArcMap, you need to save it.

h Right-click on the map image. If you don't see the Save Picture As menu option, right-click again. Click the Save Picture As option to save the image.

i In the Save Picture dialog that opens, navigate to the module 3 images folder (**C:\MapWorld9\Mod3\Data\Images**) or a location your teacher designates.

j Change the name of the image to **ABC_sstemp1297.gif**, where ABC are your initials. Click Save.

k Repeat steps 1f–1j to draw and save a map with the following parameters:
  • Sea Surface Temperature (Quarterly)
  • Date: 1999-03-31
  • Save picture as: **ABC_sstemp0399.gif** (ABC are your initials)

l Minimize your Web browser.

Step 2 Georeference the images

a Use "My Computer" or "Windows Explorer" to navigate to the Images folder (**C:\MapWorld9\Mod3\Data\Images**).

b Copy the file **master.flw** and paste it twice into the same Images folder. Now you have three copies of master.flw in the Images folder.

> *Note: If you did not save your downloaded images to this images folder, you need to paste the two copies of master.flw into the folder where the two images are saved.*

Master.flw is a master georeferencing file (also known as a world file) to use with all the large-format images you retrieve from the GLOBE home page. It will "tell" ArcMap where to place the image.

c   Rename one of the world files to match the file name of the first image you downloaded. You also need to change the file extension from .flw to .gifw. The "gif" tells ArcMap to georeference the .gif file with the same name; the "w" tells ArcMap that it is a world file. For the image named ABC_sstemp1297.gif, you need to rename the world file **ABC_sstemp1297.gifw**.

d   Change the file name of the other pasted world file to match the other downloaded image. It now reads **ABC_sstemp0399.gifw**.

You still have a copy of the original master.flw file in the Images folder. When you need to download additional images later, you can copy this file again.

Step 3   **Start ArcMap and add the georeferenced images**

a   Start ArcMap. Navigate to the module 3 folder (**C:\MapWorld9\Mod3**) and open **Adv3.mxd** (or **Adv3**).

When the map document opens, you see the following layers displayed: Latitude & Longitude, Climate, Countries, and Ocean.

b   Stretch the ArcMap window so the map is bigger.

    c   Click the Add Data button and navigate to the folder where the georeferenced images are saved (**C:\MapWorld9\Mod3\Data\Images**).

d   Add both images (**ABC_sstemp1297.gif** and **ABC_sstemp0399.gif**).

e   A message displays, warning you that the images are missing spatial reference information. In this case, the spatial reference information is not missing—it is simply stored in the "world" file instead of with the image. Click OK to dismiss the message.

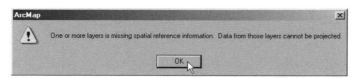

f   Drag the two images up in the table of contents and place them just below Latitude & Longitude.

The two images match up with the other layers in the map. This is because the images are in the same projection (Mollweide) as the World Climate data frame.

> *Note: If you cannot see the images display properly, it is because there was an error in georeferencing. Go back to step 2 and check that you changed the world file to match the file name of the image. Also, check that the extension is "gifw." If the file name or extension is incorrect by even one character, you will not see the image properly. After you correct the file names, replace the image layers in ArcMap.*

g   Save your map document according to your teacher's instructions.

?   *Record the new name of the map document and its location on your answer sheet.*

Step 4   Analyze characteristics of El Niño and La Niña

El Niño and La Niña are climate anomalies that affect sea surface temperature and precipitation. The scientific term for El Niño/La Niña is the "Southern Oscillation." The image maps you are exploring are Predicted Anomaly Maps. They show and describe scientists' predictions for anomalies in sea surface temperature.

a   Click the Window menu and click Magnifier. Drag the Magnification window by its title bar and center it over the legend in the upper left of the map image. (Note: The legend is the colored bar, not the zoom scale above it.) When you let go, you will see a magnified view of the legend.

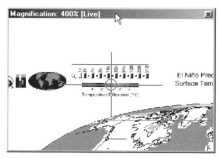
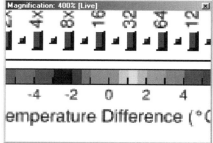

You will reduce the magnification so you can see the entire legend.

b   Right-click the Magnifier window title bar and click Properties. Change the zoom factor to 200%. Click Apply.

c   In the Magnifier Properties, click the Snapshot option. Click OK.

Now the view in the Magnification window is locked on the legend location and will not change when you drag it around or zoom the map.

> *Note: The Snapshot window displays the legend for the topmost image that is turned on in the table of contents.*

d   Zoom in on the area of the Pacific Ocean just west of the coast of South America.

e   Record your temperature observations in the table on your answer sheet, taking note of the legend in the Snapshot window.

f   Go back to the GLOBE Web site and download precipitation maps for the same time period. Follow the procedure in steps 1–3 to create and add the images to ArcMap.

Record your observations of precipitation characteristics for the same time period and location on your answer sheet.

g   Save your map document.

The impact of El Niño and La Niña is felt at a global scale; it is not limited to the small region of the Pacific where these climate anomalies are born.

h If time permits, download other GLOBE map images that you think will help you create a detailed definition for the global climate events of El Niño and La Niña. Compare these patterns with the Climate layer already in the table of contents.

? *Synthesize the information you've recorded and develop a definition of El Niño and La Niña. Write these definitions on your answer sheet.*

i Save your map document.

Step 5 Are El Niño and La Niña equal and opposite?

According to Sir Isaac Newton's third law of motion, for every action there is an equal and opposite reaction. The same holds true for the weather—which is, after all, influenced by physical elements and properties. Is La Niña the "equal and opposite" reaction to El Niño? Is one event better than the other? Your answer depends on where you live.

? *a* Look at the images you've downloaded and focus on different regions of the world. Complete the table on your answer sheet by filling in your observations for El Niño and La Niña's effect on the different regions of the world. Consult your teacher for what kinds of weather phenomena you should investigate under "Other."

? *b* Based on the data you recorded in the previous question, do you think La Niña is equal and opposite to El Niño? Explain your answer.

c Analyze the climate in your region and local area during the 1997 El Niño and the 1999 La Niña. Consult basic climate data (temperature, precipitation, and storms). Refer to the following questions to guide you, and record your answers on the answer sheet:

? *(1) Was one year better than the other for you and your community?*

? *(2) If in one year you received greater than normal rainfall, did your town have problems with flooding?*

? *(3) If weather was unseasonably warm and mild, did outdoor activities such as amusement parks have greater turnout?*

? *(4) How did these things affect the local economy?*

d As an assessment for this GIS Investigation, write an essay that answers the questions above and create ArcMap maps to support your conclusions. Check with your teacher for additional guidelines for this assignment.

Step 6 Exit ArcMap

When you are finished analyzing the El Niño and La Niña data, you may exit ArcMap.

a Save your map document.

b Exit ArcMap by choosing Exit from the File menu.

NAME _____   DATE _____

Student answer sheet
Module 3
Physical Geography II:
Ecosystems, Climate, and Vegetation

Advanced investigation: Sibling Rivalry

Step 3   Start ArcMap and add the georeferenced images

g   Write the new name you gave the map document and where you saved it.

_____   _____
(Name of map document.   **(Navigation path to where map document is saved.**
For example: ABC_Adv3.mxd)   **For example: C:\Student\ABC)**

Step 4   Analyze characteristics of El Niño and La Niña

e, f   Record your temperature observations in the table. Record your observations of precipitation characteristics for the same time period.

YEAR / SO	TEMPERATURE CHARACTERISTICS	PRECIPITATION CHARACTERISTICS
1997 / El Niño		
1999 / La Niña		

h   Synthesize the information you've recorded and develop a definition of El Niño and La Niña. Write these definitions below.

El Niño:

_____
_____
_____
_____

La Niña:

_____
_____
_____
_____

Step 5   Are El Niño and La Niña equal and opposite?

a   Complete the table below by filling in your observations for El Niño and La Niña's effect on the different regions of the world.

WORLD REGION	TEMPERATURE		PRECIPITATION		OTHER		OTHER	
	DEC. 1997 EL NIÑO	MARCH 1999 LA NIÑA	DEC. 1997 EL NIÑO	MARCH 1999 LA NIÑA	DEC. 1997 EL NIÑO	MARCH 1999 LA NIÑA	DEC. 1997 EL NIÑO	MARCH 1999 LA NIÑA
North America								
South America								
Europe								
Africa								

Step 5   Are El Niño and La Niña equal and opposite? (continued)

a   (continued)

WORLD REGION	TEMPERATURE		PRECIPITATION		OTHER		OTHER	
	DEC. 1997 EL NIÑO	MARCH 1999 LA NIÑA	DEC. 1997 EL NIÑO	MARCH 1999 LA NIÑA	DEC. 1997 EL NIÑO	MARCH 1999 LA NIÑA	DEC. 1997 EL NIÑO	MARCH 1999 LA NIÑA
Asia								
Oceania								
Pacific Ocean								

b　Based on the data you recorded in the previous question, is La Niña equal and opposite to El Niño? Explain your answer.

_____

_____

_____

_____

c-1　Was one year better than the other for you and your community?

_____

_____

_____

_____

c-2　If in one year you received greater than normal rainfall, did your town have problems with flooding?

_____

_____

_____

_____

c-3　If weather was unseasonably warm and mild, did outdoor activities such as amusement parks have greater turnout?

_____

_____

_____

_____

c-4　How did these things affect the local economy?

_____

_____

_____

_____

_____

_____

_____

module 4

Human Geography I
Population Patterns and Processes

The distribution and size of the earth's population results from the interplay between the complex forces of growth and migration.

The March of Time: A global perspective
Students will observe and analyze the location and population of the world's largest cities from the year 100 C.E. through 2000 C.E. In this investigation, students will describe spatial patterns of growth and change among the world's largest urban centers during the past two thousand years and speculate on reasons for the patterns they observe.

Growing Pains: A regional case study of Europe and Africa
In this lesson, students will compare the processes and implications of population growth in the world's fastest and slowest growing regions: Sub-Saharan Africa and Europe. Through the analysis of standard of living indicators in these two regions, students will explore some of the social and economic implications of rapid population growth.

Generation Gaps: An advanced investigation
Students will use global data to investigate variations in population age structure and the relationship of age structure to a country's rate of natural increase. After exploring global patterns, they will download and map U.S. Census data to examine the patterns of age structure in their own community.

The March of Time
A global perspective

Lesson overview

Students will observe and analyze the location and population of the world's largest cities from the year 100 C.E. through 2000 C.E. In this investigation, students will describe spatial patterns of growth and change among the world's largest urban centers during the past two thousand years and speculate as to reasons for the patterns they observe.

Estimated time   One to two 45-minute class periods

Materials   ✔ Student handouts from this lesson to be photocopied:
- GIS Investigation sheets (pages 195 to 203)
- Student answer sheets (pages 204 to 207)
- Assessment(s) (pages 208 to 213)
- Photocopy of the transparency (page 193)

Standards and objectives   *National geography standards*

	GEOGRAPHY STANDARD	MIDDLE SCHOOL	HIGH SCHOOL
1	How to use maps and other geographic representations, tools, and technologies to acquire, process, and report information from a spatial perspective	The student knows and understands how to use maps to analyze spatial distributions and patterns.	The student knows and understands how to use geographic representations and tools to analyze and explain geographic problems.
12	The processes, patterns, and functions of human settlement	The student knows and understands the spatial patterns of settlement in different regions of the world.	The student knows and understands the functions, sizes, and spatial arrangements of urban areas.
17	How to apply geography to interpret the past	The student knows and understands how the spatial organization of a society changes over time.	The student knows and understands how processes of spatial change affect events and conditions.

Objectives
The student is able to:
- Describe the location and size of the world's largest cities over time.
- Identify historical events and periods that influenced the location of cities throughout history.
- Explain the exponential pattern of growth among the world's urban populations in the past thousand years.

GIS skills and tools

 Find a specific feature in a layer

 Learn more about a selected feature

 Label selected features in a layer

 Move or unselect a graphic label

 Zoom to the full extent of the map

 Add a layer to the map

- Turn layers on and off
- Find a feature and select it
- Zoom to a layer's extent
- Open the attribute table for a layer
- Sort data in descending order
- Select a record in a table
- Clear selected features in all layers
- Zoom to a selected feature

For more on geographic inquiry and these steps, see Geographic Inquiry and GIS (pages xxiii to xxv).

Teacher notes

Lesson introduction

Begin the lesson by dividing the students into pairs. Challenge each group to name the ten most populated cities in the world today. After five minutes, each group should share their list with the rest of the class. Use the blackboard or an overhead projector to tally the cities mentioned as each group reports. Based on the tally, circle the cities that were listed most often. Tell the class that they are going to do a GIS Investigation that will use real data to identify the 10 most populated cities in the world from the last two thousand years.

Before beginning the GIS Investigation, engage students in a discussion about the cities that are circled on their list.

* What do they know about these cities?
* In what countries are these cities located?
* How many people live in these urban centers?
* Has anyone ever visited one of these cities?
* Can they think of any reasons why some cities grow to be so large?

Student activity

 Before completing this lesson with students, we recommend that you complete it as well. Doing so will allow you to modify the activity to accommodate the specific needs of your students.

After the initial discussion, have the students work on the computer component of the lesson. Ideally, each student should be at an individual computer, but the lesson can be modified to accommodate a variety of instructional settings.

Distribute the March of Time GIS Investigation to the students. Explain that in this activity, they will use GIS to observe and analyze the location and size of the world's 10 largest cities in seven different time periods from 100 C.E. to the year 2000. The activity sheets will provide them with detailed instructions for their investigations. As they investigate, they will identify changes in both location and size of the world's largest cities, and speculate on possible reasons for the patterns they describe.

In addition to instructions, the activity sheets include questions to help students focus on key concepts. Some questions will have specific answers while others will require creative thought.

Things to look for while the students are working on this activity:

* Are the students using a variety of GIS tools?
* Are the students answering the questions as they work through the procedure?
* Are students asking thoughtful questions throughout the investigation?

Conclusion

Use the March of Time transparency to compare student ideas and observations from the GIS Investigation. Summarize student observations on the transparency as they share and discuss their observations with the class. Use this discussion as a forum to elaborate on relevant themes in world history (such as the decline of the Roman Empire or the Industrial Revolution) and the value of using geography's spatial perspective to interpret the past. Ideally, this discussion should take place in the classroom with a projection device that displays the ArcMap map document as students discuss it. If this is not possible, teachers may choose to conduct the discussion while students are still working on the computer individually or in small groups.

Assessment

Middle school: Highlights skills appropriate to grades 5 through 8

The middle school March of Time assessment asks the students to create a line graph of the most populous cities for each time period studied. Students will use the graph as a reference when writing an essay that compares two major time periods they have mapped. The essay will illustrate their knowledge and understanding of the changes in spatial patterns of major population centers.

High school: Highlights skills appropriate to grades 9 through 12

The high school March of Time assessment asks the students to create a line graph of the most populous cities for each time period studied. Students will use the graph as a reference when writing an essay that compares and contrasts three or more time periods they have mapped. The essay will illustrate their knowledge and understanding of the changes in spatial patterns of major population centers. In addition, they will take the historical information from the map and their own research to make predictions about future locations of major population centers.

Extensions

- Explore the relationship between physical characteristics of the landscape and the location of the world's most populated cities by adding layers reflecting world climate data (module 3) and ecoregion data (module 7).

- Assign students to conduct research on the historic cities and time periods reflected in the project.

- Ask students what questions this activity has raised in their minds. Questions generated could serve as a springboard for further research and spatial analysis.

- Create map layouts in ArcMap of the historic time periods. Have students print these layouts and use them in reports about the ancient European cities.

- Check out the Resources by Module section of this book's companion Web site *(www.esri.com/mappingourworld)* for print, media, and Internet resources on the topics of population, historical time periods, and ancient cities.

Transparency master

YEAR C.E.	LARGEST CITY	POPULATION OF LARGEST CITY	MAJOR DIFFERENCES FROM PREVIOUS TIME PERIOD
100			
1000			
1500			
1800			
1900			
1950			
2000			

NAME _____   DATE _____

The March of Time
A GIS investigation

ACQUIRE
ASK
EXPLORE
ACT
ANALYZE

Answer all questions on the student answer sheet handout

Step 1   Start ArcMap

a   Double-click the ArcMap icon on your computer's desktop.

b   If the ArcMap start-up dialog appears, click **An existing map** and click OK. Then go
to step 2b.

Step 2 Open the Global4.mxd file

a In this exercise, a map document has been created for you. To open it, go to the File menu and choose **Open**.

b Navigate to the module folder (**C:\MapWorld9\Mod4**) and choose **Global4.mxd** (or **Global4**) from the list.

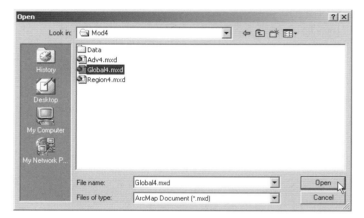

c Click Open.

When the map document opens, you see a world map. The table of contents has a data frame called March of Time.

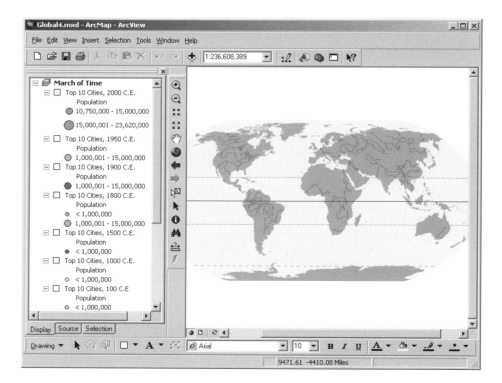

Step 3   Look at cities in 100 C.E.

a   The table of contents of the map is located on the left side of the ArcMap window. Scroll down the table of contents until you see the layer called **Top 10 Cities, 100 C.E.** Click the box to the left of the name to turn on the layer.

This places a check mark in the box and draws points on the map showing the locations of cities.

 Note: The abbreviation "C.E." stands for "Common Era."

b   On the answer sheet, write three observations about the location of the ten largest cities in the world in 100 C.E.

(1)   Where are they located on the earth's surface?

(2)   Where are they located in relation to each other?

(3)   Where are they located in relation to physical features?

c   What are possible explanations for the patterns you see on this map?

Step 4   Find historic cities and identify modern cities and countries

You will use the Find tool to locate the historic cities on the map.

a   Click the Find tool.

b   Type **Carthage** in the Find box, and then click Find.

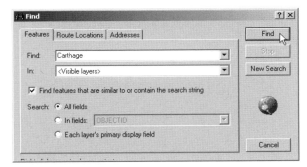

The Find dialog expands to show a results box. A row for Carthage is highlighted in the box.

c   Move the Find dialog so you can see all of the cities on your map. Then right-click the highlighted row for Carthage and click Select feature(s).

The dot that represents Carthage is highlighted blue on the map. To complete the table in the answer sheet, you need to identify the modern city located in the same place.

d Right-click the highlighted row for Carthage again, and this time click Identify feature(s).

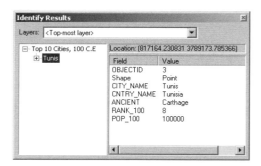

The information that you need to complete the table on the answer sheet appears in the Identify Results window.

? **e** Write the modern city name and modern country name for Carthage in the table on the answer sheet.

f Move the Identify Results window so it is off the Find dialog and the map by clicking its title bar and dragging it out of the way.

? **g** Use the Find and Identify tools to complete the information in the table on the answer sheet.

h Close the Identify Results and Find windows.

i Click Selection on the Main menu and click Clear Selected Features.

Step 5 **Find the largest city of 100 C.E. and label it**

? **a** What's your estimate of how many people lived in the world's largest city in 100 C.E.?

b Right-click the Top 10 Cities, 100 C.E. layer in the table of contents, and choose Zoom to Layer. The map zooms to the region of the world where these cities are located.

c Right-click the Top 10 Cities, 100 C.E. layer again and choose Open Attribute Table.

> *Note: If you see a lot of extra gray area under the last record, drag the bottom edge of the window up so that the table takes up less space on your screen.*

d Look at the table. Remember: each row in this table is associated with one of the city points on the map.

e   Scroll right and locate the field name POP_100. Click the field name to highlight the whole column.

f   Right-click the field name. Choose Sort Descending to list the cities from largest to smallest in terms of population.

Note: In the POP_100 column, the value –99 for Suzhou indicates that no data is available. It does not mean that there was a population of –99 in Suzhou.

g   Evaluate the table and answer the following questions:

   (1)   *What was the largest city in 100 C.E.?*

   (2)   *What was the population of the world's largest city in 100 C.E.?*

h   Click the gray box at the beginning of the row with the largest city.

The selected record turns blue in the attribute table and so does its corresponding dot on the map.

i   Close the Attributes of Top 10 Cities, 100 C.E. attribute table by clicking the ✕ in the upper right of the window.

j   Make sure the Draw toolbar is displayed. If it is not, right-click in the gray space next to the Help menu and click Draw. Dock the toolbar at the bottom of the ArcMap window.

k   Click the down arrow to the right of the New Text tool in the Draw toolbar and select the Label tool.

l   Move the Label Tool Options dialog off the map if necessary. You will accept the default options in the Label Tool Options dialog.

m　Hover your cursor over the selected city. When the MapTip displays the city name (Rome), click to add the label to the map.

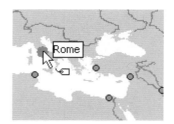

 　　:ℓ: Hints: If you accidentally label the wrong feature, click the Undo button and try again. If you want to reposition the label, use the Select Elements tool to drag the text where you want it to go. If you want to change the font, size, or style of the text, use the options on the Draw toolbar.

n　Click the Select Elements tool and click anywhere on the map away from the label to remove the selection box. The Label Tool Options box also closes.

o　Right-click the Top 10 Cities, 100 C.E. layer and click Selection, then Clear Selected Features.

p　Click the Full Extent button to see the entire world in the map.

Step 6　Look at cities in 1000 C.E. and label the most populous city

a　Turn on the Top 10 Cities, 1000 C.E. layer.

```
⊟ ☐ Top 10 Cities, 1500 C.E.
     Population
     ●  < 1,000,000
⊟ ☑ Top 10 Cities, 1000 C.E.
     Population
     ○  < 1,000,000
⊟ ☑ Top 10 Cities, 100 C.E
     Population
     ○  < 1,000,000
```

b　Right-click the Top 10 Cities, 1000 C.E. layer and choose Zoom to Layer.

c　Look for similarities and differences between these points and the cities of 100 C.E. in the location and distribution of the world's largest cities.

　　(1)　What notable changes can you see from 100 C.E. to 1000 C.E.?

　　(2)　What similarities can you see between 100 C.E. and 1000 C.E.?

Now you will find, identify, and label the most populous city of 1000 C.E.

d　Right-click the Top10 Cities, 1000 C.E. layer and choose Open Attribute Table. This shows you all the attribute data associated with the yellow dots on the map.

e　Locate the POP_1000 field. Click the field name to highlight the column.

f　Right-click the POP_1000 name and choose Sort Descending to list the cities from largest to smallest in terms of population. Evaluate the table and answer the following questions:

　　(1)　What was the largest city in 1000 C.E.?

　　(2)　What was the population of the world's largest city in 1000 C.E.?

g   Click the gray box at the beginning of the row with the world's largest city in the attribute table. The selected row turns blue in the attribute table and its corresponding dot on the map turns blue.

h   Close the Attributes of Top 10 Cities, 1000 C.E. table.

i   Click the Label tool. Click the selected city on the map to label it.

j   Click the Select Elements tool.

k   Reposition the text and use the options on the Draw toolbar to change the font, size, or style of the label text, if desired.

l   Click anywhere in the ocean with the Select Elements tool to remove the selection box and close the Label Tool Options dialog.

m   Right-click the Top 10 Cities, 1000 C.E. layer and choose Selection, then Clear Selected Features to make all the cities in 1000 C.E. yellow again.

Step 7   **Observe and compare other historic periods**

a   Click the Full Extent button.

b   One at a time, turn on each of the remaining five layers representing the years 1500, 1800, 1900, 1950, and 2000.

As you do so, explore the map to complete the table on the answer sheet. Refer back to steps 5 and 6 to get information for 100 C.E. and 1000 C.E.

> *Hint: Turn layers on and off or change the order of the layers by moving them up or down in the table of contents to see each of the largest cities from all the time periods.*

Step 8   **State a hypothesis**

a   Zoom to the full extent of the map.

b   Look back at the maps and at the table you completed in step 7. In the table on the answer sheet, complete the left column by stating which periods in history are associated with the greatest changes. In the right column, state possible explanations for the changes that you see.

Step 9   **Investigate cities in the present time**

a   As a class, before you began the GIS Investigation, you made a guess about the top 10 cities in the world today. Look at the Top 10 Cities, 2000 C.E. layer to see if any of your predictions are correct.

    (1)   How many of your original guesses are among the cities in Top 10 Cities, 2000 C.E.?

    (2)   Which cities did you successfully guess?

Most likely, there are cities on your list that are not in the Top 10 Cities, 2000 C.E. layer. In order to look at population data on these other cities, you will add a layer with the top 30 cities.

b   Click the Add Data button.

c Navigate to the module 4 layer files folder (**C:\MapWorld9\Mod4\Data\LayerFiles**). Select the Top 30 Cities, 2000.lyr layer file and click Add.

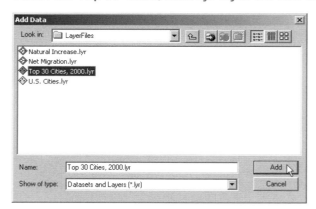

The Top 30 Cities, 2000 layer is added to your map.

 d Use the Find tool to locate the other cities that you guessed in the beginning of this activity. Make sure to select Top 30 Cities, 2000 as the layer to search in the Find dialog.

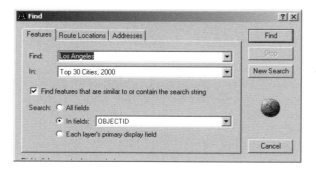

e Right-click in the results box on the row for the city you found and choose Identify feature(s).

The Identify Results dialog appears with information like that in the graphic below.

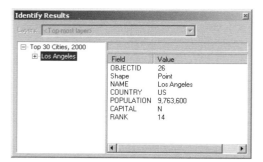

? f In the table on your answer sheet, write the city's name, population, and rank.

? g Continue to fill out the table for the other cities in your list, or cities that interest you. Use the Find and Identify tools, or open the Top 30 Cities, 2000 layer attribute table.

Note: If you have cities on your list that are not in the top 30, fill in the name, leave the Population column blank, and write >30 in the Rank column.

h   When you are finished, close the Identify Results window, and the table if it is open.

❓   *i*   In general, how far are these other cities from the top 10?

Step 10   Exit ArcMap

In this exercise, you explored population data from 100 C.E. through the year 2000. You used ArcMap to find, identify, and label the world's most populous cities during different time periods. After analyzing this data, you added data for more large cities and explored populations of your cities of interest.

a   Ask your teacher for instructions on where to save this map document and on how to rename the project.

b   If you are not going to save the project, exit ArcMap by clicking the File menu and clicking Exit. When asked if you want to save changes to Global4.mxd (or Global4), click No.

NAME _____   DATE _____

Student answer sheet

Module 4
Human Geography I: Population Patterns and Processes

Global perspective: The March of Time

Step 3   Look at cities in 100 C.E.

b-1   Where are they located on the earth's surface?

_____

_____

_____

_____

b-2   Where are they located in relation to each other?

_____

_____

_____

_____

b-3   Where are they located in relation to physical features?

_____

_____

_____

_____

c   What are possible explanations for the patterns you see on this map?

_____

_____

_____

_____

_____

_____

_____

_____

Step 4   Find historic cities and identify modern cities and countries

e, g   Use the Find and Identify tools to complete the information in the table below.

HISTORIC CITY NAME	MODERN CITY NAME	MODERN COUNTRY
Carthage		
Antioch		
Peshawar		

Step 5   Find the largest city of 100 C.E. and label it

a   What's your estimate of how many people lived in the world's largest city in 100 C.E.?

_____

g-1   What was the largest city in 100 C.E.?   _____

g-2   What was the population of the world's largest city in 100 C.E.?   _____

Step 6   Look at cities in 1000 C.E. and label the most populous city

c-1   What notable changes can you see from 100 C.E. to 1000 C.E.?

_____

_____

_____

_____

_____

c-2   What similarities can you see between 100 C.E. and 1000 C.E.?

_____

_____

_____

_____

_____

f-1   What was the largest city in 1000 C.E.?   _____

f-2   What was the population of the world's largest city in 1000 C.E.?   _____

Step 7   Observe and compare other historic periods

b   Explore the map to complete the table below. Refer back to steps 5 and 6 to get information for 100 C.E. and 1000 C.E.

YEAR C.E.	LARGEST CITY	POPULATION OF LARGEST CITY	MAJOR DIFFERENCES FROM PREVIOUS TIME PERIOD
100			
1000			
1500			
1800			
1900			
1950			
2000			

Step 8   State a hypothesis

b   In the table below, state in the left column which periods in history are associated with the greatest changes. In the right column, state possible explanations for the changes you see.

TIME PERIODS OF SIGNIFICANT CHANGE	EXPLANATION FOR CHANGE

Step 9   Investigate cities in the present time

a-1   How many of your original guesses are among the cities in Top Ten Cities, 2000 C.E.?

_____

a-2   Which cities did you successfully guess?

_____
_____
_____
_____
_____

f, g   In the table below, write the name, population, and rank for the other cities on your list. For cities that are not in the top 30, leave the population column blank and write >30 in the rank column.

CITY NAME	POPULATION	RANK

i   In general, how far are these other cities from the top 10?

_____
_____

NAME _____   DATE _____

The March of Time
Middle school assessment

Part 1

Use the information you collected in steps 5, 6, and 7 of the GIS Investigation to complete the attached graph. Plot a line graph that shows the population of the largest city in each time period. Be sure to label the points and title the graph appropriately.

Part 2

Refer to the ArcMap map document (Global4 or the one you saved) and write an essay that compares two of the time periods that you studied. You may use additional sources such as your history and geography books, encyclopedias, and the Internet to help you with your comparisons. On a separate piece of paper, address the following questions in your essay:

1   What was the primary means of transportation in each time period?
2   How did trade between various cities influence the location of the places?
3   How did advancements in technology affect the location of places?
4   How did distances between major cities change throughout time?
5   What physical features played important roles in the locations of cities (location relative to fresh water and waterways, elevation, etc.)?

Use the remainder of this page as a place to brainstorm for your essay.

The March of Time

Assessment rubric

Middle school

STANDARD	EXEMPLARY	MASTERY	INTRODUCTORY	DOES NOT MEET REQUIREMENTS
The student knows and understands how to use maps to analyze spatial distributions and patterns.	Creates an accurate and well-labeled graph of the most populous cities over time. Uses GIS to analyze population patterns and how they change throughout time, compares this with his/her original predictions, and makes predictions on future trends.	Creates an accurate graph of the most populous cities over time. Uses GIS to analyze population patterns and how they change throughout time, and compares this with his/her original predictions.	Creates a graph of the most populous cities over time that is inaccurate or mislabeled in some places. Uses GIS to view patterns of the location of major population centers.	Does not complete a graph of the most populous cities over time. Has difficulty identifying location patterns of major population centers.
The student knows and understands the spatial patterns of settlement in different regions of the world.	Describes in detail how major cities have changed and are influenced by their location relative to physical features, and elaborates on what these features are, and if they were a positive or negative force on the cities.	Identifies how major cities have changed and are influenced by their location, relative to physical features such as water and other major cities.	Identifies how the major cities (population centers) have spread throughout the various regions of the world.	Identifies the location of major cities, but does not identify a pattern of settlement.
The student knows and understands how the spatial organization of a society changes over time.	Compares the major cities of at least two time periods and makes predictions as to why changes occurred over time. Provides detailed evidence of how things such as technology and transportation influenced the societies that lived in these cities.	Compares the major cities of two time periods, and makes predictions as to why changes in these population centers occurred over time.	Identifies characteristics of the major population centers of two different time periods, but does not draw comparisons between the two eras.	Identifies a few characteristics of major population centers of two different time periods, or only describes characteristics of one time period's settlements.

This is a four-point rubric based on the National Standards for Geographic Education. The "Mastery" level meets the target objective for grades 5–8.

NAME _____   DATE _____

The March of Time
High school assessment

Part 1

Use the information you collected in steps 5, 6, and 7 of the GIS Investigation to complete the attached graph. Plot a line graph that shows the population of the largest city in each time period. Be sure to label the points and title the graph appropriately.

Part 2

Refer to the ArcMap map document (Global4 or the one you saved) and write an essay that compares two of the time periods that you studied. Use at least three additional sources such as your history and geography books, encyclopedias, and the Internet to help you with your comparisons. On a separate piece of paper, address the following questions in your essay:

1 What was the primary means of transportation in each time period?
2 How did trade between various cities influence the location of the places?
3 What physical features played important roles in the locations of cities (location relative to fresh water and waterways, elevation, etc.)?
4 Are there any unusual shifts in population centers that you see from one period to the next? What do you think caused these changes?
5 How did advancements in technology affect the location of places?
6 Predict how future advancements in technology may affect the location of population centers in the next hundred years.

Use the remainder of the page as a place to brainstorm for your essay.

The March of Time

Assessment rubric

High school

STANDARD	EXEMPLARY	MASTERY	INTRODUCTORY	DOES NOT MEET REQUIREMENTS
The student knows and understands how to use geographic representations and tools to analyze and explain geographic problems.	Creates an accurate and well-labeled graph of the most populous cities over time. Uses GIS to analyze how population patterns change throughout time, compares this with his/her original thoughts, and makes predictions on future trends. Uses GIS to create an original map to illustrate these points.	Creates an accurate graph of the most populous cities over time. Uses GIS to analyze how population patterns change throughout time, compares this with his/her original thoughts, and makes predictions on future trends.	Creates a graph of the most populous cities over time that is inaccurate or mislabeled in some places. Uses GIS to analyze population patterns and how they change throughout time.	Does not complete a graph of the most populous cities over time. Uses GIS to analyze population patterns, but has difficulty identifying how they change throughout time.
The student knows and understands the functions, sizes, and spatial arrangements of urban areas.	Elaborates, by using visual and written products, on characteristics of major cities throughout time, and how factors such as trade routes and technology influenced the growth of these places.	Identifies characteristics of major cities throughout time and how factors such as trade routes and technology influenced the growth of these places.	Identifies characteristics of major cities throughout time.	Identifies major cities and a few characteristics of these places.
The student knows and understands how processes of spatial change affect events and conditions.	Cites specific examples of how changes in politics, transportation, and so on caused major population centers to shift from one location to another throughout time.	Identifies how changes in politics, transportation, and so on influenced the location of major population centers throughout time.	Lists one or two major cities that changed from one time period to the next and provides explanation for the change.	Lists one or two major cities that changed from one time period to the next.

This is a four-point rubric based on the National Standards for Geographic Education. The "Mastery" level meets the target objective for grades 9–12.

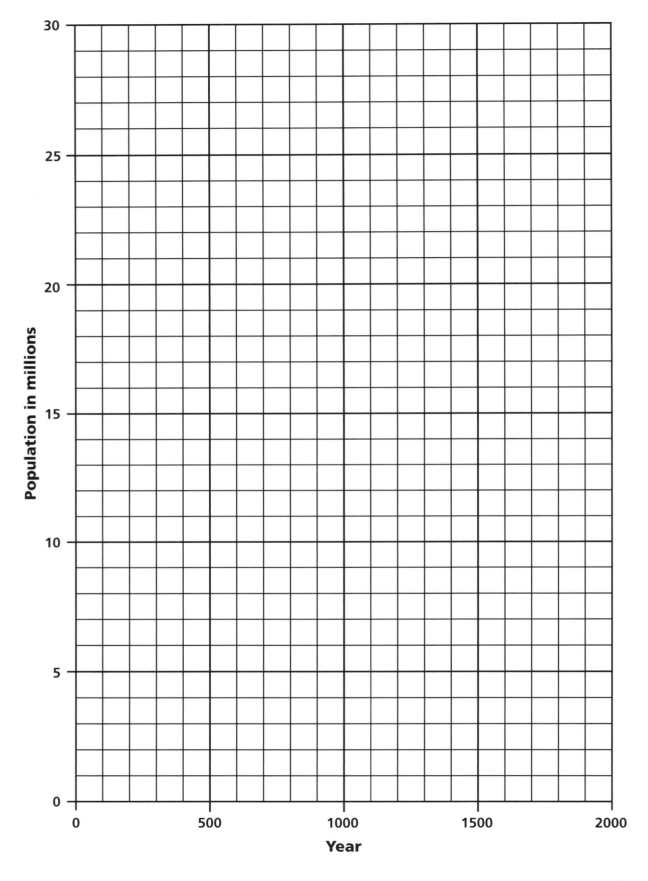

Answer key to graph

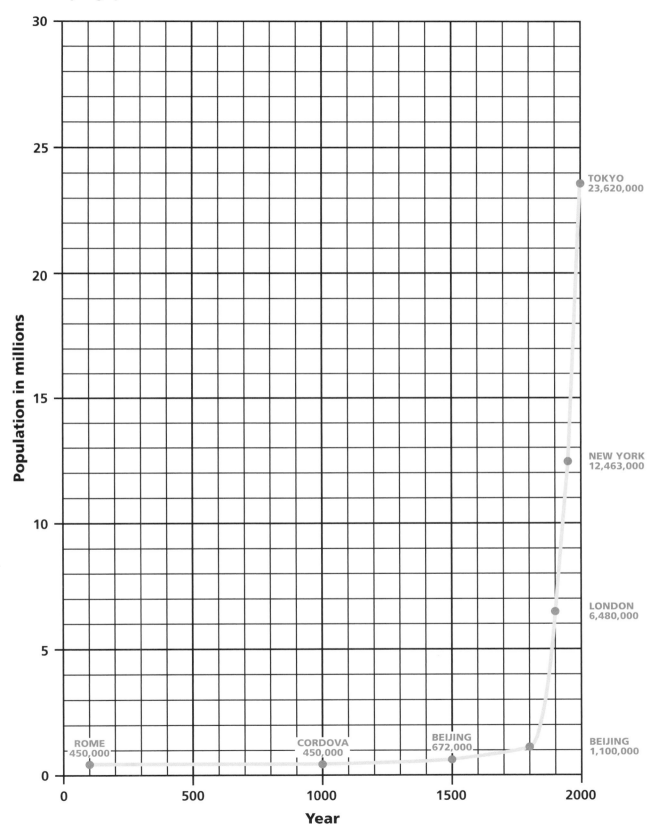

Growing Pains
A regional case study of Europe and Africa

Lesson overview

In this lesson, students will compare the processes and implications of population growth in the world's fastest and slowest growing regions: sub-Saharan Africa and Europe. Through the analysis of standard of living indicators in these two regions, students will explore some of the social and economic implications of rapid population growth.

Estimated time   Two to three 45-minute class periods

Materials   ✔ Large sheets of construction paper or drawing paper and markers—one sheet of paper and one marker per student

   ✔ Student handouts from this lesson to be copied:
   - GIS Investigation sheets (pages 219 to 230)
   - Student answer sheets (pages 231 to 235)
   - Assessment(s) (pages 236 to 239)

Standards and objectives   *National geography standards*

GEOGRAPHY STANDARD	MIDDLE SCHOOL	HIGH SCHOOL
1 How to use maps and other geographic representations, tools, and technologies to acquire, process, and report information from a spatial perspective	The student knows and understands how to use maps to analyze spatial distributions and patterns.	The student knows and understands how to use geographic representations and tools to analyze and explain geographic problems.
9 The characteristics, distribution, and migration of human populations on Earth's surface	The student knows and understands the demographic structure of a population.	The student knows and understands trends in world population numbers and patterns.
18 How to apply geography to interpret the present and plan for the future	The student knows and understands how to apply the geographic point of view to social problems by making geographically informed decisions.	The student knows and understands how to use geographic knowledge, skills, and perspectives to analyze problems and make decisions.

Standards and objectives (continued)

Objectives

The student is able to:

- Describe the fundamentals of population growth by explaining the relationship between birth rate, death rate, and natural increase.
- Identify the fastest and slowest growing regions in the world today.
- Explain the socioeconomic implications of rapid population growth.
- Explain and understand the slow population growth in Europe and how standard of living indicators are affected.
- Explain and understand the rapid population growth in Africa and how standard of living indicators are affected.

GIS skills and tools

 View attributes for a feature on the map

 Zoom in on the map

 Add a layer to the map

 Zoom to the full map extent

 Go back to the previous map extent

 View the entire layout page

 Change the layout template

 Select, move, and resize layout elements

- Turn layers on and off
- Expand and activate a data frame
- Create and print two layouts using different data frames

For more on geographic inquiry and these steps, see Geographic Inquiry and GIS (pages xxiii to xxv).

Teacher notes

Lesson introduction

Begin the lesson with a discussion of world population growth. Remind the students that the world's population reached six billion in 1999 and continues to grow.

Consider the following questions:
- Why is the earth's population growing?
- Are all regions growing at the same rate of speed?
- What does "overpopulation" mean and at what point would you characterize the world as overpopulated?

If your community or state is growing, that would be a good starting point for this discussion. How has population growth affected your community or state? Do students see this as a good thing or a bad thing?

Student activity

 Before completing this lesson with students, we recommend that you work through it yourself. Doing so will allow you to modify the activity to accommodate the specific needs of your students.

After the initial discussion, have the students work on the computer component of the lesson. Ideally, each student should be at an individual computer, but the lesson can be modified to accommodate a variety of instructional settings.

Distribute the GIS Investigation handout to the students. Explain that in this activity, they will use GIS to identify the regions of the world where population is growing fastest and where it is growing slowest. In addition, they will use GIS to compare characteristics of regions with slow and rapid population growth. As they compare, they will gather information and form a hypothesis about the relationship between a region's rate of population growth and its standard of living.

The GIS activity will provide students with detailed instructions for their investigations. In addition to the instructions, the handout includes questions to help students focus on key concepts. Some questions will have specific answers while others call for speculation and require creative thought.

Things to look for while the students are working on this activity:
- Are the students using a variety of tools?
- Are the students answering the questions as they work through the procedure?
- Do students need help with the lesson's vocabulary?

Conclusion

When the class has finished the GIS Investigation, give each student a piece of drawing paper on which to print the hypothesis they wrote in step 7 of the GIS Investigation. Students should tape their hypothesis to the wall or blackboard where they will provide the focus for a concluding discussion. Look for similarities and differences among the student hypotheses. Allow students to question each other to clarify confusing or contradictory statements. If possible, try to reach consensus about the relationship between a country's rate of natural increase and its standard of living based on evidence from the GIS Investigation.

Assessment

Middle school: Highlights skills appropriate to grades 5 through 8

The middle school assessment has the students taking the role of a special liaison to the United Nations, in charge of establishing partnerships between nations of slow and fast growth. The students will select two countries—one with fast growth and one with slow growth. They will use their work in the GIS Investigation to identify issues critical to each country and devise a way the countries can form a partnership to improve the standard of living in each place.

High school: Highlights skills appropriate to grades 9 through 12

The high school assessment has the students taking the role of a special liaison to the United Nations, in charge of establishing partnerships between nations of slow and fast growth. The students will select two groups of countries—one with fast growth and one with slow growth. They will use their work in the GIS Investigation to identify issues critical to each group and devise a way the countries can form a partnership to improve the standard of living for all countries involved.

Extensions

- Explore and map additional attributes from the module 4 data folder. Look for additional social and economic implications of rapid population growth.
- Students can test their hypothesis about the relationship between standard of living, net migration, and the rate of natural increase, by investigating countries in other parts of the world.
- Assign an African country to each student and ask them to explore the population, birth rate, death rate, and standard of living indicators of the country. Create map layouts of the individual countries and use them in a report about the African country.
- Use ArcMap attribute queries to identify countries that do not match your hypothesis. Try to figure out an explanation for these anomalies.
- Explore gender differences in standard of living in Europe and/or sub-Saharan Africa. Using the module data, map and analyze male and female life expectancy, male and female literacy rates, and male and female infant and child mortality.
- Conduct research on the impact of HIV/AIDS on death rates and population growth in sub-Saharan Africa.
- Check out the Resources by Module section of this book's companion Web site *(www.esri.com / mappingourworld)* for print, media, and Internet resources on the topics of Africa, Europe, demographics, and standard of living indicators.

NAME _____   DATE _____

Growing Pains
A GIS investigation

Answer all questions on the student answer sheet handout

In this GIS Investigation, you will observe and analyze population growth by looking at the natural increase of different countries. You will focus on Africa, the fastest growing region in the world, and Europe, the slowest growing region in the world. You will analyze the standard of living indicators for each region and form a hypothesis about the relationship between population growth and standard of living indicators.

Step 1   Start ArcMap

a   Double-click the ArcMap icon on your computer's desktop.

b   If the ArcMap start-up dialog appears, click **An existing map** and click OK. Then go to step 2b.

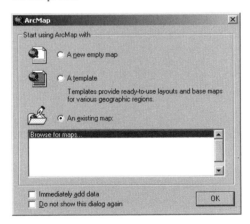

Step 2   Open the Region4.mxd file

a   In this exercise, a map document has been created for you. To open it, go to the File menu and choose **Open**.

b   Navigate to the module folder (**C:\MapWorld9\Mod4**) and choose **Region4.mxd** (or **Region4**) from the list.

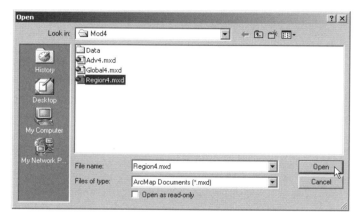

c   Click Open.

d   When the map document opens, click the plus sign next to Population Growth to expand the data frame legend in the table of contents.

You see a world map with two layers turned on (Countries and Ocean). The check mark next to a layer name tells you the layer is turned on and visible on the map. Two layers, Birth Rate and Death Rate, are listed in the table of contents, but are not turned on.

Step 3   Compare birth rate and death rate data

The world's population is growing because there are more births than deaths each year. This fact can be expressed as a simple formula:

Birth rate   =   BR   BR – DR = NIR
Death rate   = – DR

Natural increase   =   NIR

In the first part of this activity, you will compare birth rates and death rates around the world to see if you can identify the regions that are growing fastest and slowest.

World Vital Events

World Vital Events Per Time Unit: 2004

(Figures may not add to totals due to rounding)

Time unit	Births	Deaths	Natural increase
Year	129,108,390	56,540,896	72,567,494
Month	10,759,033	4,711,741	6,047,291
Day	352,755	154,483	198,272
Hour	14,698	6,437	8,261
Minute	245	107	138
Second	4.1	1.8	2.3

Source: U.S. Bureau of the Census, International Data Base

This table from the U.S. Census Bureau shows the number of births, deaths, and rate of natural increase for the world population for every year, month, day, hour, and second in 2004.

a   **Click the box to the left of the Countries layer in the table of contents to turn the layer off.**

b   **Click the box next to Birth Rate to turn on the layer. This layer shows the number of births for every one thousand people in a country.**

☐ ☑ Birth Rate
   Births/1000
   ☐ 7.80 - 15.12
   ☐ 15.13 - 22.60
   ■ 22.61 - 29.81
   ■ 29.82 - 39.63
   ■ 39.64 - 51.45
   ☐ No Data
☐ ☐ Countries
   ☐

(1)   Which world region or regions have the highest birth rates?

(2)   Which world region or regions have the lowest birth rates?

c   **Turn on the Death Rate layer by clicking the box to the left of its name.**

(1)   Which world region or regions have the highest death rates?

(2)   Which world region or regions have the lowest death rates?

d   Turn these layers on and off to compare them easily.

?   *(1)   If the overall rate of growth is based on the formula BR – DR = NI, which world regions do you think are growing the fastest?*

?   *(2)   Which world regions do you think are growing the slowest?*

e   Turn off the Death Rate layer.

f   You can use the Identify tool to learn more about the birth and death rates of specific countries. Click the Identify tool and choose Birth Rate from the Layers list in the Identify Results window.

g   Move the Identify Results window so you can see the map. (Hint: To move the window, click the window's title bar and drag the window to the desired location.)

h   Move your cursor over an African country where the birth rate is very high. Click the country. Your Identify Results window should look similar to the one below:

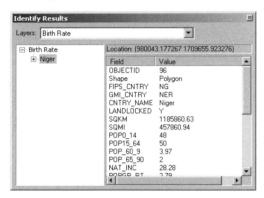

i   Scroll down the list.

This layer contains a lot of information about each country. As you scroll down the list, you see attribute field names in one column and attribute values in another. The birth rate field is abbreviated as "BRTHRATE" and the death rate field is abbreviated as "DTHRATE." In the example above, Niger has a birth rate of 51.45 births per 1,000 living people, and a death rate of 23.17 deaths per 1,000 living people.

j   Click the × in the upper right corner of the Identify Results window to close it.

k   Click the Zoom In tool and draw a box around Africa and Europe.

?   *l*   Choose two European countries and two African countries and record their birth and death rates in the table on the answer sheet. Use the Identify tool, as described above, to find information on your chosen countries.

?   *m*   List three questions that the Birth Rate and Death Rate maps raise in your mind.

Step 4   Add the Natural Increase layer

You can test your predictions of the fastest and slowest growing regions (step 3d) by adding the Natural Increase layer to your map. This layer shows the yearly increase in population that results from the difference between births and deaths in each country.

a   Click the Add Data button.

b   Navigate to the LayerFiles folder within the module 4 Data folder (**C:\MapWorld9 \Mod4\Data\LayerFiles**). Select **Natural Increase.lyr** from the list and click Add.

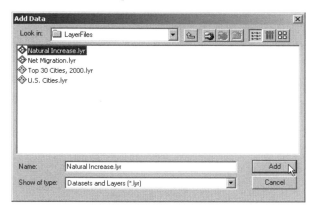

c   Click the Full Extent button to return to the view of the entire world. (Hint: If your map doesn't draw completely, click the Refresh View button below the map.)

Like the birth and death rates, the numbers for natural increase are expressed as a rate per 1,000 population. This means that in all the countries colored dark blue on the map, there are between 25 and 39 people being added to the population each year for every 1,000 people already there.

> *Note: The actual growth rate of an individual country is based on its natural increase plus the net migration of people into or out of that country each year.*

(1)   What is happening to the population in the countries that are red?

(2)   Which world region is growing the fastest?

(3)   Which world region is growing the slowest?

(4)   Think about what it would mean for a country to have a population that is growing rapidly or one that is growing slowly. Which of these two possibilities (fast growth or slow growth) do you think would cause more problems within the country? On the answer sheet, briefly list some of the problems you would expect to see.

d   Click the Go Back to Previous Extent button to return the view to Africa and Europe.

Step 5   **Look at standard of living indicators for Europe and Africa**

Geographers look at certain key statistics when they want to compare the standard of living in different countries. They refer to these statistics as "indicators" because they typically reveal or provide some information about the quality of life in that country. The indicators that you will look at in this activity are the following:

- Population 60 years old and over: The percent of the total population that is in this age group.
- GDP per capita: The GDP (gross domestic product) divided by the total population for that year, where the GDP is expressed as a per-person figure.
- Infant mortality rate: The number of deaths of infants under one year old in a given year per 1,000 live births in the same year.
- Life expectancy: The number of years a newborn infant would live if prevailing conditions of mortality at the time of birth continue.
- Literacy rate: The percentage of the adult population that can read and write.
- Services: The percentage of the workforce that is employed in the service sector of the economy.

a   Click the minus sign next to Population Growth in the table of contents to collapse the data frame legend.

b   Click the plus sign next to Standard of Living Indicators to expand this data frame legend.

c   Right-click the Standard of Living Indicators data frame name and choose Activate.

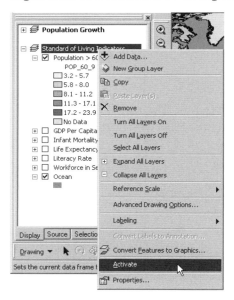

d   Click View on the Main menu, point to Bookmarks, and click Europe and Africa.

The Standard of Living Indicators map is focused on Europe and Africa. Europe is the slowest growing region and sub-Saharan Africa is the fastest growing region in the world.

e   Look down the table of contents to see each of the six standard of living indicators layer names. The first indicator, Population >60 years, shows the percentage of each country's population over 60 years old.

In order to complete the chart on the answer sheet, you need to explore each of the six standard of living indicators. Remember:

- A layer will cover the one beneath it when it is turned on. You will need to turn layers on and off to see all the indicators.
- You can also reposition the layers by dragging them to a new position in the table of contents.
- You can expand or collapse the layers in the table of contents to show or hide the legends as you examine different layers.

?   *f*   **Now, complete the table on the answer sheet.**

Step 6   Add the Net Migration layer

The net rate of migration is a statistic that indicates the number of people per 1,000 gained or lost each year as a result of migration. A negative number indicates that more people are leaving the country than coming in. A positive number means more people are coming to the country than leaving it.

?   *a*   **In step 5 you compared standard of living indicators in Europe and sub-Saharan Africa. Based on your observations of those indicators, which region would you expect to have a negative net migration? A positive net migration? Explain your answer.**

+   *b*   **Click the Add Data button.**

            c   **Navigate to the LayerFiles folder (C:\MapWorld9\Mod4\Data) and add Net Migration.lyr.**

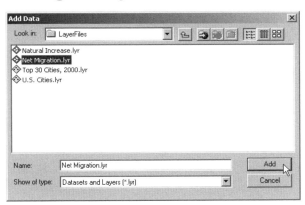

?   *d*   **Summarize the overall patterns of net migration in Europe and sub-Saharan Africa in the table on the answer sheet.**

?   *e*   **What are possible political or social conditions or events that could explain any of the migration patterns you see on the map?**

Step 7   Draw conclusions

    a   Click the Layout View button at the bottom of the map area to switch from Data View to Layout View.

The two data frames are displayed side by side on a layout so you can easily compare the two maps.

b   The Layout toolbar automatically becomes active. If your Layout toolbar is floating, dock it above the map.

c   Expand the Population Growth data frame in the table of contents to display its legend.

    d   Enlarge your ArcMap window so that it fills your screen. Click the Zoom Whole Page button on the Layout toolbar.

e   In the next steps, you will be comparing the Natural Increase and Net Migration layers. Make sure that both these layers are visible (not hidden beneath other layers) and that their legends are visible in the Table of Contents.

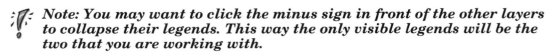

> *Note: You may want to click the minus sign in front of the other layers to collapse their legends. This way the only visible legends will be the two that you are working with.*

    f   Click the Zoom In or Zoom Out tool on the Tools toolbar and zoom as needed to focus both maps on Europe and Africa once again. (Hint: If you accidentally use the Zoom In or Zoom Out tool from the Layout toolbar, click the Zoom Whole Page button.)

g   Look at the two maps and compare the rate of Natural Increase for each country to its Net Migration.

Think about the following question: What correlation, if any, exists between a country's standard of living, its rate of natural increase, and its rate of net migration?

    h   Based on your map investigations, write a hypothesis about how a country's rate of natural increase affects its standard of living and its net rate of migration.

?   *i*   In the table on the answer sheet, illustrate your hypothesis with data from one European country and one sub-Saharan African country. Use the Identify tool to see the data for an individual country.

j   Close the Identify Results window.

Step 8   Design a layout

You will make some changes to the layout's design before you print it. You will use a template to add a title, legend, and other map elements.

a   Turn off all of the layers in the Population Growth data frame except Natural Increase and Ocean.

    b   Click the Change Layout button.

c   Click the General tab in the Select Template dialog. If your window looks different from the picture below, click the List button in the lower left corner. Select **LetterLandscape.mxt** in the list.

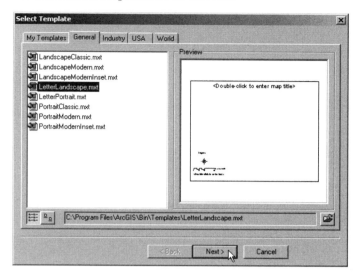

d   Click Next. Make sure the Population Growth data frame is number 1 in the list and click Finish.

The Population Growth data frame fills the layout. You see text marking the place for a title, a legend, a north arrow, a scale bar, and another place for text.

⚠ *Notice that the Standard of Living Indicators data frame has not been deleted—it has been moved off the layout. You can see part of it in the lower left corner of the layout view.*

e   Click the Select Elements tool if it is not already selected. Click on the map to display the blue squares at the border of the data frame.

f   Place your cursor over the top middle blue square. When it changes to a double-headed arrow, click and drag the top of the data frame down below the title.

g   Repeat the procedure to drag the left side of the data frame over to the right of the legend and other map elements.

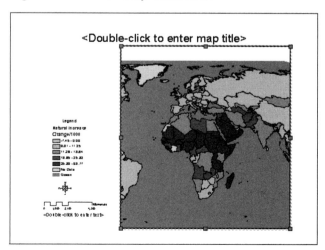

h   Double-click the text that reads **<Double-click to enter map title>**. Delete the text in the Properties dialog and type **Population Growth**. Click OK.

i   Click anywhere on the legend. Click one of the blue corner squares and drag it to make the legend larger.

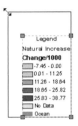

j   Place your cursor in the middle of the legend. Click the legend and hold the mouse button down. Drag the legend up to a suitable location closer to the top of the map.

k   Double-click the compass rose that's below the legend to open the North Arrow Properties dialog.

l   Click the North Arrow Style button. Choose a different north arrow that you like from the North Arrow Selector. Click OK in both of the dialog boxes to apply your change.

The new north arrow displays automatically in the layout.

m   Resize the north arrow using the blue squares.

n   Double-click the scale bar to see its properties.

❓   *What are the units of measurement?*

By default, ArcMap assigns the units of measurement and intervals based on the coordinate system of the data frame.

o   Click the Division Units drop-down list and select Miles. Click OK to update the scale bar's units of measurement.

p   Enlarge the scale bar slightly so that it is easier to read.

Step 9   Label your map and print it

When making maps, it's important to include the cartographer's name and date that the map was created. In order to make room for that in the bottom right corner of the map, you may need to move the north arrow and scale bar up slightly.

a   Adjust the position of the north arrow and scale bar. Click and drag each one to the desired position so they are centered beneath the legend and in the white space to the left of the map.

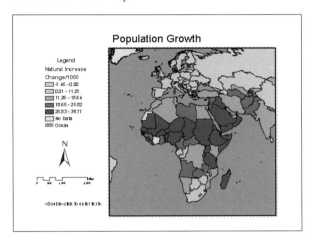

b   To add the name of the cartographer (that's you!) and date, double-click the text below the scale bar.

A Text Properties dialog appears.

c   In the dialog, type your name. Press Enter to move to the next line and then type the date of the project. Click OK.

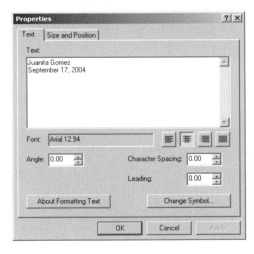

d   Click and drag the text box to the bottom right corner of the layout. Then click any blank area on the layout to clear the box around the text element.

Your layout of Population Growth is now ready to print. If you need to make any final adjustments to your map, make them now. Otherwise, proceed to the next step.

e   From the File menu, click Print. Click the Setup button.

f   Use the following settings in the Page and Print Setup dialog.

- Make sure the printer you want to use is the one shown.
- Under Paper, set the paper size to Letter and the orientation to Landscape.
- Under Map Page Size, check the box to Use Printer Paper Settings.
- At the bottom of the dialog, check the box to "Scale Map Elements proportionally to changes in page size."

g   Click OK on both the Page and Print Setup window and the Print window.

Your map should print after a few moments.

Step 10   Save your map document

In the next steps, you will change the layout. If you want to save your work on this layout, for example to print it or modify it at a later time, you need to save a copy of the map document in its current state.

a   If you wish to save the map document at this point, click the File menu and choose Save As. Ask your teacher for instructions on where to save this map document. Name it **ABC_PopGrowth**, where ABC are your initials.

b   Choose Save As again and save another copy of the map document for continuing your work on the next layout. Name this copy **ABC_StdLiving**, where ABC are your intitials.

Step 11   Make a standard of living indicators map and print it

a   Turn off all layers in the Standard of Living Indicators data frame except Net Migration and Ocean.

b   Click the Change Layout button. Make sure the LetterLandscape.mxt template is selected, and click Next.

c   Click Standard of Living Indicators and then click the Move Up button to move the data frame to the number 1 position. Click Finish.

The data frames switch positions in the layout.

d   Follow the procedures outlined in steps 8e–8p and step 9 to design your Standard of Living presentation map and print it.

Step 12   Save your map document and exit ArcMap

In this exercise, you explored world population growth and analyzed standard of living indicators in the fastest and slowest growing regions of the world (Africa and Europe). You added layers and worked in two data frames. You created a layout for each data frame and printed your maps.

a   If you already saved this map document in step 10b, click the Save button to save your work on the second layout. Otherwise, ask your teacher for instructions on where to save this map document and on how to rename it. Save it according to your teacher's instructions.

b   If you are not going to save the map document, exit ArcMap by choosing Exit from the File menu. When asked if you want to save changes to the map document, click No.

PHOTOCOPY

NAME _____   DATE _____

Student answer sheet
Module 4
Human Geography I: Population Patterns and Processes

Regional case study: Growing Pains

Step 3   Compare birth rate and death rate data

b-1   Which world region or regions have the highest birth rates?

_____

b-2   Which world region or regions have the lowest birth rates?

_____

c-1   Which world region or regions have the highest death rates?

_____

c-2   Which world region or regions have the lowest death rates?

_____

d-1   If the overall rate of growth is based on the formula BR – DR = NI, which world regions do you think are growing the fastest?

_____

d-2   Which world regions do you think are growing the slowest?

_____

l   Choose two European and two African countries and record their birth and death rates in the table below.

COUNTRY AND CONTINENT NAME	BIRTH RATE/1,000	DEATH RATE/1,000
Niger (Africa)	51.45	23.17

m   List three questions that the Birth Rate and Death Rate maps raise in your mind.

_____
_____
_____
_____
_____

Step 4   Add the Natural Increase layer

c-1   What is happening to the population in the countries that are red?

_____

c-2   Which world region is growing the fastest?

_____

c-3   Which world region is growing the slowest?

_____

c-4   Think about what it would mean for a country to have a population that is growing rapidly or one that is growing slowly. Which of these two possibilities (fast growth or slow growth) do you think would cause more problems within the country?

_____

Briefly list some of the problems you would expect to see.

_____

_____

_____

_____

_____

_____

Step 5   Look at standard of living indicators for Europe and Africa

f   Complete the table below:

INDICATOR	COMPARE SUB-SAHARAN AFRICA AND EUROPE	WHAT DOES THIS "INDICATE" ABOUT THE STANDARD OF LIVING IN THESE REGIONS?
Population > 60 years		
GDP per capita		
Infant mortality rate		
Life expectancy		
Literacy rate		
Percent of workforce in service sector		

Step 6   Add the Net Migration layer

a   In step 5 you compared standard of living indicators in Europe and sub-Saharan Africa. Based on your observations of those indicators, which region would you expect to have a negative net migration?

_____

A positive net migration?

_____

Explain your answer.

_____

_____

_____

_____

_____

d   Summarize the overall patterns of net migration in Europe and sub-Saharan Africa in the table below.

NET MIGRATION IN SUB-SAHARAN AFRICA	NET MIGRATION IN EUROPE

e   What are possible political or social conditions or events that could explain any of the migration patterns you see on the map?

_____

_____

_____

_____

_____

PHOTOCOPY

Step 7   Draw conclusions

h   Based on your map investigations, write a hypothesis below about how a country's rate of natural increase affects its standard of living and its net rate of migration.

_____

_____

_____

_____

_____

i   In the table below, illustrate your hypothesis with data from one European country and one sub-Saharan African country.

EUROPE	DATA	AFRICA
	Country name	
	Natural increase	
	Net migration	

Step 8   Design a layout

n   What are the units of measurement?

_____

NAME _____   DATE _____

Growing Pains
Middle school assessment

You have been selected by the United Nations to establish a model partnership between nations. Your job is to select two countries—one country with a slow growth rate from Europe and one with a fast growth rate from Africa. Refer to your initial GIS Investigation if you need help in identifying growth rate in a country. Select the countries from the list below:

Belgium	Italy	Norway
Germany	Nigeria	Poland
Mozambique	Madagascar	Tanzania
Somalia	Zambia	
United Kingdom	Denmark	

Once you have selected your two countries, you need to write a report that addresses the following points:

1 Identify issues critical to each country in regard to growth and standard of living.

2 How can these countries work together to help improve the standard of living in their countries and address their respective problems?

3 Do current relationships (such as trade agreements) exist between these countries and are these positive or negative in nature? How can these current partnerships be improved upon?

Support this report with maps and data from the GIS Investigation. You may also want to refer to additional sources such as textbooks, encyclopedias, and the Internet to find out additional details about your countries.

Growing Pains

Assessment rubric

Middle school

STANDARD	EXEMPLARY	MASTERY	INTRODUCTORY	DOES NOT MEET REQUIREMENTS
The student knows and understands how to use maps to analyze spatial distributions and patterns.	Uses GIS to compare and analyze growth and demographic trends in countries throughout the world. Makes predictions from the data on future population trends. Creates original printed/electronic maps to support findings.	Uses GIS to compare and analyze growth and demographic trends in countries throughout the world. Creates views that isolate regions of slow and fast growth. Creates printed/electronic maps to support findings.	Uses GIS to identify regions and countries with slow and fast growth rates. Printed maps do not support findings.	Has difficulty identifying patterns in population and demographics. No printed maps are included.
The student knows and understands the demographic structure of a population.	Analyzes demographic data of countries with fast and slow rates of natural increase and draws conclusions on how the structure of these populations came to exist.	Identifies and analyzes data that illustrates the demographic makeup of countries with slow and fast rates of natural increase.	Identifies demographic characteristics of countries with slow and fast rates of natural increase.	Has difficulty distinguishing differences between countries with slow and fast rates of natural increase.
The student knows and understands how to apply the geographic point of view to social problems by making geographically informed decisions.	Identifies critical growth issues for a country of fast growth and a country of slow growth. Creates a detailed program by which the countries could establish a partnership of nations to solve growth-related issues. Elaborates on how this program could be replicated by other countries.	Identifies critical growth issues for a country of fast growth and a country of slow growth. Determines several ways that the countries could form a partnership for the mutual benefit of all in regard to growth issues.	Identifies critical growth issues for a country of fast growth and another of slow growth. Determines one or two ways the countries could form a partnership.	Identifies critical growth issues for two countries, but does not address the issue of establishing a partnership between nations.

This is a four-point rubric based on the National Standards for Geographic Education. The "Mastery" level meets the target objective for grades 5–8.

NAME _____   DATE _____

Growing Pains
High school assessment

You have been selected by the United Nations to establish a model partnership among nations. Your job is to select two groups of countries—one group with a slow growth rate and one group with a fast growth rate. Each group of countries should consist of four to five nations from Europe or sub-Saharan Africa. The countries within each group need to share similar growth characteristics. Refer to your initial GIS investigation if you need help identifying growth rate in a particular country.

Once you have selected your two groups, you will write a report that addresses the following points:

1   Identify issues related to growth and standard of living that are critical to each group's welfare.
2   How can these countries work together to help improve the standard of living in their countries and address their respective problems?
3   Do current relationships (such as trade agreements) exist between these countries and are these positive or negative in nature? How can these current partnerships be improved upon?

Support this report with maps and data from the GIS Investigation. You may also want to refer to additional sources such as textbooks, encyclopedias, and the Internet to find out additional details about your countries.

Growing Pains

Assessment rubric

High school

STANDARD	EXEMPLARY	MASTERY	INTRODUCTORY	DOES NOT MEET REQUIREMENTS
The student knows and understands how to use geographic representations and tools to analyze and explain geographic problems.	Uses GIS to compare and analyze growth and demographic trends in countries throughout the world. Makes predictions from the data provided and additional sources on future population trends. Creates original printed/electronic maps to support findings.	Uses GIS to compare and analyze growth and demographic trends in countries throughout the world. Makes predictions from the data on future population trends. Creates printed/electronic maps to support findings.	Uses GIS to compare and analyze growth and demographic trends in countries throughout the world. Makes predictions from the data on future population trends. Printed maps do not support findings.	Attempts to make comparisons between countries using demographic data in a GIS. No printed maps are included.
The student knows and understands trends in world population numbers and patterns.	Creates two groups of countries, each with similar demographic trends, including standard of living indicators. One group represents fast-growth countries and one represents slow-growth. Makes predictions on how these trends will change through time.	Creates two groups of countries, each with similar demographic trends, including standard of living indicators. One group represents fast-growth countries and one represents slow-growth.	Creates two groups of countries with similar growth characteristics. One group represents fast-growth countries and one represents slow-growth.	Identifies one or two countries that represent either slow or fast growth.
The student knows and understands how to use geographic knowledge, skills, and perspectives to analyze problems and make decisions.	Creates a report that establishes a coalition of slow- and fast-growth countries that work together for the mutual benefit of all involved. The report takes into account issues critical to the countries that participate and elaborates on possible solutions.	Creates a report that establishes a coalition of slow- and fast-growth countries that work together for the mutual benefit of all involved. The report takes into account issues critical to the countries that participate.	Lists several issues that are critical to fast- and slow-growth countries. Attempts to find ways in which each group of countries can partner.	Identifies one or two issues critical to countries with slow growth and those with fast growth.

This is a four-point rubric based on the National Standards for Geographic Education. The "Mastery" level meets the target objective for grades 9–12.

Generation Gaps
An advanced investigation

Lesson overview

Students will use global data to investigate variations in population age structure and the relationship of age structure to a country's rate of natural increase. After exploring global patterns, they will download and map U.S. census data to examine the patterns of age structure in their own community.

Estimated time Two to three 45-minute class periods

Materials ✔ Student handouts from this lesson to be copied:
- GIS Investigation sheets (pages 245 to 252)
- Student answer sheets (pages 253 to 254)

Standards and objectives

National geography standards

	GEOGRAPHY STANDARD	MIDDLE SCHOOL	HIGH SCHOOL
1	How to use maps and other geographic representations, tools, and technologies to acquire, process, and report information from a spatial perspective	The student knows and understands how to use maps to analyze spatial distributions and patterns.	The student knows and understands how to use geographic representations and tools to analyze and explain geographic problems.
9	The characteristics, distribution, and migration of human populations on the earth's surface	The student knows and understands the demographic structure of a population.	The student knows and understands trends in world population numbers and patterns.
18	How to apply geography to interpret the present and plan for the future	The student knows and understands how to apply the geographic point of view to social problems by making geographically informed decisions.	The student knows and understands how to use geographic knowledge, skills, and perspectives to analyze problems and make decisions.

Objectives

The student is able to:
- Explain how the age structure of a population is affected by its rate of natural increase.
- Describe age structure trends in different world regions.
- Use the Internet to locate U.S. Census Bureau data for their own community.
- Use GIS to map acquired data from the Internet.
- Describe patterns of age structure in their local area.

GIS skills and tools
- Obtain census data from the Internet and prepare database tables for GIS implementation
- Add data, including a table, to a map document
- Copy layers
- Join tables in a map document
- Edit a layer's legend to map thematic data
- Create a new data frame

For more on geographic inquiry and these steps, see Geographic Inquiry and GIS (pages xxiii to xxv).

Teacher notes

Lesson introduction

Introduce the concept of population age structure by displaying and discussing several examples of population pyramids. You can obtain population pyramids for any world country at the U.S. Census Bureau's International Database (IDB) (www.census.gov/ipc/www/idbpyr.html). Compare a population pyramid for a country that is growing rapidly with one for a country that is growing very slowly. Initially, just display the pyramids without telling the students what countries they represent—refer to them as Country A and Country B. Ask students to consider the following questions:

- How is the makeup of population different in these two countries?
- One of these countries is growing rapidly and one is growing slowly—can you determine which is which? How do you know?
- What specific issues does each country face because of its age structure?

Explain to students that in this lesson they will investigate age structure trends around the world and then download U.S. Census data to examine patterns of age structure in their own community.

Student activity   Distribute the Generation Gaps GIS Investigation. Students should follow the GIS Investigation guidelines to analyze global and local patterns of age structure within specific populations. After looking at global patterns, they will download census data for their own county and examine local age distribution patterns.

 Note: When downloading the census data from the Internet, each student needs to have a folder or a computer directory that they can save the data to. Students are instructed to consult their teacher on how to name the data that's downloaded and where to save it. Without this information, students will have difficulty accessing the data at a later time.

 Teacher Tip: In step 5b, students are directed to unzip the census data files. If this is a task that your students are not familiar with, you will need to explain the process to them.

Conclusion   To conclude this lesson, engage the class in a discussion of the implications of population age structure. Be sure to bring out the concept that the age structure of a population in a country or a community affects its key socioeconomic issues. Countries and communities with young populations (high percentage under age 15) need to invest more in schools while countries and communities with older populations (high percentage ages 60 and over) might need to invest more in the health sector. The age structure can also be used to help predict potential political issues. For example, the rapid growth of a young adult population unable to find employment can lead to unrest. Large numbers of the elderly can lead to political concerns about health benefits and pension plans.

Challenge the class to identify local social and political issues that are related to the age structure of the community. Bring in newspaper clippings about stories related to these issues for further discussion and display in the classroom.

Assessment   Students will create a layout that illustrates the distribution of one particular age group in their county. They will submit a written report summarizing their observations and explaining their data analysis. The report should address all or some of these points:

- Summarize the spatial distribution of the selected age group in your county.
- Analyze the patterns that you observe by relating those patterns to other features of the county such as:
  - Location of age-based institutions (colleges, retirement communities, etc.)
  - Proximity to major roads and transportation systems
  - Economic activities
  - Housing characteristics
  - Income and educational attainment of the population

 Note: Due to the independent nature of this lesson, there is no supplied assessment rubric. You are free to design assessment rubrics that meet the needs of your specific adaptation.

Extensions
- Do an Internet search on "population pyramid" to find directions for using a spreadsheet program to create population pyramids. Make a pyramid of your own community and incorporate it into an ArcMap map document, displaying demographic data about your community.
- Download TIGER data from the U.S. Census Bureau for a college town and a town with a large retirement population. Map and compare the age distributions in these communities with your own community.
- Analyze your county data and make a prediction on population growth for the future. Create map layouts that support your prediction. Download Census 2000 data for your county and enhance your claim by using that data to support your prediction.
- Check out additional resources available online at *www.esri.com/ mappingourworld*.

NAME _____   DATE _____

Generation Gaps
An advanced investigation

> ⚡ *Note: Due to the dynamic nature of the Internet, the URLs listed in this lesson may have changed, and the graphics shown below may be out of date. If the URLs do not work, refer to this book's Web site for an updated link: www.esri.com/mappingourworld.*

Answer all questions on the student answer sheet handout

Step 1   Start ArcMap and open the Adv4.mxd file

a   Start ArcMap.

b   Open the **Adv4.mxd** (or **Adv4**) map document from your module folder (**C:\MapWorld9\Mod4**).

When the map document opens, you see a world map with two layers turned on (Natural Increase and Ocean).

Step 2   Create new layers

The world's population is growing because there are more births than deaths each year. This fact can be expressed as a simple formula, BR (birth rate) – DR (death rate) = NIR (natural increase rate). The spatial patterns displayed on the Natural Increase map indicate where population is growing rapidly and slowly in the world today. In this part of the activity, you will see if there are comparable spatial patterns in the world's age structure.

World Vital Events

World Vital Events Per Time Unit: 2004

(Figures may not add to totals due to rounding)

Time unit	Births	Deaths	Natural increase
Year	129,108,390	56,540,896	72,567,494
Month	10,759,033	4,711,741	6,047,291
Day	352,755	154,483	198,272
Hour	14,698	6,437	8,261
Minute	245	107	138
Second	4.1	1.8	2.3

Source: U.S. Bureau of the Census, International Data Base

---

a Look at the map and answer the following questions:

    (1)   What world regions have the highest rates of natural increase today?

    (2)   What world regions have the lowest rates of natural increase today?

b Right-click the Natural Increase layer and choose Copy.

c Click the Paste button twice to paste two copies of the layer into the Generation Gaps data frame.

d Click the highlighted Natural Increase layer name to activate the text cursor. Rename the layer **Population <15 Yrs** and press Enter.

e Slowly click twice on the other copied layer and rename it **Population 60+ Yrs**.

Step 3 Thematically map world age structure

a Double-click the Population <15 Yrs layer name to open the Layer Properties dialog.

b Click the Symbology tab. For the Value Field, scroll up and select **POP0_14**.

The class ranges begin with a value of –99, which is a code indicating that no data is available for some countries. You need to exclude this value from the classification.

c Click the Classify button. In the Classification dialog, click the Exclusion button.

d Click the Query tab in the Data Exclusion Properties dialog. A query statement already exists for the layer for the natural increase field.

e Highlight **[NAT_INC]** and then in the Fields list above, double-click **[POP0_14]**. The query statement is updated.

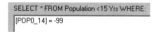
```
SELECT * FROM Population <15 Yrs WHERE:
[POP0_14] = -99
```

f Click Verify to make sure the statement has no errors and then click OK on the message box. Fix any errors if necessary, and then click OK in the Data Exclusion Properties and Classification dialogs.

g From the Color Ramp drop-down list, choose Green Light to Dark. (Hint: You can toggle the color ramp menu to show names instead of colors. To do so, right-click the menu and click to uncheck Graphic View.)

Color Ramp: Green Light to Dark

h Click OK to close the Layer Properties dialog and update the map.

i In the table of contents, click slowly two times on the legend heading (POP0_14) to activate the text cursor. Change the heading to **Percent** and press Enter.

Population <15 Yrs
Percent
0
1 - 24
25 - 33
34 - 41
42 - 51
No Data

j   Turn off the Population <15 Yrs layer. Repeat steps 3a–3i to classify and symbolize the Population 60+ Yrs layer. Use the following parameters:

●   For the value field, select **POP_60_9**.

●   For the color ramp, choose Purple-Red Bright.

●   Before closing the Properties dialog, right-click any white space in the symbol box and choose Format Labels. Reduce the number of decimal places to 2 and click OK.

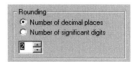

k   Change the POP_60_9 heading in the legend to **Percent**.

l   Turn off the Natural Increase layer. Toggle back and forth between the Population 60+ Yrs layer and the Population <15 Yrs layer to compare them.

?   *(1)   What regions of the world have populations with a high percentage below age 15?*

?   *(2)   What regions of the world have populations with a relatively low percentage below age 15?*

?   *(3)   Why do you think there is a much higher percentage of children in some populations than in others?*

?   *(4)   Why are there far more people 60 years of age and older in some populations than in others?*

?   *(5)   Based on the layers in the Generation Gaps data frame, how do you think natural increase is related to age structure?*

m   Save a copy of this map document in a location specified by your teacher. Name it **ABC_Adv4** where ABC are your initials.

n   Minimize the ArcMap window.

Step 4   **Download census data for your county**

a   Before beginning this part of the exercise, create a folder to store the census data you will download later. Ask your teacher where to put the folder and how to name it. One suggestion is to put the folder in your Mod4/Data folder and name it ABC_censusdata, where ABC are your initials.

b   Use your Web browser to go to the ESRI site at *www.esri.com/data/download/census2000_tigerline*. (At this ESRI site, you will be able to download 2000 U.S. Census data for your county.)

c   Click Download Data. A new page opens.

d Select your state from the drop-down list and click Submit Selection, or click on your state on the map.

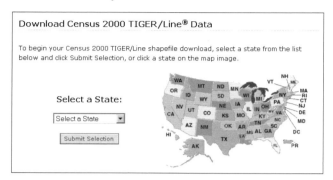

e Select your county from the Select by County drop-down list. Click Submit Selection.

A list of available data for your county appears.

f Check off all of the following data layers.

Available data layers:
- Block Groups 2000
- Line Features - Roads
- Water Polygons

Available statewide layers:
- Census Block Group Demographics (SF1)

g Scroll to the bottom of the page and click Proceed to Download.

A page with the message "Your Data File is Ready" appears.

h Click the Download File button.

i Follow your computer's prompts to save the file to disk. Navigate to the folder that you created in step 4a. Save the file as **ST_Countyname.zip**, where ST is your state's abbreviation and countyname is your county's name (example: CT_Hartford for Hartford County, Connecticut).

j Click Save. When the download is complete, close any open dialogs and quit your Web Browser.

Step 5   Unzip your data and add it to the map document

a   Navigate to your census data in the designated folder.

The downloaded file (ST_Countyname.zip) contains the data layers you selected. Each of those layers is, in turn, a zipped file.

b   Unzip the individual layers and save the unzipped data in a location that your teacher specifies. Open and read the readme.html file that's included with the data.

> *Note: It is very important that your data is saved to a folder that you can access later. If you intend to save this map document and open it later, you must have access to the data.*

c   Restore the ArcMap window. Collapse the Generation Gaps data frame legend in the table of contents.

d   Click the Insert menu and choose Data Frame.

A new data frame is added to the map document. It automatically becomes the active data frame.

e   Rename the new data frame **Generation Gaps 2**.

 f   Click the Add Data button. Navigate to the location of the census data you down-loaded and add all four of the files. They end with the following suffixes: sf1grp.dbf, grp00.shp, lkA.shp, and wat.shp.

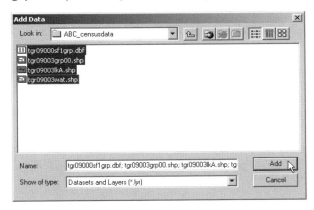

> *Note: The number included in the name of the file is the census number for your county.*

g   Notice that the table of contents switched to the Source tab when you added the data. This is because the sf1grp.dbf file is a table. Tables do not appear on the Display tab because they cannot be shown on a map unless you associate them with a layer. (You will do this in step 6.)

h Click the Display tab. Arrange the layers so that the streets layer (lkA) is on top, followed by water (wat), and then census block groups (grp00) on the bottom. (Note: Click the layer you want to move to deselect the other two before you try to drag it to a new location.)

i Rename the layers Streets, Water, and Block Groups, respectively.

j Choose appropriate symbol colors for Streets and Water.

Step 6 **Join the census table to the map**

In this step you will join the census block group demographics table to the Block Group layer so that you can map the demographic attributes. Both tables contain a field called STFID. You will use this common field to join the tables.

a Right-click the Block Groups layer and choose Open Attribute Table. Notice that all of the fields contain various identification numbers.

b Move the table so that you can see the map and table of contents. In the table of contents, right-click Block Groups. Point to Joins and Relates and click Join. A Join Data dialog appears.

c Use the graphic below as a guide for completing the Join Data dialog.

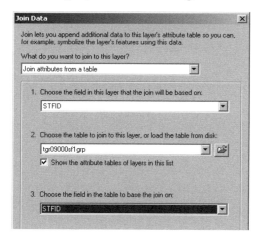

d Click OK to join the data. Click Yes if the Create Index message box appears.

e Scroll through the Attributes of Block Groups table and look at the fields. The fields containing demographic data have been appended to this table. Each field name is now preceded by the name of the source table it comes from.

f Close the attribute table.

Step 7   Analyze census data for your county

a   Double-click the Block Groups layer in the table of contents to open the Layer Properties. Click the Symbology tab.

b   In the Show box click Quantities, Graduated colors.

c   Click the Value field drop-down list. Scroll down the list and select **tgrxxxxxsf1grp.AGE_UNDER5** for the Value field. (Hint: The field name may be cut off in the list. After you choose the field you think is the correct one, pause your mouse over the Value box to see a tip displaying the complete name.)

d   Choose an appropriate color ramp, and then click OK.

The map displays the distribution of children between birth and five years of age in your county.

e   Change the Block Groups legend heading to **Age < 5 yrs**.

⊟ ☑ Block Groups
     Age < 5 Yrs

You will add a layer showing the main cities of the United States to use as a reference. (Note: This layer may have no cities in your county if it is highly rural.)

f   Click the Add Data button and navigate to the module 4 layer files folder (**C:MapWorld9/Mod4/Data/LayerFiles**). Add U.S. Cities.lyr.

ArcMap displays a warning that the U.S. Cities layer uses a different coordinate system than the other layers in the data frame. For the purposes of this exercise you will accept ArcMap's attempt to align the layer correctly.

g   Click OK in the Warning window.

You can view the city names using MapTips or by turning labels on for this layer.

h   If it's hard to see the color patterns with the Streets layer on, turn it off.

    Are there identifiable concentrations of children between birth and five years old in your county? How do you explain these concentrations?

i   Copy and paste the Block Groups layer. Move it below the Water layer if necessary. Use the Layer Properties to map the value field AGE_65_UP. Choose a different color ramp.

The map displays the distribution of adults over the age of 65 in your county.

    Where is the greatest number of this age group found?

j   Change the legend heading to Age 65+ yrs. Repeat the process of copying a layer and mapping a new age group to explore other age group populations.

    (1)   Do you have any colleges in your county? Any retirement communities? Are these institutions reflected in the census data?

    (2)   On the answer sheet, identify patterns of age distribution in your county and suggest explanations for those patterns.

Step 8   Save your map document and exit ArcMap

You can further explore population in your county by investigating the TIGER data you downloaded. For information on how to read all the abbreviations and interpret the assigned codes, visit the Census Bureau's Web site on TIGER data at www.census.gov/geo/www/tiger or read the readme file that was downloaded from the ESRI Web site.

In this exercise, you analyzed the natural increase and population by age of different countries. You searched for your county data on the Internet, downloaded it, and incorporated it into an ArcMap map document. Now the data is available for analysis and presentation.

    a   Save the changes you made to your map document.

b   Exit ArcMap.

NAME _____   DATE _____

Student answer sheet
Module 4
Human Geography I: Population Patterns and Processes

Advanced investigation: Generation Gaps

Step 2   Create new layers

a-1   What world regions have the highest rates of natural increase today?

_____

_____

_____

a-2   What world regions have the lowest rates of natural increase today?

_____

_____

_____

Step 3   Thematically map world age structure

l-1   What regions of the world have populations with a high percentage below age 15?

_____

_____

_____

l-2   What regions of the world have populations with a relatively low percentage below age 15?

_____

_____

_____

l-3   Why do you think there is a much higher percentage of children in some populations than in others?

_____

_____

_____

_____

l-4   Why are there far more people 60 years of age and older in some populations than in others?

_____

_____

_____

_____

_____

l-5   Based on the layers in the Generation Gaps data frame, how do you think natural increase is related to age structure?

_____

_____

_____

_____

_____

Step 7   Analyze census data for your county

h   Are there identifiable concentrations of children between birth and five years old in your county? How do you explain these concentrations?

_____

_____

_____

_____

_____

i   Where is the greatest number of this age group found?   _____

j-1   Do you have any colleges in your county? Any retirement communities? Are these institutions reflected in the census data?

_____

_____

_____

_____

_____

j-2   In the space below, identify patterns of age distribution in your county and suggest explanations for those patterns.

_____

_____

_____

_____

_____

_____

_____

_____

_____

_____

module 5

Human Geography II
Political Geography

Invisible boundary lines on the earth's surface divide our world into discrete political entities and have significant influence on the stability and cohesiveness of the countries they define.

Crossing the Line: A global perspective
Students will explore the nature and significance of international political boundaries. Through an investigation of contemporary political boundaries, they will identify boundary forms, compare patterns of size and shape, and explore the influence of boundaries on national cohesiveness and economic potential. By comparing world political boundaries in 1992 and 2000, students will observe the evolution of boundary characteristics over time.

A Line in the Sand: A regional case study of Saudi Arabia and Yemen
Students will study the creation of a new border between Yemen and Saudi Arabia on the Arabian Peninsula. Using data included in the June 2000 Treaty of Jeddah, they will draw the new boundary described in the treaty and analyze the underlying physiographic and cultural forces that influenced the location of that boundary. In the process they will come to understand how any map of the world must be considered a tentative one, as nations struggle and cooperate with each other.

Starting from Scratch: An advanced investigation
Students will use physiographic (physical features) and anthropographic (cultural features) data to redraw some of the world's international boundaries, thereby creating states characterized by internal cohesiveness and economic parity. By comparing their maps to ones reflecting contemporary political boundaries, students will identify world regions where political boundaries are in conflict with physical and cultural imperatives.

Crossing the Line
A global perspective

Lesson overview

Students will explore the nature and significance of international political boundaries. Through an investigation of contemporary political boundaries, they will identify boundary forms, compare patterns of size and shape, and explore the influence of boundaries on national cohesiveness and economic potential. By comparing world political boundaries in 1992 and 2000, students will observe the evolution of boundaries over time.

Estimated time

Two to three 45-minute class periods

Materials

✔ A world atlas or map of Europe showing mountain ranges and their names

✔ Student handouts from this lesson to be copied:
  - GIS Investigation sheets (pages 265 to 273)
  - Student answer sheets (pages 274 to 279)
  - Assessment(s) (pages 280 to 283)

Standards and objectives

National geography standards

GEOGRAPHY STANDARD	MIDDLE SCHOOL	HIGH SCHOOL
3 How to analyze the spatial organization of people, places, and environments on Earth's surface	The student knows and understands how to use the elements of space to describe spatial patterns.	The student knows and understands how to apply concepts and models of spatial organization to make decisions.
13 How the forces of cooperation and conflict among people influence the division and control of Earth's surface	The student knows and understands the multiple territorial divisions of the student's own world.	The student knows and understands why and how cooperation and conflict are involved in shaping the distribution of social, political, and economic spaces on Earth at different scales.
18 How to apply geography to interpret the present and plan for the future	The student knows and understands how varying points of view on geographic context influence plans for change.	The student knows and understands contemporary issues in the context of spatial and environmental perspectives.

Objectives

The student is able to:

- Define and give examples of physiographic, geometric, and anthropographic boundaries.
- Describe the political and economic implications of a country's size and shape.
- Explain the relationship between boundary characteristics and national cohesiveness.
- Give examples and explain the nature of international boundary changes in the late twentieth century.

GIS skills and tools

 Zoom to a specific area on the map

 Draw a line

 Change the line color

 Identify a feature on the map

 Select a graphic

 Pan to a different section of the map

 Zoom to full extent

 Add layers to the map

 Find a feature in a layer, select it, and zoom to it

 Save the map document

- Turn layers on and off
- Expand and collapse legends in the table of contents

For more on geographic inquiry and these steps, see Geographic Inquiry and GIS (pages xxiii to xxv).

Teacher notes

Lesson introduction

Write the following quotation on the board or on a transparency:

"When you go around the Earth in an hour and a half . . . you look down there and you can't imagine how many borders and boundaries you cross, again and again and again, and you don't even see them . . . from where you see it, the thing is a whole, and it's so beautiful."

Russell L. Schweickart
Apollo 9, March 3–13, 1969
(As published in *The Overview Effect,* 1998)

Use this quotation as a springboard to discuss the following questions:

- What are boundaries?
- Who draws the boundary lines?
- What purpose do boundaries serve?
- If boundaries are invisible lines, how do you know when you've crossed one?
- Once you've crossed one of these invisible lines, what has changed?
- Can you think of any problems that boundaries may cause?

Throughout the discussion, emphasize that although most boundaries are unmarked and invisible, they determine our perception of spaces and places on earth. Boundaries between countries help maintain order in the world because they define internationally recognized and sovereign political entities. Conflict can result when boundary lines are disputed. Boundaries are also potential sources of conflict because they are the point of contact between neighboring people.

Challenge students to identify places in the world where international boundaries have changed or are in conflict. What do students know about the reasons for those boundary changes and conflicts? Tell the class that they are going to do a GIS Investigation that will explore the characteristics of modern international boundaries and investigate recent boundary changes.

Student activity

 Before completing this lesson with students, we recommend that you complete it as well. Doing so will allow you to modify the activity to accommodate the specific needs of your students.

After the initial discussion, have the students work on the computer component of the lesson. Ideally, each student should be at an individual computer, but the lesson can be modified to accommodate a variety of instructional settings.

Distribute the GIS Investigation sheets to the students. Explain that in this activity, they will use GIS to investigate different types of international boundaries, explore the implications of various boundary configurations, and observe boundary changes in recent years. The activity sheets will provide them with detailed instructions for their investigations.

In addition to instructions, the handout includes questions to help students focus on key concepts. Some questions have specific answers while others call for speculation and have a range of possible responses. In addition, answers to many questions will vary with student knowledge of current events and contemporary world political issues.

Things to look for while the students are working on this activity:

- Are the students using a variety of GIS tools?
- Are the students answering the questions as they work through the procedure?
- Are students asking thoughtful questions throughout the investigation?

Teacher Tip: This GIS Investigation contains instructions for students to periodically stop and save their work. These are good spots to stop the class for the day and to pick up the investigation the next day. Be sure to inform your students as to how they should rename their map document and where to save it.

Conclusion

Use a projection device to display the global5.apr in the classroom. As a group, compare student observations and conclusions from the lesson. Students can take turns being the "driver" on the computer to highlight boundaries and observations that are identified by members of the class. Focus on the following aspects of the GIS Investigation in your discussion:

- Where did students observe the coincidence of political and physiographic boundaries?
- What examples of territorial morphology (countries representing categories of different shapes and sizes) did the students find? Ask students to speculate on ways that a country's size and shape could influence its sense of unity or cohesiveness.
- How can a country's boundaries influence its economic strength and advantage?
- What kinds of problems are likely to arise when political and anthropographic (cultural) boundaries do not coincide?
- What is the nature of the boundary changes that occurred between 1992 and 2000? Based on student response in step 8, which of the new countries do students believe are in the strongest position today in terms of cohesiveness?

Assessment

Middle school: Highlights skills appropriate to grades 5 through 8

In the middle school assessment, students are asked to identify an international boundary in 2000 that they predict could change in the next 25 years. They are asked to prepare a map of the projected boundary change and describe its impact in terms of the lesson's concepts.

- What types of boundaries are involved in the projected change?
- How will the territorial morphology of the countries involved be affected by the projected change?
- What will the economic impact of the projected change be?
- How will the projected change affect the internal cohesiveness of all countries involved?

High school: Highlights skills appropriate to grades 9 through 12

In the high school assessment, students are asked to identify two international boundaries in 2000 that they predict could change in the next 25 years. One of their predictions should involve splitting a current country into two or more smaller countries, and one of their predictions should involve merging two or more countries into one larger one. They are asked to prepare a map of the projected boundary changes and compare them in terms of the lesson's concepts.

- What types of boundaries are involved in the changes?
- How will the territorial morphology of the countries involved be affected by the projected changes?
- What will the economic impact of the projected changes be?
- How will the projected changes affect the internal cohesiveness of all countries involved?

Extensions

- Assign students to conduct research on world boundary changes during the twentieth century. Use ArcView to prepare a sequence of layouts reflecting those changes.

- Explore the nature of political boundaries in your own community and state. What kind of boundaries are they, how does the shape of your town or state affect its cohesiveness, what are the economic advantages and disadvantages of your town or state's boundary configuration, how have your town or state's boundaries changed over time?

- Search newspapers and magazines (both on- and offline) for coverage of border conflicts and related issues around the world. Use ArcView to illustrate the nature of these conflicts.

- Check out the Resources by Module section of this book's companion Web site *(www.esri.com/mappingourworld)* for print, media, and Internet resources on the topic of political boundaries.

NAME _____   DATE _____

Crossing the Line
A GIS investigation

Answer all questions on the student answer sheet handout

Boundaries are invisible lines on the earth's surface. They divide the surface area into distinct separate political entities. In this activity, you will use GIS to investigate different types of international boundaries, explore the implications of various boundary configurations, and observe boundary changes in recent years. When you have completed the activity, you will use your knowledge of boundary dynamics to speculate on world boundaries that are likely to change in the future.

Step 1   Start ArcMap

a   Double-click the ArcMap icon on your computer's desktop.

b   If the ArcMap start-up dialog appears, click **An existing map** and click OK. Then go to step 2b.

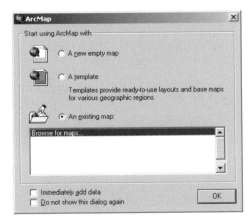

Step 2   Open the Global5.mxd file

a   In this exercise, a map document has been created for you. To open it, go to the File menu and choose **Open**.

b   Navigate to the module 5 folder (**C:\MapWorld9\Mod5**) and choose **Global5.mxd** (or **Global5**) from the list.

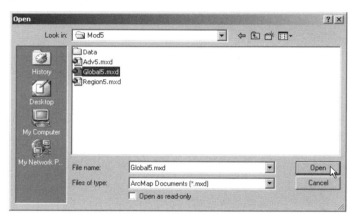

c   Click Open.

The map document opens and you see a composite satellite image of the world. The check mark next to the layer name tells you the layer is turned on and visible in the data frame.

Step 3   Explore mountain ranges as physiographic boundaries

As astronaut Russell L. Schweickart said, if you could view the world from space, you would see no boundary lines. Boundaries are human-made lines that define the world's political entities.

There are several types of boundaries between countries. One type of boundary is called a physiographic boundary. They are based upon natural features on the land-scape such as mountain ranges or rivers.

a   Click the box next to the Boundaries 2004 layer to turn it on. A check mark appears and the red lines show the international boundaries for the year 2004.

    b   Click the Zoom In tool. On the map, click and drag a box around Europe. The view is now centered on Europe.

c Drag another box around Europe to zoom in more so you can see the physical features in greater detail.

d Turn off Boundaries 2004 by clicking the check mark next to the layer name.

Locate Europe's mountain ranges in the satellite image. Notice that ranges such as the Pyrenees Mountains in northeastern Spain form a natural boundary. You will use the Draw Line tool to draw lines where you see a mountain range forming a natural boundary between different parts of the continent. First, you will select a symbol type and color for drawing.

e On the Draw toolbar click the drop-down arrow to the right of the New Rectangle tool and select the New Line tool.

  f On the Draw toolbar click the down arrow key to the right of the Line Color button and change the color to yellow.

Now you are ready to draw a physiographic boundary in Europe.

g Click the westernmost edge of the Pyrenees Mountains to start your line. Continue clicking along the path of the mountain range until you reach its easternmost edge. Double-click to end the line.

h The yellow line is displayed in the map. Click on the map away from the yellow line to make the blue selection box disappear.

i Turn on the Boundaries 2004 layer. You see that your line corresponds to a border between two countries.

  j Click the Identify button. Move the Identify Results window so you can see the map.

k Click the country that borders the Pyrenees Mountains to the north.

The left side of the window shows the layer name and below it the country name. You can also see all the attributes for that country that are in the attribute table.

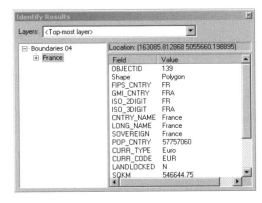

l Click on the country on the other side of the border.

The Pyrenees Mountains are the border between which two countries?

m Click the Select Elements tool. Click and drag a box over the boundary line you previously drew on the map to select the line (the blue box should appear). Press the Delete key on your keyboard to delete the line.

n Use the Identify tool to find other Western European countries where physiographic boundaries created by mountains correspond to actual political boundaries. Click a country to see its information in the Identify Results window.

Complete the table on the answer sheet.

Step 4 **Explore bodies of water as physiographic boundaries**

a Turn off the Satellite Image layer and turn on the Rivers, Lakes, Countries 2004, and Ocean layers.

Wherever countries have physiographic boundaries based on rivers, the red boundary line disappears beneath the blue river on the map. Look closely at Europe to see if you can find any boundaries that are rivers. The different colored countries will help you find these places.

b In the Identify Results window click the Layers box and make Countries 2004 the active layer. (If the Identify Results window is closed click the Identify button to open it.)

c Click a country that has a river as all or part of a boundary. You may need to zoom and pan to see the rivers that are aligned with the country boundaries. Make sure you click the Identify tool again when you are in a location that you like.

> *Note: Refer to the ArcMap Toolbar Quick Reference for a brief explanation of the Zoom and Pan tools.*

In the table on the answer sheet, record the names of three sets of countries that share a boundary that is a river.

> *Note: In order to identify the names of the rivers, you must make Rivers the active layer in the Identify Results window.*

Coastlines are also physiographic boundaries. Countries that do not have a coastline are said to be landlocked.

d Zoom or pan so that Western Europe is in your full view.

Name three landlocked countries in Western Europe. Use the Identify tool if you don't know the name of a specific country. (Remember to set Countries 2004 as the active layer in the Identify Results window.)

Step 5 **Explore geometric boundaries**

Another type of boundary is a geometric boundary. Geometric boundaries consist of straight or curved lines that do not correspond to physical features on the earth's surface.

a Click the Full Extent button to see all the continents.

b Click the Zoom In tool. Use it to zoom in on Africa.

You see many rivers that overlap boundaries throughout the African continent.

c Look at the map and locate countries that have geometric boundaries.

d Click the Identify button. Make Countries 2004 the active layer in the Identify Results window.

e Use the Identify tool to identify African countries that are separated by geometric boundaries.

Record three sets of countries in the table on the answer sheet.

f Close the Identify Results window.

g Click Full Extent to see the entire world.

h Ask your teacher if you should stop here and save this map document. Follow your teacher's instructions on how to rename the map document and where to save it. If you do not need to save the map document, proceed to step 6.

Write the new name you gave the map document and where you saved it.

Step 6 **Explore anthropographic boundaries based on language and religion**

A third type of boundary is an *anthropographic* boundary. This boundary marks the transition between cultural characteristics on the landscape. Anthropographic boundaries are based on characteristics such as language, religion, or ethnicity.

a Turn off Rivers, Lakes, Boundaries 2004, and Countries 2004.

b Click the Add Data button.

c Navigate to the LayerFiles folder within the module 5 Data folder (**C:\MapWorld9\Mod5\Data\LayerFiles**).

d Select **Language.lyr**. Hold down the Ctrl key and click **Religion.lyr**. Click Add.

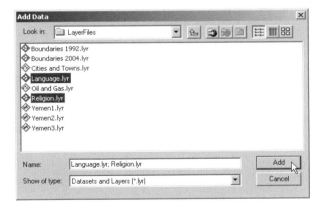

e Drag the Boundaries 2004 layer above the Language and Religion layers in the table of contents.

f   Turn on Language and click the plus sign next to it to expand the legend. The distribution of major language groups in the world is displayed. Drag the right edge of the table of contents to widen it if you need to.

g   Observe the pattern of anthropographic boundaries based on language in the world.

h   Use the Identify tool to determine the principal language groups in South America and Western Europe. (Don't forget to make Language the active layer in the Identify Results window.) Record them on the answer sheet.

i   Turn on Boundaries 2004.

j   Use the Identify and Zoom tools to locate three examples in the world where political boundaries coincide with anthropographic boundaries based on language. Record them on the answer sheet.

k   Click Full Extent. Turn off Boundaries 2004 and Language.

l   Turn on the Religion layer and expand its legend. The distribution of major religions is displayed.

m   Observe the pattern of anthropographic boundaries based on religion throughout the world.

n   Use the Identify and Zoom tools to determine the principal religions in North America and Africa. Record them on the answer sheet.

o   Turn on Boundaries 2004.

p   Use the Identify and Zoom tools to locate three examples in the world where political boundaries coincide with anthropographic boundaries based on religion. Record them on the answer sheet.

Step 7   **Review physiographic, geometric, and anthropographic boundaries**

a   Find additional examples of physiographic, geometric, and anthropographic boundaries between countries. Record your findings in the table on the answer sheet.

b   Close the Identify Results window when you are finished.

Step 8   **Explore the impact of boundary shape, cultural diversity, and access to natural resources**

Boundaries determine the size and shape, or territorial morphology, of countries. Size and shape can exert a powerful influence on the cohesiveness of a country. Small compact nations or ones that are circular or hexagonal, for example, are more easily united than ones that are elongated or fragmented.

a   Click the Full Extent button to see the whole world again. Turn off all layers except for Countries 2004 and Ocean and collapse the Languages and Religions legends in the table of contents (click the minus sign next to the layer title).

b   Click the Find button.

c In the Find box type **Chile**. Select Countries 2004 from the drop-down list.

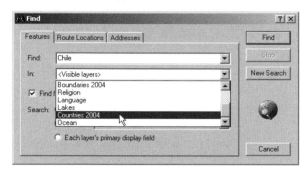

d Click Find. A results box appears at the bottom of the Find dialog and Chile is listed. Move the Find dialog to the side so you can see the map.

e Right-click on the highlighted row for Chile and click Zoom to feature(s). Then right-click again and click Flash feature to locate Chile in the map.

The table in the answer sheet illustrates six types of countries based on shape and gives an example of each. Chile is listed as an example of an elongated country.

f Use the Find, Zoom, and Pan tools to locate another example of each type of country. Record them in the table on the answer sheet in the Example 2 column. Remember, you can use the Identify tool to find the names of countries that you do not know.

Another factor that influences cohesiveness is the extent of cultural diversity.

g Click the Full Extent button. Turn off the Countries 2004 layer. Turn on Boundaries 2004 and Language.

(1) *By using language groups as an indicator of cultural uniformity, identify three countries that reflect cultural uniformity.*

(2) *By using language groups as an indicator of cultural diversity, identify three countries that reflect cultural diversity.*

Boundaries also influence economic activities. Earlier in this GIS Investigation, you identified landlocked countries in Western Europe. Historically, these countries were limited in their ability to trade directly with other nations because imports and exports had to pass through other countries en route to their destination.

h Click the Full Extent button. Turn off Boundaries 2004 and Language. Turn on Countries 2004.

i Use the ArcMap tools and buttons you've learned in this investigation to find an example of a landlocked country on each continent listed in the table on the answer sheet. For a continent that does not have a landlocked country, write "none."

j Close the Find dialog if it is still open.

Boundaries also influence economic activities by establishing a country's access to natural resources.

k Click the Full Extent button.

l Click the Add Data button. Navigate to the LayerFiles folder within the module 5 Data folder (**C:\MapWorld\Mod5\Data\LayerFiles**). Double-click **Oil and Gas.lyr**.

m The locations of oil and natural gas sources around the world are displayed on the map.

 n Use the Zoom In tool to focus on Southeast Asia.

o Use the Pan, Zoom, and Identify tools to help you answer the following questions.

? *(1)   Name two Southeast Asian countries that do not have any oil and gas resources within their borders.*

? *(2)   Name two Southeast Asian countries that have oil and gas resources within their borders.*

p Turn off the Oil and Gas layer.

 q Click the Full Extent button. Turn off Countries 2004 and Ocean, and turn on Boundaries 2004 and Satellite Image.

Step 9   Explore boundary changes in the 1990s

Political boundaries can change in many ways. Large countries may split into several smaller ones, small countries may combine to produce larger ones, territories that were once part of one country may be incorporated into another.

 a Click the Add Data button. Navigate to the LayerFiles folder within the module 5 Data folder (**C:\MapWorld9\Mod5\Data\LayerFiles**) and add **Boundaries 1992.lyr**.

b The international boundaries from 1992 display as yellow lines.

> *Note: Because the 1992 boundaries cover the 2004 boundaries on the map, the 2004 boundary lines are not visible when they are in the same location as they were in 1992. The only 2004 boundary lines (red) that are visible are those that did not exist in 1992.*

c Observe the map closely to see the difference between 1992 and 2004. What kind of changes to you see? (Use your Zoom and Pan tools to get a good look at these changes.)

? *(1)   Describe three political boundary changes you see between 1992 and 2004.*

? *(2)   Name two countries that existed in 1992 but do not exist in 2004.*

d If you have already saved this map document at the end of step 5, click the Save button to save your work. Otherwise, ask your teacher for instructions on where to save this map document and how to rename it. If you do not need to save the map document, continue to the next step.

? *Write the new name you gave the map document and where you saved it.*

Step 10   Compare new countries

Countries in groups A and B below are new countries that have emerged since 1992.

Group A	**Group B**
Czech Republic	Russia
Slovakia	Belarus
Slovenia	Ukraine
Croatia	Moldova
Bosnia and Herzegovina	Armenia
Serbia and Montenegro	Azerbaijan
Macedonia	Georgia
Eritrea	Kazakhstan
	Uzbekistan
	Tajikistan
	Turkmenistan
	Kyrgyzstan

? *a* Select three countries from group A and three from group B and complete the table on the answer sheet. Use the information and GIS skills you learned in this investigation to answer the questions.

Step 11   Exit ArcMap

In this exercise, you used ArcGIS to explore the various types of political boundaries and their impact on the countries they define. You added layers and used the Find, Identify, Zoom, and Pan tools to investigate the maps. You observed and analyzed boundary changes between 1992 and 2004.

a Click the File menu and click Exit. When asked if you want to save changes to the map document, click No.

NAME _____   DATE _____

Student answer sheet

Module 5
Human Geography II: Political Geography

Global perspective: Crossing the Line

Step 3   Explore mountain ranges as physiographic boundaries

l   The Pyrenees Mountains are the border between which two countries?

_____ **and** _____

n   Complete the table below (consult an atlas to find the names of unknown mountain ranges):

COUNTRIES THAT HAVE MOUNTAIN RANGES AS POLITICAL BOUNDARIES	MOUNTAINS THAT FORM THE BOUNDARY
and	
and	
and	

Step 4   Explore bodies of water as physiographic boundaries

c   In the table below, record the names of three sets of countries that share a boundary that's a river:

COUNTRIES THAT HAVE RIVERS AS BOUNDARIES	RIVER THAT FORMS THE BOUNDARY
and	
and	
and	

d   Name three landlocked countries in Western Europe. Use the Identify tool if you don't know the name of a specific country.

_____

_____

_____

Step 5   Explore geometric boundaries

e   Record three sets of countries in the table below:

COUNTRIES THAT ARE SEPARATED BY GEOMETRIC BOUNDARIES
and
and
and

h   Write the new name you gave the map document and where you saved it.

(Name of map document. For example: ABC_Global5.mxd)	(Navigation path to where map document is saved. For example: C:\Student\ABC)

Step 6   Explore anthropographic boundaries based on language and religion

h   Determine the principal language groups in the regions listed below.

South America: _____

Western Europe: _____

j   Locate three examples in the world where political boundaries coincide with anthropographic boundaries based on language.

ANTHROPOGRAPHIC BOUNDARIES BASED ON LANGUAGE COINCIDE WITH POLITICAL BOUNDARIES BETWEEN
and
and
and

n   Determine the principal religions in the following regions:

North America: _____

Africa: _____

p   Locate three examples in the world where political boundaries coincide with anthropographic boundaries based on religion.

ANTHROPOGRAPHIC BOUNDARIES BASED ON RELIGION COINCIDE WITH POLITICAL BOUNDARIES BETWEEN
and
and
and

Step 7   Review physiographic, geometric, and anthropographic boundaries

a   Find additional examples of physiographic, geometric, and anthropographic boundaries between countries. Record your findings in the following table:

CONTINENT	PHYSIOGRAPHIC BOUNDARIES SEPARATE THE FOLLOWING COUNTRIES	GEOMETRIC BOUNDARIES SEPARATE THE FOLLOWING COUNTRIES	ANTHROPOGRAPHIC BOUNDARIES SEPARATE THE FOLLOWING COUNTRIES
North and Central America	_____ and _____ Boundary formed by (circle one): Mountains Rivers Lakes	_____ and _____	_____ and _____
South America and the Caribbean	_____ and _____ Boundary formed by (circle one): Mountains Rivers Lakes	_____ and _____	_____ and _____
Europe	_____ and _____ Boundary formed by (circle one): Mountains Rivers Lakes	_____ and _____	_____ and _____
Africa	_____ and _____ Boundary formed by (circle one): Mountains Rivers Lakes	_____ and _____	_____ and _____
Asia	_____ and _____ Boundary formed by (circle one): Mountains Rivers Lakes	_____ and _____	_____ and _____

Step 8   Explore the impact of boundary shape, cultural diversity, and access to natural resources

f   Locate another example of each type of country. Record them in the following table in the Example 2 column. Remember, you can use the Identify tool to find the names of countries that you do not know.

TYPE OF COUNTRY	EXAMPLE	EXAMPLE 2
Elongated	Chile	
Fragmented	Philippines	
Circular/Hexagonal	France	
Small/Compact	Bulgaria	
Perforated	South Africa	
Prorupted	Namibia	

g-1 By using language groups as an indicator of cultural uniformity, identify three countries that reflect cultural uniformity.

_____

_____

_____

g-2 By using language groups as an indicator of cultural diversity, identify three countries that reflect cultural diversity.

_____

_____

_____

i   Use the ArcMap tools and buttons you've learned in this investigation to find an example of a land-locked country on each of the following continents. For a continent that does not have a landlocked country, write "none."

CONTINENT	LANDLOCKED COUNTRY
North America (including Central America)	
South America	
Africa	
Asia	

o-1 Name two Southeast Asian countries that do not have any oil and gas resources within their borders.

_____

o-2 Name two Southeast Asian countries that have oil and gas resources within their borders.

_____

Step 9   Explore boundary changes in the 1990s

c-1   Describe three political boundary changes you see between 1992 and 2004.

_____

_____

_____

c-2   Name two countries that existed in 1992 but do not exist in 2004.

_____

_____

d   Write the new name you gave the map document and where you saved it.

_____   _____

(Name of map document.   **(Navigation path to where map document is saved.**
For example: ABC_Global5.mxd)   **For example: C:\Student\ABC)**

Step 10   Compare new countries

a   See next page.

10a   Select three countries from group A and three from group B and complete the following table. Use the information and skills you learned in this GIS investigation to answer the questions.

GROUP	COUNTRY	WHAT TYPE OF BOUNDARIES DOES IT HAVE?	HOW WOULD YOU CHARACTERIZE ITS SHAPE?	WHAT ECONOMIC ADVANTAGES OR DISADVANTAGES DO YOU SEE?	DO THE CURRENT CHARACTERISTICS OF THE COUNTRY PROMOTE COHESIVENESS OR SPLITTING APART?
A					
B					

NAME _____   DATE _____

Crossing the Line
Middle school assessment

In this assessment activity, you must predict the future! Instead of using a crystal ball, you will use a GIS to see 25 years into the future of the world.

1 Use the information you learned in this GIS Investigation to identify a current international boundary that you think will change in the next 25 years.

2 Draw a map (in ArcMap or with paper and pencil) to illustrate what this boundary will look like in 25 years. Be sure to include a legend, north arrow, scale, and date of creation on the map.

3 Write an essay that describes the consequences of the change you predict. Address the following questions in your essay:

• What types of boundaries are involved in the projected change?

• How will the territorial morphology of the countries involved be affected by the projected change?

• What will the economic impact of the projected change be?

• How will the projected change affect the internal cohesiveness of all countries involved?

Crossing the Line

Assessment rubric

Middle school

STANDARD	EXEMPLARY	MASTERY	INTRODUCTORY	DOES NOT MEET REQUIREMENTS
The student knows and understands how to use the elements of space to describe spatial patterns.	Creates a digital map using GIS to illustrate a predicted international boundary change. The student takes into account a variety of data when developing his/her new map.	Creates a map to illustrate a predicted international boundary change. Takes into account a variety of data when developing his/her new map.	Creates a map to illustrate a predicted international boundary change. Provides some data to support ideas.	Describes a change in a border, but does not provide a map and has little or no data to support ideas.
How cooperation and conflict among people contribute to economic and social divisions of Earth's surface.	Uses maps and written description to illustrate how the proposed boundary changes will affect the countries involved, including their role in the global economy. Provides sufficient data to support ideas.	Describes how the proposed boundary changes will affect the countries involved, including their role in the global economy. Provides sufficient data to support ideas.	Attempts to describe how the proposed boundary changes will affect the countries involved, but does not address effect on global economy. Provides some data to support ideas.	Does not address economic issues for the countries involved in the boundary changes.
The student knows and understands how varying points of view on geographic context influence plans for change.	Describes the perspectives of the countries involved in the proposed boundary change. Addresses a variety of issues including political, cultural, and so on. Also addresses how this will affect the global community.	Describes the perspectives of the countries involved in the proposed boundary change. Addresses a variety of issues including political, cultural, and so on.	Attempts to describe the perspectives of the countries involved in the proposed boundary change. Addresses only one issue in their description.	Describes only one country's perspective on the proposed boundary change.

This is a four-point rubric based on the National Standards for Geographic Education. The "Mastery" level meets the target objective for grades 5–8.

NAME _____   DATE _____

Crossing the Line
High school assessment

In this assessment activity, you must predict the future! Instead of using a crystal ball, you will use a GIS to see 25 years into the future of the world.

1   Use the information you learned in this GIS Investigation to identify two current international boundaries that you think will change in the next 25 years. One prediction must involve splitting a current country into two or more smaller countries, and the other must merge two or more countries into one larger one.

2   Draw a map (in ArcMap or with paper and pencil) to illustrate what these boundaries will look like in 25 years. Be sure to include a legend, north arrow, scale, and date of creation on the map.

3   Write an essay that describes potential consequences of the change you predict. Address the following questions in your essay:

•   What types of boundaries are involved in the changes?

•   How will the territorial morphology of the countries involved be affected by the projected changes?

•   What will the economic impact of the projected changes be?

•   How will the projected changes affect the internal cohesiveness of all countries involved?

Crossing the Line

Assessment rubric

High school

STANDARD	EXEMPLARY	MASTERY	INTRODUCTORY	DOES NOT MEET REQUIREMENTS
The student knows and understands how to apply concepts and models of spatial organization to make decisions.	Creates a digital map using GIS to illustrate a predicted international boundary change. Takes into account a variety of data when developing his/her new map.	Creates a map to illustrate a predicted international boundary change. The map illustrates two types of changes: the merging of countries together, and the splitting of countries apart. Takes into account a variety of data when developing his/her new map.	Creates a map to illustrate a predicted international boundary change. The map shows only one type of change. Provides some data to support ideas.	Describes a change in a border, but does not provide a map and has little or no data to support ideas.
The student knows and understands why and how cooperation and conflict are involved in shaping the distribution of social, political, and economic spaces on Earth at different scales.	Uses maps and written description to illustrate how the proposed boundary changes will affect the countries involved, including their role in the global network. Provides sufficient data to support ideas.	Describes how the proposed boundary changes will affect the countries involved and their role in the global network. Provides sufficient data to support ideas.	Attempts to describe how the proposed boundary changes will affect the countries involved, but does not address effect on global network. Provides some data to support ideas.	Does not address any specific issues for the countries involved in the boundary changes.
The student knows and understands contemporary issues in the context of spatial and environmental perspectives.	Describes the perspectives of the countries involved in the proposed boundary change. Addresses a variety of issues including political, cultural, and so on. Also addresses how this will affect the perspective of the global community.	Describes the perspectives of the countries involved in the proposed boundary change. Addresses a variety of issues including political, cultural, and so on.	Attempts to describe the perspectives of the countries involved in the proposed boundary change. Addresses only one issue in their description.	Describes only one country's perspective on the proposed boundary change.

This is a four-point rubric based on the National Standards for Geographic Education. The "Mastery" level meets the target objective for grades 9–12.

A Line in the Sand
A regional case study of Saudi Arabia and Yemen

Lesson overview

Students will study the creation of a new border between Saudi Arabia and Yemen on the Arabian Peninsula. Using data included in the June 2000 Treaty of Jeddah, they will draw the new boundary described in the treaty and analyze the underlying physiographic and cultural forces that influenced the location of that boundary. In the process they will come to understand how any map of the world must be considered a tentative one, as nations struggle and cooperate with each other.

Estimated time   Three to four 45-minute class periods

Materials   ✔ Student handouts from this lesson to be copied:
- GIS Investigation sheets (pages 291 to 304)
- Student answer sheets (pages 305 to 309)
- Assessment(s) (pages 310 to 313)

Standards and objectives

National geography standards

GEOGRAPHY STANDARD	MIDDLE SCHOOL	HIGH SCHOOL
1 How to use maps and other geographic representations, tools, and technologies to acquire, process, and report information from a spatial perspective	The student understands how to use maps to analyze spatial distributions and patterns.	The student understands how to use geographic representations and tools to analyze and explain geographic problems.
4 The physical and human characteristics of places	The student understands how physical processes shape places and how different human groups change places.	The student understands the changing human and physical characteristics of places.
13 How the forces of cooperation and conflict among people influence the division and control of Earth's surface	The student understands the multiple territorial divisions of the student's own world.	The student understands why and how cooperation and conflict are involved in shaping the distribution of social, political, and economic spaces on Earth at different scales.
18 How to apply geography to interpret the present and plan for the future	The student understands how various points of view on geographic context influence plans for change.	The student understands contemporary issues in the context of spatial and environmental perspectives.

Standards and objectives (continued)

Objectives

The student is able to:

- Describe the physical and human characteristics of the Arabian Peninsula.
- Define and describe the Empty Quarter.
- Explain major elements of the Treaty of Jeddah boundary agreement between Saudi Arabia and Yemen.
- Identify physical and cultural characteristics of the Arabian Peninsula that are reflected in the new Saudi–Yemeni border agreement.

GIS skills and tools

 Zoom in on the map

 Zoom to the previous extent

 Zoom in on the center of the map

 Add layers to the map

 Draw a line

 Select a graphic

 Specify a location to save work

 Create a line feature by adding points at specific latitude/longitude coordinates

 Save the map document

 Select a map feature from a specific layer

 Show the ArcToolbox window

- Use a bookmark
- Use MapTips to identify features
- Turn layers on and off and expand and collapse their legends
- Activate a data frame
- Export a layer to a shapefile
- Display and work with the Edit toolbar
- Set selectable layers
- Create a buffer
- Print a map

*GIS skills and tools
(continued)*

For more on geographic inquiry and these steps, see Geographic Inquiry and GIS (pages xxiii to xxv).

Teacher notes

Lesson introduction

Divide your class into small groups. Explain to your students that the lesson they will begin today is about drawing boundary lines. In order to identify some of the important considerations in the demarcation of boundaries, each group will take five minutes to consider the following hypothetical scenario:

Size limitations in the school building require that their classroom be divided to create two new, smaller classrooms. Other than the wall dividing the two classrooms, there will be no new construction. Each group is charged with two tasks:

- Identify the features of the present classroom which are valuable to the teachers and students who use that room (windows, for example).
- Suggest a possible boundary line to divide the present classroom and identify the features from the previous step that each of the new classrooms will have.

When five minutes have passed, make a list on the blackboard or an overhead projector of the valuable classroom features that students identified in the first step. Let each group report on the boundary they propose. Use this activity as a springboard for a discussion of the issues involved in the creation of national boundaries. Be sure to include the following points in the discussion:

- Certain features of the physical environment have greater value than others to the people who will occupy and use that space.
- The human uses of a place influence the perceived value of its physical features.
- When a boundary line is drawn, it may not be possible to divide the valuable features evenly between the parties involved.

Tell the class that this activity will explore a twenty-first-century case of the demarcation of a boundary between two countries. Although at a much different scale, this decision involved many of the same issues they faced in drawing a hypothetical boundary in their classroom.

Student activity

 Before completing this lesson with students, we recommend that you work through it yourself. Doing so will allow you to modify the activity to accommodate the specific needs of your students.

After the initial discussion, have the students work on the computer component of the lesson. Ideally each student should be at an individual computer, but the lesson can be modified to accommodate a variety of instructional settings.

Distribute the GIS Investigation sheets to your students. Explain that in this activity they will use GIS to explore a region of the world where a recent boundary dispute has been settled after 65 years of conflict: the Arabian Peninsula. They will explore alternatives for boundaries between the countries involved and analyze the underlying physiographic and cultural considerations that played a part in the resolution of that conflict.

The GIS activity will provide students with detailed instructions for their investigations. In addition to the instructions, the handout includes questions to help them focus on key concepts. Some questions will have specific answers while others require creative thought.

Things to look for while the students are working on this activity:

- Are the students using a variety of tools?
- Are they answering the questions as they work through the procedure?
- Are they experiencing any difficulty managing the display of information in their map as they turn layers on and off?
- Are students experiencing any difficulty plotting latitude/longitude points or finishing their sketch when creating the Saudi–Yemeni boundary line feature?

 Teacher Tip: Decide ahead of time where you want students to save their boundary line data in step 8f. Students can export a feature class to the MiddleEast geodatabase if they have their own copy of the module 5 folder. Otherwise, you may want students to export a shapefile to another location.

Conclusion

Refer your students to the activity that introduced this lesson: the creation of a hypothetical boundary line that divides their classroom into two new rooms. Review their conclusions and ask them to identify parallel issues in the settlement of the boundary dispute between Saudi Arabia and Yemen.

- Certain features of the physical environment have greater value than others to the people who will use that space. On the Arabian Peninsula, areas that get enough precipitation for agriculture, areas of grassland for grazing, sources of water, and areas with the strategic advantage of mountain peaks have greater value.
- The human uses of a place influence the perceived value of its physical features. On the Arabian Peninsula, livestock herding and farming are examples of traditional human uses.
- When a boundary line is drawn, it may not be possible to divide the valuable features evenly between the parties involved. On the Arabian Peninsula, most of the areas that get enough precipitation for agriculture, areas of grassland for grazing, and sources of water went to Yemen.

Ask students to identify issues that played a role in the Saudi–Yemeni border conflict that were not present in the classroom boundary scenario. For example:

- Historic boundaries and patterns of political control in the region played an important part in the Saudi–Yemeni border conflict. Discuss important historical events and periods such as the Ottoman Empire, the consequences of World War I in this region, international interest in the region during the twentieth century, and the British Protectorate of Aden.

- Nomadism and strong identity with regional tribal traditions are at odds with the delineation of a fixed boundary in this region. Discuss the various factors that influence a community's or a region's sense of itself.

 Note: Students may wonder why the Saudis were willing to yield so much territory to Yemen. Ask them to speculate on possible reasons for this apparent generosity. The Treaty of Jeddah states that the two countries will negotiate if sources of "shared natural wealth" are discovered in the border region. This means that Saudi Arabia reserves the right to reopen negotiations in the event that something that they value very highly—oil or gas, for example—is discovered near the new boundary. Also, Saudi Arabia has long been interested in constructing a pipeline to the Arabian Sea across the southern part of the peninsula. The Saudis may have been hoping that a generous settlement with Yemen on the border issue could make the Yemenis more willing to agree to a Saudi pipeline across their territory.

Assessment

Middle school: Highlights skills appropriate to grades 5 through 8

In the middle school assessment, students will write a newspaper article reporting on the settlement of the Saudi–Yemeni border dispute by the Treaty of Jeddah, which was agreed to in June 2000. The article, written from either a Saudi or a Yemeni perspective, should describe the new boundary established by the treaty and analyze underlying physiographic and cultural considerations that influenced the location of that boundary. They will also prepare a map to go with the article.

High school: Highlights skills appropriate to grades 9 through 12

In the high school assessment, students will write a newspaper article reporting on the settlement of the Saudi–Yemeni border dispute by the Treaty of Jeddah, which was agreed to in June 2000. The article, written from either a Saudi or a Yemeni perspective, should describe the new boundary established by the treaty and analyze underlying physiographic and cultural considerations that influenced the location of that boundary. The article should also include information about historical factors that contributed to this 65-year-old boundary conflict. A map should accompany the article.

Extensions

- In the introductory activity, have student groups negotiate with each other to arrive at a mutually satisfactory boundary for the two new classrooms.
- Use the Internet to identify other areas of the world where international boundaries are in dispute.
- Research the events of World War I on the Arabian Peninsula. Create an ArcMap map document illustrating these events.
- Use ArcMap to compare the countries of the Arabian Peninsula by mapping and analyzing relevant economic and demographic data.
- Check out the Resources by Module section of this book's companion Web site *(www.esri.com/mappingourworld)* for print, media, and Internet resources on the topics of Saudi Arabia, Yemen, and the Treaty of Jeddah.

NAME _____   DATE _____

A Line in the Sand
A GIS investigation

ACQUIRE

ASK

EXPLORE

ACT

ANALYZE

Answer all questions on the student answer sheet handout

The ever-changing map of the world reflects the forces of conflict and cooperation among nations and peoples of the world. In this GIS Investigation, you will explore one of the first boundary changes of the twenty-first century—the creation of a new border between Yemen and Saudi Arabia on the Arabian Peninsula. After more than 60 years of conflict, the two nations signed the historic boundary agreement in June 2000. Using data provided in the Treaty of Jeddah, you will create a map reflecting the treaty's territory, and analyze underlying physiographic and cultural considerations that influenced the location of the boundary.

Step 1   Start ArcMap

a   Double-click the ArcMap icon on your computer's desktop.

b   If the ArcMap start-up dialog appears, click **An existing map** and click OK. Then go to step 2b.

Step 2   **Open the Region5.mxd file**

a   In this exercise, a map document has been created for you. To open it, go to the File menu and choose **Open**.

b   Navigate to the module folder (**C:\MapWorld9\Mod5**) and choose **Region5.mxd** (or **Region5**) from the list.

c   Click Open.

When the map document opens, you will see a shaded relief map of the Middle East and northeast Africa. A red outline locates the Arabian Peninsula.

d   Stretch your ArcMap window so that it fills most of your screen.

Step 3   **Identify countries that border the Arabian Peninsula**

a   Click the View menu and choose Bookmarks, Arabian Peninsula. Now the Arabian Peninsula fills the view.

b   Look in the table of contents for a layer called Neighbors - outline. Click the box to the left of the layer name to turn it on.

c   Slide the mouse pointer over the map to display the country names.

❓   *What countries border the Arabian Peninsula to the north?*

Step 4   **Investigate the physical characteristics of the Arabian Peninsula**

a   The map on your screen is a shaded relief map. It depicts landforms such as mountain ranges, valleys, plateaus, and plains.

❓   *(1)   Is any part of the Arabian Peninsula mountainous?*

❓   *(2)   If so, where are the mountains located?*

b   Click the plus sign next to Water in the table of contents to expand this group of layers.

c   Turn on the Bodies of Water and Streams layers. Then display them by checking the box next to Water.

Most of the streams you see on your map are intermittent, which means that they are dry during some parts of the year.

❓   *(1)   Are there any parts of the Arabian Peninsula that do not have any water at all? If so, where are these regions?*

❓   *(2)   Do you see any relationship between landforms and the availability of water?*

d   Turn off the Streams layer and observe the distribution of permanent bodies of water on the Arabian Peninsula.

e   Click the Zoom In tool. Click and drag a small box around an area of blue dots.

Now you can see the bodies of water more closely.

Describe the bodies of water on your answer sheet.

f   Click the Previous Extent button to return to your view of the entire peninsula.

g   Turn off the following layers: Water, Arabian Peninsula - outline, Neighbors - outline, and Shaded Relief. (You may need to scroll down in the table of contents.)

Your map display should look like this:

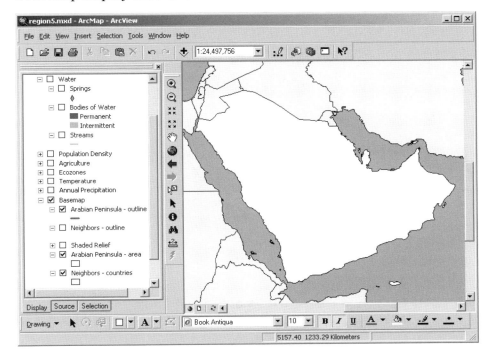

h   Collapse the Water group. Expand the Annual Precipitation layer and turn it on.

Amounts of rainfall are given in millimeters. Here is a conversion table that compares millimeters to inches (25.4 mm. = 1 in.).

MM.	100	200	300	400	500	600	700
IN.	3.9	7.9	11.8	15.7	19.7	23.6	27.6

(1)   A desert is defined as a place that gets less than 10 inches of rain per year. How many millimeters equal 10 inches?

(2)   Based on the amounts of rainfall displayed on the map, do you think there is much farming on the Arabian Peninsula? Explain.

(3)   Approximately what percentage of the Arabian Peninsula is desert?

i   Turn off Annual Precipitation. Turn on and expand the Temperature layer group. Turn on Temp: Sept. – Nov.

Use this conversion table to help you answer the next questions.

°C	5°	10°	15°	20°	25°	30°	35°	40°
°F	41°	50°	59°	68°	77°	87°	95°	104°

?   *What is the approximate range of temperatures across the Arabian Peninsula during this period?*

The three layers below Temp: Sept.- Nov. display temperature information for the periods December–February, March–May, and June–August.

j   Turn the temperature layers on and off one at a time to see the change of temperatures on the Arabian Peninsula through the four seasons.

?   *(1)   Which season is the hottest?*

?   *(2)   What is the approximate range of temperatures across the Arabian Peninsula during this period?*

k   Turn off the Temperature group. Scroll up and turn on the Ecozones layer. Expand its legend.

?   *(1)   What relationship do you see between the Arabian Peninsula's ecozones as displayed on this map and its patterns of landforms, precipitation, and temperature?*

?   *(2)   Use your answers from previous questions and turn different layers on and off to complete the Physical Characteristics of the Arabian Peninsula table on your answer sheet. List three observations for each physical characteristic.*

?   *(3)   In your opinion, which of the region's physical characteristics would be considered "valuable" in a boundary decision? Explain.*

Step 5   **Investigate the human characteristics of the Arabian Peninsula**

The population of the Arabian Peninsula is approximately 45 million. The majority of this population lives in Saudi Arabia (22 million) and Yemen (17.5 million). The remaining 5.5 million can be found in Oman, the United Arab Emirates, and Qatar.

a   Turn off Ecozones. Turn on the Arabian Peninsula - names layer. This layer locates the countries by name, but does not show their borders. You will explore the borders of these countries in the second part of the investigation.

b   Turn on the Major Cities and Agriculture layers. Expand the Agriculture legend to see a list of the types of agricultural activity on the Arabian Peninsula.

MODULE 5 • HUMAN GEOGRAPHY II: POLITICAL GEOGRAPHY

?   **(1)**   *What is the principal agricultural activity on the peninsula?*

?   **(2)**   *Based on what you now know about the physical characteristics of the region, why do you think the agricultural activity is so limited?*

c   Turn on and expand the Population Density layer. The human population around major cities and throughout the Arabian Peninsula is displayed as number of people per square kilometer.

?   **(1)**   *How does Yemen compare to the rest of the Arabian Peninsula in population density?*

?   **(2)**   *Describe the overall population density of the Arabian Peninsula.*

d   Turn off Population Density. Make sure Agriculture is still turned on.

e   Turn on Water, then expand it and turn on Springs. Turn off Bodies of Water.

The Springs layer shows the location of springs and water holes.

f   Look at the map.

?   **(1)**   *On your answer sheet, speculate about the ways water is most commonly and frequently used at these springs and water holes.*

?   **(2)**   *Use your answers from previous step 5 questions and analysis of the maps to complete the table on your answer sheet. List three observations for each human characteristic.*

?   **(3)**   *If an international boundary were to be drawn across some part of the Arabian Peninsula, how would these characteristics influence the perception of certain regions as being more valuable than others?*

Step 6   **Locate and describe the Empty Quarter**

a   Turn off Springs and turn on Roads. Expand the Roads legend.

Take note of the large area with practically no roads in the south-central part of the peninsula. This region is called the Rub´ al-Khali and is also known as the Empty Quarter. The Empty Quarter is important to this lesson because most Saudi Arabian borders with its southern neighbors cross this region.

b   Turn the following layers on and off so you can observe the characteristics of the Empty Quarter: Streams, Population Density, Agriculture, Ecozones, Temperature, and Annual Precipitation.

?   **(1)**   *Complete the table on your answer sheet. List three observations in each column.*

?   **(2)**   *What difficulties would an area like this present if an international boundary must cross it?*

c   Right-click the Middle East data frame in the table of contents and choose Collapse All Layers.

d   Right-click Middle East again and choose Turn All Layers Off. Then turn the following layers back on: Major Cities, Arabian Peninsula - names, and Basemap.

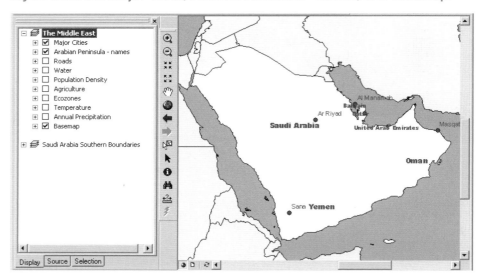

?   *e*   Ask your teacher for instructions on where to save this map document and on how to rename it. Record the map document's new name and where you saved it on your answer sheet.

Follow the steps below to exit ArcMap if you will be doing steps 7–12 at a later time. If you will be continuing to work now, skip to step 7b.

f   From the File menu, click Exit.

Step 7   **Explore Saudi Arabia's southern boundaries**

a   Start ArcMap. Navigate to the folder where you renamed and saved Region5. Open the map document.

b   Right-click the Saudi Arabia Southern Boundaries data frame and click Activate. Expand the data frame legend.

This map looks very similar to the Middle East map.

c   Turn on the 20th Century Boundary layer.

This layer reflects the boundary agreements Saudi Arabia made with most of its southern neighbors at the end of the twentieth century.

?   (1)   *Are the boundaries what you expected them to be?*

?   (2)   *Which boundary remained unsettled?*

According to international boundary expert Richard Schofield, this boundary was "the last missing fence in the desert." The only part of the boundary that was mutually agreed upon was the western area adjacent to the Red Sea. Over the years, the boundary has shifted. Now you will add layers that reflect some of the major boundary changes.

d Click the Add Data button. Navigate to the module 5 data folder and look in the LayerFiles folder (**C:\MapWorld9\Mod5\Data\LayerFiles**). Hold down the Shift key and click once on each of the following file names: **Yemen1.lyr**, **Yemen2.lyr**, and **Yemen3.lyr**. Click Add.

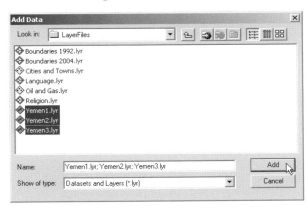

e Turn off Yemen2 and Yemen3 for the moment and look at the red line of Yemen 1.

This line represents the boundary between Yemen and Saudi Arabia established by the Treaty of Ta'if in 1934. It is the only part of the boundary that both countries recognized at the turn of the twenty-first century.

f Turn on Yemen2.

The green line of Yemen2 represents the Saudi–Yemeni border recognized by Yemen at the end of the century. It is based on lines established when Yemen (then the Aden Protectorate) was under British control in the early twentieth century. Most maps used these lines to delineate the extent of Yemen prior to 2000. This boundary was not recognized by Saudi Arabia.

g Turn on Yemen3.

This purple line represents the Saudi–Yemeni border claimed by Saudi Arabia at the end of the twentieth century. It is based on lines established by the Saudis in the mid-1930s. This line was still being used on Saudi Arabian maps to represent the boundary in the 1990s.

h Click the Zoom In tool. On your map, click the label Yemen. Now the map is centered on the country of Yemen.

i Click the Fixed Zoom In button two or three times until the country of Yemen fills the view.

? *What does the area between the green and purple lines represent?*

j Turn on Agriculture and expand its legend.

? *What is the principal economic activity of the regions in dispute?*

k Turn off Agriculture and turn on and expand Population Density.

? *Describe the population distribution in the disputed territory.*

If you were asked to settle the disputed boundary between Saudi Arabia and Yemen, where would you draw the line? In this next step, you will draw a proposed boundary line between Saudi Arabia and Yemen, using the Draw Line tool.

l Make sure the Draw toolbar is displayed. If you don't see it, right-click in the gray space to the right of the Help menu to display the toolbar list and click Draw. A good place to dock the Draw toolbar is at the bottom of the ArcMap window.

m In the Draw toolbar, click the down arrow next to the New Rectangle tool and choose the New Line tool.

n On your map, click the eastern end point of the red boundary line. (This is the boundary that both countries agree on.) Proceed eastward (to your right) and click a proposed boundary line. Double-click when you get to the end of your boundary. Now you have an additional black line that extends from the red line to Oman.

o Make sure the Select Elements tool is now active. Click anywhere on the map away from the line you drew to make the blue selection box disappear.

In the next step, you will view the new boundary actually agreed upon by Saudi Arabia and Yemen in 2000.

Step 8   Draw the Saudi–Yemeni boundary

In June of 2000, Saudi Arabia and Yemen signed the Treaty of Jeddah, which settled their 65-year-long boundary dispute. The boundary agreement had three parts. The first part of the treaty reaffirmed agreement on the 1934 Ta'if line (Yemen1.shp). The agreement did say, however, that the line would be amended in any place where it cuts through villages.

a Click the Zoom In tool. Zoom to the area of the Ta'if line (red line) by dragging a box around it.

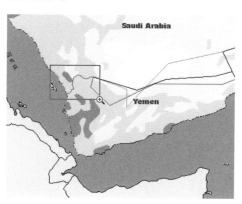

b   Click the Add Data button. Navigate to the module 5 Data folder and look in the LayerFiles folder (**C:\MapWorld9\Mod5\Data\LayerFiles**). Select **Cities and Towns.lyr**. Click Add.

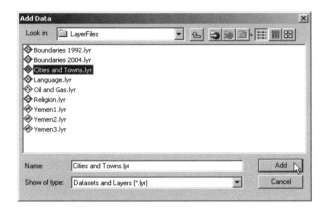

? **(1)** *Does the red line go through any cities or towns? (Hint: You may need to zoom in again to answer the question.) If yes, approximately how many does the boundary pass through?*

? **(2)** *How would you decide which side of the town to put the boundary on? Remember, this decision would determine whether the residents of that village would be citizens of Saudi Arabia or Yemen.*

The second part of the Treaty of Jeddah determined the new boundary from the end of the red line to the border with Oman, 500 miles to the east. The treaty did not actually draw the line, but gave its starting and ending points and points in between as latitude/longitude grid coordinates. You will now plot these points on your map to locate the new boundary line.

c   Turn off Cities and Towns and Population Density.

d   Click the View menu, point to Bookmarks, and choose Yemen to zoom out to the entire country.

You will need your own layer for holding the data you will plot. You will export a copy of the BoundaryTemplate layer, which has no features, for this purpose.

e   Right-click the BoundaryTemplate layer. Point to Data, then click Export Data.

f   Click the Browse button in the Export Data dialog. Ask your teacher what type of file you should save and where you should save it. If you will be saving a feature class, choose Personal Geodatabase feature classes from the Save as type drop-down list and navigate to the MiddleEast geodatabase. Otherwise, choose Shapefile and navigate to the location your teacher directed you.

g   Name the file **ABC_Yemen4** where **ABC** are your initials. Click Save, and then click OK in the Export Data dialog.

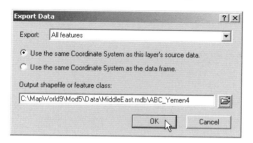

h   Click Yes to add the exported data to the map. Check the box to turn on ABC_Yemen4.

i   Right-click the BoundaryTemplate layer and click Remove to remove it from the map.

You will use tools on the Editor toolbar to plot the latitude and longitude coordinates for the new Saudi–Yemeni boundary.

j   Click the Editor Toolbar button to turn on the Editor toolbar. Dock the toolbar above the map.

k   Click the Editor menu on the Editor toolbar and choose Start Editing. Click the source for your ABC_Yemen4 layer in the dialog that appears and click OK.

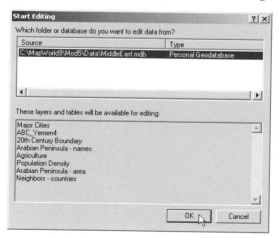

l   On the Editor toolbar make sure the Task is set to Create New Feature. Make sure the Target is your layer: ABC_Yemen4.

m   Click the Sketch Tool on the Editor toolbar.

n   Right-click anywhere in the map and choose Absolute X,Y.

The first point you will enter is 52 degrees longitude, 19 degrees latitude.

o   Type **52** for X. Press the Tab key and type **19** for Y. Press the Enter key.

You see a red square called a vertex appear at the border with Oman. Your cursor is attached to the vertex with an elastic line. As you move your mouse around the map without clicking, the line changes. The elastic line tells you that the new line feature you are creating is not yet finished.

p   To enter the remaining longitude and latitude points determined in 2000, consult the table below (points 2–17). For each point, you must right-click in the map and choose Absolute X,Y and then enter the coordinates from the table. Enter all the points now.

Point	Longitude	Latitude
1	52.00	19.00
2	50.78	18.78
3	49.12	18.61
4	48.18	18.17
5	47.60	17.45
6	47.47	17.12
7	47.18	16.95
8	47.00	16.95
9	46.75	17.28

Point	Longitude	Latitude
10	46.37	17.23
11	46.10	17.25
12	45.40	17.33
13	45.22	17.43
14	44.65	17.43
15	44.57	17.40
16	44.47	17.43
17	44.37	17.43

      Hint: If you make a mistake entering a point, click the Undo button to delete it. Then enter the coordinates for that point again. If you want to delete all the points you entered and start over, double-click to complete the polygon and then press the Delete key. Enter the points again beginning with point number 1.

q   When you are satisfied that you have entered all of the points correctly, right-click anywhere in the map and choose Finish Sketch. The completed line is highlighted in blue.

r   Click the Editor menu on the Editor toolbar and choose Save Edits.

   Does the new line seem to favor Yemen or Saudi Arabia? Explain.

Step 9   **Enter the maritime part of the boundary**

The third and final part of the Treaty of Jeddah clarified the maritime boundary between Saudi Arabia and Yemen. A maritime boundary defines the offshore limits of a country. It too was defined by a series of latitude/longitude grid coordinates.

a   Follow the procedure outlined in steps 8n–8q to map the maritime boundary between Saudi Arabia and Yemen.

Point	Longitude	Latitude
1	42.77	16.40
2	42.15	16.40
3	41.78	16.29

b   When you have finished the sketch, click the Editor menu and choose Stop Editing. Click Yes to save your edits.

c   Click the Editor Toolbar button to dismiss the toolbar.

d   Click the Fixed Zoom Out button twice. Look at your map.

❓   *What body of water does the maritime boundary traverse?*

Because ArcMap randomly selects a color for a new layer, you need to change it. You will also give the layer a more descriptive name.

e   Click the name of the ABC_Yemen4 layer two times slowly and change the name to **2000 Boundary**. Right-click the line symbol and choose a dark blue color from the color picker.

❓   *How does the actual boundary established by the Treaty of Jeddah compare with the boundary you drew earlier (black line)?*

f   Turn on Agriculture and Population Density as needed to answer the following question.

❓   *Write three observations about the boundary line created by the Treaty of Jeddah.*

g   Save your map document.

Step 10   **Define the pastoral area**

The Treaty of Jeddah included additional provisions about the new Saudi–Yemeni boundary. One of these was the creation of a "pastoral area" on either side of the boundary. Shepherds from either Yemen or Saudi Arabia are allowed to use the pastoral area and water sources on both sides of the border according to tribal traditions. The treaty declared that the pastoral area extends 20 kilometers on either side of the border. In this step, you will map the 20-kilometer pastoral area.

❓   *a*   How many miles is 20 kilometers? (Hint: 1 kilometer = .6214 miles)

b   Click the Selection menu and click Set Selectable Layers.

c Click the Clear All button in the Set Selectable layers dialog to uncheck all of the layers. Then click the box for 2000 Boundary to make it the only layer that is checked on. Click Close.

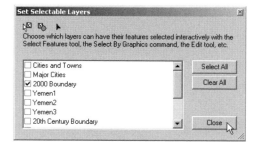

d Click the Select Features tool in the Tools toolbar.

e Click the blue 2000 Boundary line that crosses the land. It becomes highlighted blue on the map.

f Click the Show/Hide ArcToolbox Window button to open the ArcToolbox window.

ArcToolbox is where you can access many ArcGIS tools that work on your data. You will use the Buffer tool to draw the 20-kilometer zone around the boundary line.

g Expand the Analysis Tools toolbox and then expand the Proximity toolbox. Double-click the Buffer tool to open the Buffer dialog.

h Click the down-arrow to show the Input Features drop-down list and choose 2000 Boundary.

i The default Output Feature Class will be the same location and type (geodatabase feature class or shapefile) as the 2000 Boundary layer (ABC_Yemen4). Keep the default unless your teacher asks you to change it.

j For Distance, choose Kilometers from the Linear unit drop-down list, and then type **20** in the box on the left as the buffer distance.

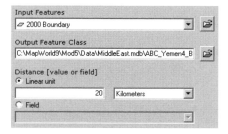

k Click OK. After the buffer is completed, close the Buffer window if necessary. Close the ArcToolbox window by clicking the small × in the upper right corner.

l Drag the ABC_Yemen4_Buffer layer below the 2000 Boundary layer in the table of contents.

m Right-click the 2000 Boundary layer, point to Selection, and click Clear Selected Features to clear the selection.

? *(1) In which part of the Saudi–Yemeni border will the pastoral area be most significant? Explain.*

? *(2) Why do you think the Treaty of Jeddah created a pastoral area?*

Step 11   Create a map of the Arabian Peninsula

Before you print a map of the Arabian Peninsula, you need to clean up the map.

 a   Click the Select Elements button. Click the boundary line you first drew (it's black and doesn't have the buffer around it). A dashed box appears to show the line is selected. Press the Delete key. Your line disappears from the map.

b   Go to the Arabian Peninsula bookmark.

c   Decide what layers you want to display on your map. Include the following layers:
- Major Cities
- 2000 Boundary
- Yemen1
- 20th Century Boundary
- Arabian Peninsula - names
- Arabian Peninsula - area
- Neighbors - countries

d   From the File menu, click Page and Print Setup. Follow your teacher's instructions to select the correct printer name. Be sure to check the Use Printer Paper Settings box and the Scale Map Elements proportionally to changes in Page Size box. In the Paper section, choose Landscape paper orientation. Click OK.

e   From the File menu, click Print. Click OK to print a copy of your map.

Step 12   Exit ArcMap

In this exercise, you explored physical and human characteristics of the Arabian Peninsula. After analyzing this data, you explored boundary issues in this region and plotted the new Yemeni–Saudi boundary established by the 2000 Treaty of Jeddah.

a   Save your map document.

b   From the File menu, click Exit. When asked if you want to save your changes, click No.

NAME _____   DATE _____

Student answer sheet
Module 5
Human Geography II: Political Geography

Regional case study: A Line in the Sand

Step 3   Identify countries that border the Arabian Peninsula

c   Record the names of the countries on the map that border the Arabian Peninsula to the north.

_____   _____   _____

Step 4   Investigate the physical characteristics of the Arabian Peninsula

a-1   Is any part of the Arabian Peninsula mountainous? _____

a-2   If so, where are the mountains located?

_____

_____

c-1   Are there any parts of the Arabian Peninsula that do not have any water at all? If so, where are these regions?

_____

_____

c-2   Do you see any relationship between landforms and the availability of water?

_____

_____

e   Describe the bodies of water.

_____

_____

h-1   How many millimeters equal 10 inches? _____

h-2   Based on the amounts of rainfall displayed on the map, do you think there is much farming on the Arabian Peninsula? Explain.

_____

_____

_____

h-3   Approximately what percentage of the Arabian Peninsula is desert? _____

i   What is the approximate range of temperatures across the Arabian Peninsula during this period?

       ° C: _____   ° F: _____

j-1   Which season is the hottest? _____

j-2   What is the approximate range of temperatures across the Arabian Peninsula during this period?
       _____°C to _____ °C   _____ °F to _____ °F

k-1   What relationship do you see between the Arabian Peninsula's ecozones as displayed on this map and
       its patterns of landforms, precipitation, and temperature?
       _____
       _____

k-2   Complete the table. List three observations for each physical characteristic.

PHYSICAL CHARACTERISTICS OF THE ARABIAN PENINSULA	
Landforms and bodies of water	
Climate	
Ecozones	

k-3   In your opinion, which of the region's physical characteristics would be considered "valuable" in a
       boundary decision? Explain.
       _____
       _____
       _____

Step 5   Investigate the human characteristics of the Arabian Peninsula

b-1   What is the principal agricultural activity on the peninsula?
       _____

b-2   Based on what you now know about the physical characteristics of the region, why do you think the
       agricultural activity is so limited?
       _____
       _____
       _____

c-1   How does Yemen compare to the rest of the Arabian Peninsula in population density?
       _____
       _____
       _____

c-2   Describe the overall population density of the Arabian Peninsula.

_____

_____

_____

f-1   Speculate on the most frequent use of the water at these springs and water holes.

_____

_____

_____

f-2   Use your answers from previous step 5 questions and analysis of the maps to complete the table. List three observations for each human characteristic.

HUMAN CHARACTERISTICS OF THE ARABIAN PENINSULA	
Agricultural activities	
Population density and distribution	

f-3   If an international boundary were to be drawn across some part of the Arabian Peninsula, how would these characteristics influence the perception of certain regions as being more valuable than others?

_____

_____

_____

Step 6   Locate and describe the Empty Quarter

b-1   Complete the table. List three observations in each column.

THE EMPTY QUARTER	
PHYSICAL CHARACTERISTICS	HUMAN CHARACTERISTICS

b-2   What difficulties would an area like this present if an international boundary must cross it?

_____

_____

_____

MODULE 5 • HUMAN GEOGRAPHY II: POLITICAL GEOGRAPHY

e   If you are going to save the project, record the map document's new name and where you saved it.

_____   _____
(Name of map document.   **(Navigation path to where map document is saved.**
For example: ABC_Region5.mxd)   **For example: C:\Student\ABC)**

Step 7   Explore Saudi Arabia's southern boundaries

c-1   Are the boundaries what you expected them to be?

_____

c-2   Which boundary remained unsettled?

_____

i   What does the area between the green and purple lines represent?

_____
_____

j   What is the principal economic activity of the regions in dispute?

_____
_____

k   Describe the population distribution in the disputed territory.

_____
_____
_____

Step 8   Draw the Saudi–Yemeni boundary

b-1   Does the red line go through any cities or towns? (Hint: You may need to zoom in again to answer the question.) If yes, approximately how many does the boundary pass through?

_____
_____
_____

b-2   How would you decide which side of the town to put the boundary on? Remember, this decision would determine whether the residents of that village would be citizens of Saudi Arabia or Yemen.

_____
_____
_____

r   Does the new line seem to favor Yemen or Saudi Arabia? Explain.

_____
_____
_____

Step 9   Enter the maritime part of the boundary

d   What body of water does the maritime boundary traverse?

_____

e　How does the actual boundary established by the Treaty of Jeddah compare with the boundary you drew earlier?

_____

_____

_____

_____

f　Write three observations about the boundary line created by the Treaty of Jeddah.

_____

_____

_____

_____

_____

_____

Step 10　Define the pastoral area

a　How many miles is 20 kilometers? (Hint: 1 kilometer = .6214 miles) _____

m-1　In which part of the Saudi–Yemeni border will the pastoral area be most significant? Explain.

_____

_____

_____

_____

_____

m-2　Why do you think the Treaty of Jeddah created a pastoral area?

_____

_____

_____

_____

_____

NAME _____   DATE _____

A Line in the Sand
Middle school assessment

You are a newspaper reporter assigned to cover the Treaty of Jeddah, signed on June 12, 2000, which settled the 65-year-old border dispute between Saudi Arabia and Yemen. You must choose to be a reporter for a newspaper in either Saudi Arabia or Yemen and write your article from that country's perspective. In preparing your article, you may use the Line in the Sand ArcMap map document as well as additional resources such as your history and geography books, encyclopedias, and the Internet. Your article should include the following:

- A map showing the new boundary line and a relevant physical or cultural characteristic discussed in your article
- A description of the physical and cultural characteristics of the region affected by the boundary change
- A description of the new boundary established by the treaty and its implications for people living in the affected areas

Use the remainder of the page as a place to brainstorm for your article.

A Line in the Sand

Assessment rubric

Middle school

STANDARD	EXEMPLARY	MASTERY	INTRODUCTORY	DOES NOT MEET REQUIREMENTS
The student knows and understands how to use maps to analyze spatial distributions and patterns.	Creates a detailed map using GIS that shows the new boundary lines and relevant physical or cultural characteristics.	Creates a map showing the new boundary line and most relevant physical or cultural characteristics.	Creates a map showing the new boundary line, but does not include any relevant physical or cultural characteristics.	Creates a map of the Middle East, but does not focus on the boundary issue between Saudi Arabia and Yemen.
The student knows and understands how physical processes shape places and how different human groups change places.	Writes a detailed description of the physical and cultural characteristics that will be affected by the boundary change and explains any possible ramifications.	Writes a description of the physical and cultural characteristics that will be affected by the boundary change.	Writes a description of physical or cultural characteristics that will be affected by the boundary change.	Describes physical or cultural characteristics of the region, but does not explain how these things are affected by the boundary change.
The student knows and understands the multiple territorial divisions of the student's own world.	Describes the implications of the new boundary for the people living in the affected areas. Includes quotes or stories from individuals in the area (these could be real or fictional and derived from research).	Describes the implications of the new boundary for the people living in the affected areas.	Describes characteristics of people living in the affected region, but does not relate it specifically to the boundary change.	Gives little description on the characteristics of the people living in the affected areas.
The student knows and understands how various points of view on geographic context influence plans for change.	Writes two clear and coherent news articles, one from the perspective of Saudi Arabia and one from the perspective of Yemen on the boundary issue.	Writes a clear and coherent news article from the perspective of either country involved in the boundary issue.	Writes a news article on the boundary issue, but does not offer the perspective of either country on the boundary issue.	Writes an essay on the boundary issue, but does not offer any geographic perspective, and it is not in the form of a news article.

This is a four-point rubric based on the National Standards for Geographic Education. The "Mastery" level meets the target objective for grades 5–8.

NAME _____   DATE _____

A Line in the Sand
High school assessment

You are a newspaper reporter assigned to cover the Treaty of Jeddah, signed on June 12, 2000, which settled the 65-year-old border dispute between Saudi Arabia and Yemen. You must choose to be a reporter for a newspaper in either Saudi Arabia or Yemen and write your article from that country's perspective. In preparing your article, you may use the Line in the Sand ArcMap map document as well as additional resources such as your history and geography books, encyclopedias, and the Internet. Your article should include the following:

- A map showing the new boundary line, the boundaries claimed by Yemen and Saudi Arabia prior to the settlement, and a relevant physical or cultural characteristic discussed in your article
- A description of the physical and cultural characteristics of the region affected by the boundary change
- A description of the historical factors that contributed to this long-standing conflict
- A description of the new boundary established by the treaty and its implications for people living in the affected areas

Use the remainder of the page as a place to brainstorm for your article.

Assessment *rubric*

High school

A Line in the Sand

STANDARD	EXEMPLARY	MASTERY	INTRODUCTORY	DOES NOT MEET REQUIREMENTS
The student knows and understands how to use geographic representations and tools to analyze and explain geographic problems.	Creates a detailed map using GIS showing the new boundary line and relevant physical and cultural characteristics.	Creates a map showing the new boundary line and most relevant physical or cultural characteristics.	Creates a map showing the new boundary line and relevant physical or cultural characteristics.	Creates a map showing the new boundary line, but does not include any relevant physical or cultural characteristics.
The student knows and understands the changing human and physical characteristics of places.	Writes a detailed description of the physical and cultural characteristics that will be affected by the boundary change and explains any possible ramifications.	Writes a description of the physical and cultural characteristics that will be affected by the boundary change.	Writes a description of physical or cultural characteristics that will be affected by the boundary change.	Describes physical or cultural characteristics of the region, but does not explain how these things are affected by the boundary change.
The student knows and understands why and how cooperation and conflict are involved in shaping the distribution of social, political, and economic spaces on Earth at different scales.	Describes the implications of the new boundary for the people living in the affected areas in relationship to social issues, politics, and the economy. Includes quotes or stories from individuals in the area (these could be real or fictional and derived from research).	Describes the implications of the new boundary for the people living in the affected areas in relationship to social issues, politics, and the economy.	Describes the implications of the new boundary for the people living in the affected areas.	Describes characteristics of people living in the affected region, but does not relate it specifically to the boundary change.
The student knows and understands contemporary issues in the context of spatial and environmental perspectives.	Writes two clear and coherent news articles, one from the perspective of Saudi Arabia and one from the perspective of Yemen, on the boundary issue. It includes historical factors that contributed to the conflict.	Writes a clear and coherent news article from the perspective of either country involved in the boundary issue. It includes historical factors that contributed to the conflict.	Writes a news article on the boundary issue, but does not offer the perspective of either country on the boundary issue. The article may include one or two historical factors.	Writes an essay on the boundary issue, but does not offer any geographic perspective, and it is not in the form of a news article.

This is a four-point rubric based on the National Standards for Geographic Education. The "Mastery" level meets the target objective for grades 9–12.

Starting from Scratch
An advanced investigation

Lesson overview

Students will use physiographic (physical features) and anthropographic (cultural features) data to redraw some of the world's international boundaries, thereby creating states characterized by internal cohesiveness and economic parity. By comparing their maps to ones reflecting contemporary political boundaries, students will identify world regions where political boundaries are in conflict with physical and cultural imperatives.

Estimated time   Three to four 45-minute class periods

Materials   ✔ Student handouts from this lesson to be copied:
- GIS Investigation sheets (pages 319 to 324)
- Student answer sheet (page 325)

Standards and objectives   *National geography standards*

GEOGRAPHY STANDARD	MIDDLE SCHOOL	HIGH SCHOOL
1 How to use maps and other geographic representations, tools, and technologies to acquire, process, and report information from a spatial perspective	The student knows and understands how to use maps to analyze spatial distributions and patterns.	The student knows and understands how to use technologies to represent and interpret Earth's physical and human systems.
3 How to analyze the spatial organization of people, places, and environments on Earth's surface	The student knows and understands how to use the elements of space to describe spatial patterns.	The student knows and understands how to apply concepts and models of spatial organization to make decisions.
10 The characteristics, distribution, and complexity of Earth's cultural mosaics	The student knows and understands how to read elements of the landscape as a mirror of culture.	The student knows and understands how cultures shape the character of a region.
13 How the forces of cooperation and conflict among people influence the division and control of Earth's surface	The student knows and understands the multiple territorial divisions of the student's own world.	The student knows and understands the impact of multiple spatial divisions on people's daily lives.

Objectives
The student is able to:
- Describe the physical features that form natural boundaries between major regions on earth.
- Describe the distribution of major language groups and religions on earth.
- Explain how political and cultural boundaries influence a country's internal cohesiveness and opportunity for economic parity with other nations.

GIS skills and tools
- Change layer transparency
- Change map symbolization with the layer properties
- Change a layer's label properties
- Export a layer and add new fields to an attribute table
- Digitize new polygons and enter their attribute data
- Design a presentation layout and print it
- Analyze data and make decisions based on spatial patterns

For more on geographic inquiry and these steps, see Geographic Inquiry and GIS (pages xxiii to xxv).

Teacher notes

Lesson introduction

Introduce the lesson by challenging students to identify regions in the world that are characterized by instability or long-term economic hardship. After generating a list, ask students if such problems in these regions are the result of boundary issues. Use this discussion to raise questions about boundary issues that influence stability, cohesiveness, and economic opportunity in a country.

- How can the size and shape of a country promote cohesiveness or instability?
- How do cultural (anthropographic) boundaries differ from political boundaries?
- How do cultural (anthropographic) boundaries contribute to cohesiveness or instability?
- How could the political boundaries of a country foster or hinder economic advantage?

Explain that in this lesson, students will have an opportunity to redraw the boundaries of the world. The purpose in this exercise will be to create a more stable and peaceful world by drawing boundaries that foster cohesiveness and economic parity among nations.

Student activity

 Before completing this lesson with students, we recommend that you complete it as well. Doing so will allow you to modify the activity to accommodate the specific needs of your students.

After the initial discussion, have the students work on the computer component of the lesson. This lesson is particularly well suited to students working in pairs, but the lesson can be modified to accommodate a variety of instructional settings.

Distribute and explain the Starting from Scratch GIS Investigation. It is important that students understand that the ultimate purpose of the activity is to create a more stable and peaceful world.

Explain that in this activity they will use GIS to observe and analyze cultural and physiographic data in order to determine the world's new boundaries. The GIS Investigation sheets will provide them with detailed instructions for the analysis of data and drawing of borders for their assigned continent.

Things to look for while students are working on this investigation:
- Are your students taking both physiographic and anthropographic boundaries into account as they draw their boundaries?
- Do they understand that there are many more factors that influence boundary decisions and that this lesson only uses simple datasets?
- Are they applying concepts of territorial morphology (size and shape) as they draw their boundaries?

 Teacher Tip: Students will need access to a computer and ArcMap for the equivalent of two class periods to draw their new boundaries. Decide ahead of time how and where you want your students to save their map documents. Encourage them to save their work frequently.

 Teacher Tip: Decide where you want students to save their country boundary data in step 4e. Students can export a feature class to the World5 geodatabase if they will have their own copy of the module 5 folder. Otherwise, you may want students to export a shapefile to another location.

Conclusion

Each student will present a New World map to the class, describing all boundary changes, and explaining how the decisions to make those changes were arrived at. Ideally, students will use a projection device to display their maps to the class; printed copies from an ArcMap layout will also work. Be sure that presenters provide cultural or physiographic justifications for their boundary decisions. Use these presentations as a springboard to highlight present-day areas of instability, characterized by boundaries that conflict with physical and cultural imperatives in the area. Engage the students in a conversation that analyzes the data they used. Guide the students toward understanding that although language and religion are important in determining boundaries, there are many other factors that influence boundaries: access to natural resources, economics, and infrastructure are just a few. Using a GIS can help people deal with these complicated issues.

Assessments for middle- and high-school students

In the assessment, students will prepare a map and a report focusing on a country where political boundaries are causing cultural, political, or economic instability. The map will compare the country's present boundaries with new boundaries proposed by the student. The report will explain how boundary issues contribute to instability in the country today and why the new boundaries will foster cohesiveness and economic stability in the future.

In preparing their report, students should consider the influence that each of the following factors has on cohesiveness and economic strength:
- Size, shape, and relative location of countries
- Cultural characteristics (language and religion)
- Distribution of natural resources
- Physiographic connections and barriers between places

Extensions

- Assign students to use the Internet to collect and map additional data about the regions they have characterized as unstable.

- Assign students to use the Internet to locate historic maps of these regions and to prepare an ArcMap layout reflecting boundary changes over time.

- Add climate, land-use, population density, and natural resource data to the project and revise boundaries based on this data.

- Have students create boundaries for all the continents and compare them to present-day boundaries.

- Ask students to use the graphs function to present religious and linguistic data that supports their boundary decisions.

- Check out the Resources by Module section of this book's companion Web site *(www.esri.com/mappingourworld)* for print, media, and Internet resources on the topic of international boundaries.

NAME _____   DATE _____

Starting from Scratch
An advanced investigation

You have been selected to serve on a special commission charged with redrawing the boundaries of world nations. The goal of this commission is to defuse threats to world peace by creating political boundaries that foster stability, harmony, and economic parity among nations. You will create countries on one continent.

You may create anywhere from three to 15 countries on your continent. You will use this GIS Investigation to identify and analyze key variables that have a bearing on this important decision and to create a map reflecting your New World boundaries.

Step 1 Start ArcMap

    a   Double-click the ArcMap icon on your computer's desktop.

    b   Open the **Adv5.mxd** (or **Adv5**) map document from the module 5 folder (**C:\MapWorld9\Mod5**).

       When the map document opens, you see a composite satellite image of the world. The table of contents also includes religion, language, rivers, lakes, continents, and ocean layers.

Step 2 Explore map layers

    As you prepare to draw new country boundaries, you will use the Satellite Image and Rivers layers to evaluate physiographic boundaries, and you will use the Religion and Language layers to determine anthropographic boundaries.

      Note: Religion and language are not the only anthropographic factors that influence boundary decisions. For the purposes of this investigation you will use a simple dataset with only two factors.

    a   List other important factors that influence boundary decisions.

    b   Zoom to South Asia. Turn on Religion.

      Note: Click the Refresh View button at the bottom of the map area if the Religion or Satellite Image layers do not draw completely.

    c   Use the Religion legend and MapTips to determine three principal religions of South Asia. Record them on the answer sheet.

d Turn on the Effects toolbar and dock the toolbar above the map.

  e Select the Religion layer in the Effects toolbar. Click the Adjust Transparency button and move the Transparency slider bar up to approximately 60%.

Now you are able to see both the religion boundaries and the physical features of the earth.

  The boundary between which two religions corresponds to a physiographic boundary visible in the satellite image? (Hint: Use MapTips to help you identify religion areas.)

? *f* Turn off Religion and turn on Language. Identify the principal language groups in South Asia.

g Use the Effects toolbar to make the Language layer approximately 50% transparent.

? *The boundary between which two language groups corresponds to a physiographic boundary visible in the satellite image?*

h Turn off the Effects toolbar.

Step 3 Create a new layer for new world boundaries

a Zoom to Full Extent.

b Change the name of the data frame to My New World.

  c Click the Add Data button. Navigate to the World5 geodatabase in the module 5 data folder (**C:\MapWorld9\Mod5\Data\World5.mdb**). Add **CountryTemplate**.

You will export your own copy of this feature class to hold the new country features you will create.

d Right-click the CountryTemplate layer and choose Data, Export Data.

e Click the Browse button in the Export Data dialog. Ask your teacher what type of file you should save. If you will be saving a feature class, choose Personal geodatabase feature class from the Save as type drop-down list and navigate to the World5 geo-database. Otherwise, choose Shapefile and navigate to the location your teacher directed you.

f Name your exported data **ABC_NewWorld** where **ABC** are your initials.

g   Click OK in the Export Data dialog. Click Yes to add the exported data to the map.

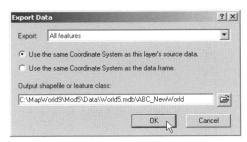

h   Right-click the CountryTemplate layer and remove it from the data frame.

i   Open the properties for the new layer you created. Click the General tab in the Layer Properties dialog and change the layer name to New World.

j   Click the Symbology tab and click the symbol. Choose Hollow with a bright red outline for the symbol. Click OK to apply your changes and close the Layer Properties dialog.

Step 4   Add language and religion fields to the New World table

a   Open the attribute table for New World.

b   Click the Options button in the table and click Add Field. Complete the Add Field dialog to create a Language field with the following parameters:

FIELD NAME	TYPE	LENGTH
Language	Text	16
Religion	Text	16

c   Repeat step b to add a Religion field. Scroll to the right to see the two fields you have just added.

d   Close the attribute table.

?   *e*   Save the map document and rename it according to your teacher's instructions. Record the new name of the map document and where you saved it.

Step 5   Draw the boundaries of new countries

It is now time to begin creating your new map of the world. You will select one continent and use the ArcMap editing tools to draw the boundaries (outlines) of the countries you create on that continent. It will be helpful to turn layers on and off so you can see the physiographic and anthropographic boundaries that exist. As you create new countries, identify the principal religion and language group of each in the attribute table.

Remember that your goal is to stabilize hot spots and foster world peace with your new boundary configuration. Consider the following points:

• Countries that are culturally uniform are generally more stable.

• Countries that have natural or physiographic boundaries as political boundaries tend to be more stable.

• Landlocked countries are at an economic disadvantage if they do not have some access to the sea.

a   Choose the continent you will be working on and record it on the answer sheet.

b   Zoom to your continent so you can see major landforms.

c   Display the Editor toolbar.

> *Note: Refer to the ArcMap Toolbar Quick Reference for a listing of the tools on the Editor toolbar.*

d   Click the Editor menu and click Start Editing. In the top pane of the Start Editing window, choose the source that contains your feature class. (Hint: You can widen the Source column by dragging the right edge of the column header to the right.) The New World layer will be listed in the bottom pane. Click OK.

e   On the Editor toolbar make sure the task is Create New Feature and the target is New World.

f   Click the Sketch tool.

g   When you have chosen the first boundary to draw, click the cursor along the proposed boundary. The vertexes are displayed as small green squares. Completely encircle your proposed country, including any coastlines. You may need to click many times to create a curved line. Double-click to complete the polygon.

> *Note: If you make a mistake, click the Undo button to delete the last vertex you entered. Or, delete the entire sketch by double-clicking to complete the polygon and then press the Delete key. Then start again.*

Next you need to add language and religion attribute data to the new country.

h   Click the Attributes button on the Editor toolbar.

i   In the Value column next to the Id field, change the number from zero to 1 and press Enter.

j   For Language, type the name of the major language group in the country you just created and press Enter. If there isn't one major language group, type **Mixed**.

k   For Religion, type the principal religion for the country you just created and press Enter. If there isn't one major religion, type **Mixed**.

l   Close the Attributes window.

m   From the Editor menu, choose Save Edits.

Look at your boundary and determine whether there are any parts of it you need to edit. If you don't need to edit the boundary, proceed to step 6q. Otherwise, continue with the editing instructions below.

n   Zoom to the section of the boundary that you would like to edit.

o   Change the Task to Modify Feature in the Editor toolbar. The edit sketch is displayed showing the vertexes.

p   Click the Edit tool. Drag the vertices to modify your boundary line. Add new vertices by right-clicking on the green line and choosing Insert Vertex. Delete vertices by right-clicking on a green square and choosing Delete Vertex.

q   Repeat the process of making a new boundary polygon, adding its attribute data, and editing it to create countries in the remainder of your continent. Give each country a unique number in the Id field. Save your edits periodically.

r   From the Editor menu, click Stop Editing and save your edits.

s   Click the Editor Toolbar button to dismiss the toolbar. Save your map document.

Step 6   Label your map

a   Open the New World layer properties.

b   Click the Labels tab and check the box for Label features in this layer. From the Label Field drop-down list, choose Id.

c   In the Text Symbol section, set the following parameters:

Font	Arial
Size	14
Style	bold
Color	red

d   Click OK. The labels appear on the map.

Step 7   Create and print a layout of your New World boundaries

a   Turn off all layers except New World, Continents, and Ocean.

b   Zoom out so you can see your continent in its entirety.

c   Change the map to Layout View. Move or dock the Layout toolbar in your preferred location.

d   Click the Change Layout button and choose a landscape layout template from the General tab.

The template you chose may be designed for paper that is larger than Letter size (8.5 × 11 inches). If so, you may need to change it to a paper size that is appropriate for your printer.

e   From the File menu, open the Page and Print Setup dialog. Make sure the box is checked to Use Printer Paper Settings.

f   In the Paper section, choose Letter or another desired paper size that your printer uses. Make sure the landscape orientation is selected. Click OK.

g   Use the Select Elements tool to edit the properties for any titles, scale bars, or other elements that appear on the layout.

h   Update an existing text placeholder or insert new text to include your name and the date on the layout.

i   Reposition any graphics or text until you are satisfied with how the layout looks.

j   Print your map.

You should be prepared to show your New World map to your classmates and explain the cultural and physical features upon which you based your decisions.

Step 8 Exit ArcMap

In this exercise, you used ArcMap to explore patterns in the world's physiographic and anthropographic borders. Based on your observations, you drew new international boundaries to create countries that you feel would be more culturally and physiographically unified than those in the real world today.

a **Save your map document.**

b **From the File menu, click Exit.**

NAME _____   DATE _____

Student answer sheet
Module 5
Human Geography II: Political Geography

Advanced investigation: Starting from Scratch

Step 2   Explore map layers
a   List other important factors that influence boundary decisions.

   _____
   _____
   _____
   _____

c   Use the Religion legend and MapTips to determine three principal religions of South Asia. Record them
       here.

   _____
   _____
   _____

e   The boundary between which two religions corresponds to a physiographic boundary visible in the
       satellite image?

   _____
   _____

f   Identify the principal language groups in South Asia.

   _____
   _____
   _____

g   The boundary between which two language groups corresponds to a physiographic boundary visible in
       the satellite image?

   _____
   _____

Step 4   Add language and religion fields to the New World table
e   Save the map document and rename it according to your teacher's instructions. Record the new name
       of the project and where you saved it.

   _____   _____
   (Name of map document.   **(Navigation path to where map document is saved.**
   For example: ABC_Adv5.mxd)   **For example: C:\Student\ABC)**

Step 5   Draw the boundaries of new countries
a   Choose the continent you will be working on and record it here.

   _____

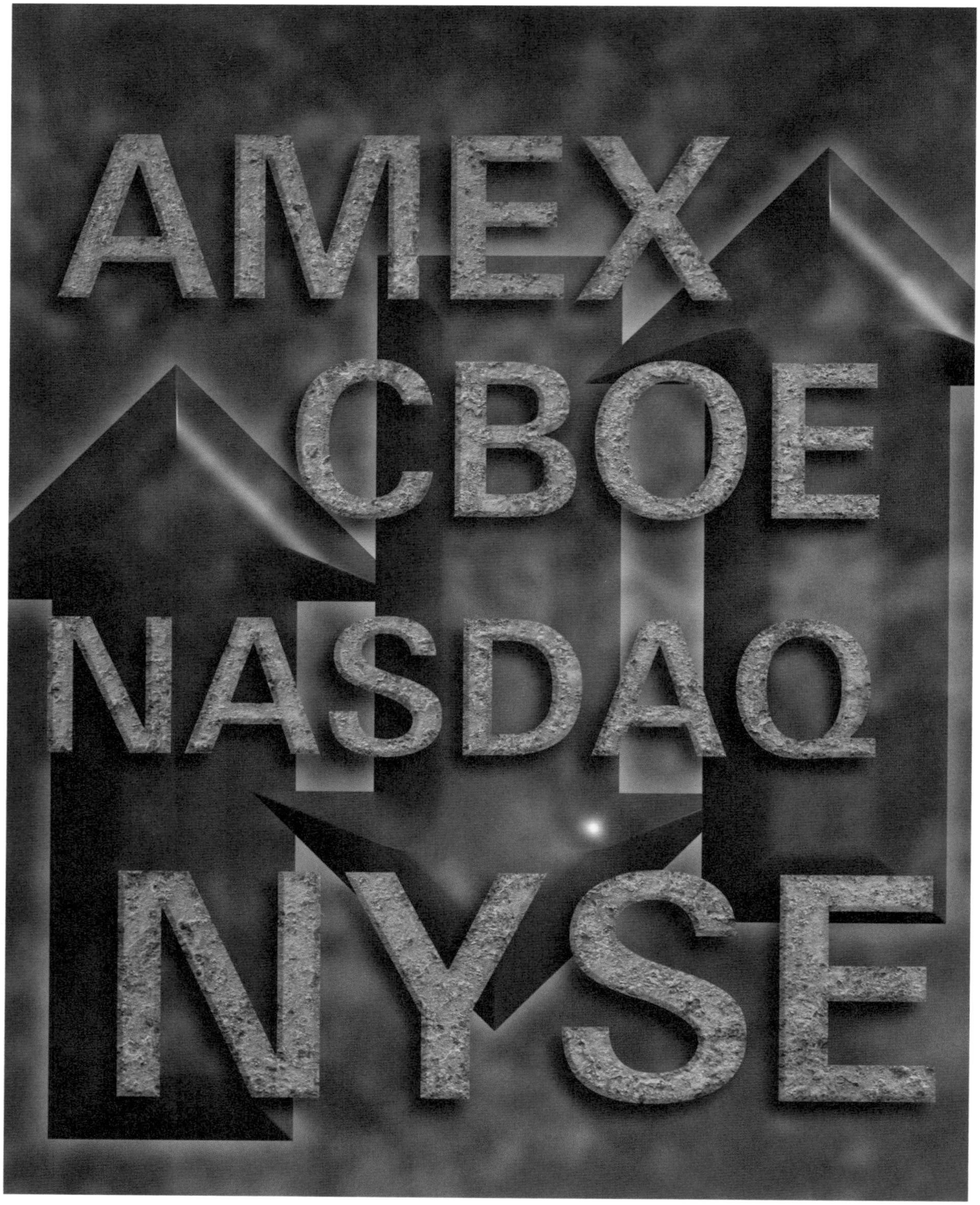
AMEX
CBOE
NASDAQ
NYSE

Human Geography III
Economic Geography

Economic development, modernization, and trade illustrate the interrelatedness of the global community.

The Wealth of Nations: A global perspective
In this lesson students will be presented with the three modes of economic production—agriculture, industry, and services—as the initial criteria for a country's developed or developing status. They will select layers of data from a group of economic indicators to determine patterns in developed and developing countries. They will be challenged to draw their own conclusions as to a country's economic development status and support their conclusions with data.

Share and Share Alike: A regional case study of North America and NAFTA
Students will explore trade in North America focusing on the three trading partners in the North American Free Trade Agreement (NAFTA)—Canada, Mexico, and the United States. They will study export data for the past 10 years from each of the NAFTA countries, then use this information to identify trading trends before and after NAFTA, and to assess its effectiveness. Finally, students will make an ArcMap layout containing a map and charts that support their opinions.

Live, Work, and Play: An advanced investigation
Students will use the Internet to acquire the most recent data on job classifications in California. By exploring this data through a GIS, students will construct a thematic map for California counties that displays population, number of jobs in a given field, and average wage data. They will perform a complex query and create a map that illustrates their results.

The Wealth of Nations
A global perspective

Lesson overview

In this lesson students will be presented with the three modes of economic production—agriculture, industry, and services—as the initial criteria for a country's developed or developing status. They will select layers of data from a group of economic indicators to determine patterns in developed and developing countries. They will then be asked to draw their own conclusions as to a country's economic status, and to support those conclusions with data.

Estimated time   Two 45-minute class periods

Materials   ✔ Student handouts from this lesson to be copied:
- GIS Investigation sheets (pages 333 to 342)
- Student answer sheet (pages 343 to 345)
- Assessment(s) (pages 346 to 350)

Standards and objectives

National geography standards

GEOGRAPHY STANDARD	MIDDLE SCHOOL	HIGH SCHOOL
1 How to use maps and other geographic representations, tools, and technologies to acquire, process, and report information from a spatial perspective	The student understands how to make and use maps, globes, graphs, charts, models, and databases to analyze spatial distributions and patterns.	The student understands how to use geographic representations and tools to analyze, explain, and solve geographic problems.
11 The patterns and networks of economic interdependence on Earth's surface	The student understands ways to classify economic activity.	The student understands the classification, characteristics, and spatial distribution of economic systems.
18 How to apply geography to interpret the present and plan for the future	The student understands how varying points of view about geographic context influence plans for change.	The student understands how to use geographic knowledge, skills, and perspectives to analyze problems and make decisions.

Standards and objectives (continued)

Objectives

The student is able to:

- Define the three economic production criteria traditionally used to determine economic development status.
- Compare and contrast these criteria.
- Evaluate them as suitable measurements for developed or developing status.
- Understand additional economic indicators used to classify a country as developed or developing.
- Develop a definition of developed and developing.
- Predict a country's or region's economic status within the next 20 years.

GIS skills and tools

 Zoom the layout to view the whole page

 Zoom in to a specific area of the map

 Activate a data frame in a layout

 Find a specific feature on the map and identify it

 Zoom to the full map extent

 Add a layer to the map

- Add a new data frame to a layout
- Symbolize data using graduated colors
- Build a query to exclude specific attribute values from classification
- Display excluded values in a No Data class

For more on geographic inquiry and these steps, see Geographic Inquiry and GIS (pages xxiii to xxv).

Teacher notes

Lesson introduction
Introduce this lesson to your students with a discussion of the three economic production criteria: agriculture, industry, and services. Ask them what they know about these three methods of earning a living. Explain to them that economists generally rate a country's economic status as developing or developed by how much of its workforce is engaged in agriculture, industry, and services. Generally, countries with a high percentage of the workforce in agriculture (whether it be subsistence, commercial, or another type) are placed in the category of "developing."

Have the class begin by using the three production criteria maps of the world to determine whether a country is classified as developed or developing.

Before starting the exercise on the computer, ask the following questions to elicit knowledge, beliefs, or ideas your students may already have about countries around the world:

- Name several countries that have a high percentage of their workforce participating in agriculture.
- Name a country that has a high percentage of its workforce participating in services.
- Which countries do we generally think of as highly industrialized (have high percentages of the workforce participating in industry)?
- Can they think of other factors that might be helpful in determining whether a country is developed or developing?

Student activity
 Before completing this lesson with students, we recommend that you complete it as well. Doing so will allow you to modify the activity to accommodate the specific needs of your students.

After the initial discussion, have the students work on the computer component of the lesson individually or in groups of no more than two or three. Distribute the GIS Investigation sheets to the students. Explain to them that they will learn and use GIS skills to enable them to gather data for the assessment worksheet.

In addition to instructions, the handout includes questions to help students focus on key concepts.

 Teacher Tip: In order for students to complete the assessment, they must save their project. Be sure to have a suggested naming convention and location on a computer or computer network for your students.

Things to look for while students are working on this activity:

- Are all students in each group participating in the activity, taking turns using the computer and writing information on the chart?
- Are the students using a variety of tools to obtain the information they need?
- Are they experiencing any difficulty symbolizing the economic data layers in step 7?

Conclusion   After the students complete the lesson and the assessment, discuss their find-ings. If you have time, you can have each group share findings on an overhead projector and explain how they came to their conclusions. Students can also take turns presenting the thematic maps they have created, either in printed format or on a computer projection device from the front of the room. Conclude the les-son by asking the students to explain which factors they feel are most important in deciding if a country is developed or developing, and to provide support for their evaluation. Do they feel this two-class system ("developed" and "developing") is adequate? What alternative or additional classes can they suggest to describe a country's economic status?

Assessment   *Middle school and high school: Highlights skills appropriate to grades 5 through 12*

The middle- and high-school assessments ask students to choose one country they believe is developed and another they believe is developing. Students will use the thematic mapping skills they have learned in this lesson to map additional indica-tors and draw conclusions about the two countries they have chosen. They will individually write an essay that answers specific questions. The middle- and high-school assessments differ in the types of questions asked.

Extensions   • Ask students to look in local newspaper employment advertisements to classify the jobs in their own community as agriculture, industry, or service. How would they classify the types of employment in their own community?

• Research employment classifications of two different counties, and compare and contrast the percentages of jobs in each category.

• Research the GDP for the top 10 trading partners of the United States, compute the GDP per capita, and map it.

• Have students choose a country and analyze all of the standard of living and economic indicator data, and present their findings orally to the rest of the class.

• Check out the Resources by Module section of this book's companion Web site *(www.esri.com/mappingourworld)* for print, media, and Internet resources on the topics of economic indicators, developing countries, and economic production of countries.

NAME _____   DATE _____

The Wealth of Nations
A GIS investigation

Answer all questions on the student answer sheet handout

Economists generally classify a country's economic status as developing or developed by determining the percentage of its workforce engaged in each of three sectors of the economy— agriculture, industry, and services. In general, a country with a high percentage of its work- force in agriculture is considered to be "developing," while a country with a high percentage of its workforce in services and industry is considered to be "developed."

In this GIS Investigation, you will use world maps of the workforce in the three employment sectors to explore patterns of development around the world. You will also examine two other economic indicators—energy use and GDP per capita—and compare the maps of employment sectors to the maps of GDP and energy use. You will evaluate whether or not the employment criteria are good indicators of a country's economic status.

Step 1   Start ArcMap

a   Double-click the ArcMap icon on your computer's desktop.

b   If the ArcMap start-up dialog appears, click **An existing map** and click OK. Then go to step 2b.

Step 2   Open the Global6.mxd file

a   In this exercise, a map document has been created for you. To open it, go to the File menu and choose **Open**.

b   Navigate to the module folder (**C:\MapWorld9\Mod6**) and choose **Global6.mxd** (or **Global6**) from the list.

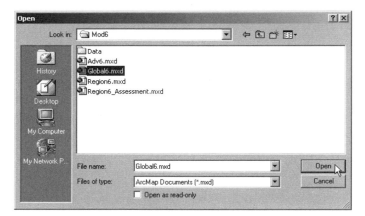

c   Click Open.

When the map document opens, you see a world map of employment in the agricultural sector. The table of contents lists three data frames in all, one for each of the employment sectors.

d   Click the Layout View button below the map.

The layout shows all three data frames side by side on the layout so you can compare them. Also, the Layout Toolbar appears or becomes active.

e   If your Layout toolbar is floating, dock it above the table of contents.

f   Maximize your ArcMap window by clicking the Maximize button in the top right corner of the window.

g   Click the Zoom Whole Page button on the Layout toolbar to make the layout fill the view.

Step 3   Evaluate the legends and patterns of the maps

    a   Look at all of the legends in the table of contents.

        (1)   What do the darkest colors represent?

        (2)   What do the lightest colors represent?

    b   Study the Workforce Involved in Agriculture map.

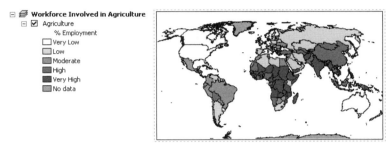

        (1)   What does the description "Very High" refer to in this map?

        (2)   Where are the countries with a high percentage of agricultural workers generally located?

        (3)   Where are the countries with a low percentage of agricultural workers generally located?

    c   Study the Workforce Involved in Services map.

        (1)   Where are the countries with a high percentage of service workers generally located?

        (2)   Where are the countries with a low percentage of service workers generally located?

        (3)   What relationship, if any, do you see between the agriculture and services workforce maps?

    d   Study the Workforce Involved in Industry map.

        What patterns do you see on the map?

    e   Using the workforce information in all three maps, in what part or parts of the world do you find the greatest number of developing countries?

Step 4   Analyze data on Bolivia

    a   Click the Zoom In tool on the Tools toolbar.

    b   Click and drag a box around South America in the Workforce Involved in Industry map.

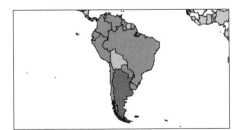

    c   Zoom in to South America on the other two maps.

d   Click the Select Elements tool and click the Agriculture map to activate the data frame. The data frame is outlined with a colored dashed line with blue squares to indicate it is active.

e   Click the Find tool.

f   Type **Bolivia** in the Find box on the Features tab.

g   Click the circle next to **In fields** in the Search area. Then select CNTRY_NAME from the drop-down list as the field to search.

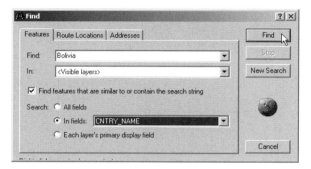

h   Click Find.

Bolivia is listed in the results box at the bottom of the Find window. Now you will use ArcMap to gather and record data about Bolivia.

i   Move the Find window so you can see the Agriculture map.

j   Right-click the Bolivia row in the results box and choose Flash feature so that you can pick out Bolivia on the map.

k   Right-click Bolivia again, and choose Identify feature(s). Close the Find dialog.

The Identify Results window shows data on Bolivia.

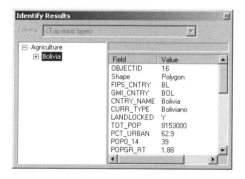

l   Scroll down the right side of the window and complete the following questions:

   (1)   What percentage of workers in Bolivia are involved in agriculture?

   (2)   What percentage of workers in Bolivia are involved in industry?

   (3)   What percentage of workers in Bolivia are involved in service?

   (4)   Assume that a developing country has a high percentage of its workforce in agriculture and lower percentages of its workforce in industrial-and service-related occupations and analyze the data on Bolivia. Based on these criteria, would you classify Bolivia as a developed or developing country? Explain.

m   Close the Identify Results window.

?   *n*   On your answer sheet, you see a table that has some information completed for Bolivia. Record the classification (developed or developing) you gave Bolivia in the previous question by making a check mark in the appropriate column.

Step 5   Analyze data on other countries

In order to complete the table on your answer sheet, you will need to find each country on the maps, interpret the legends for each map, write your answers in the columns under Agriculture, Industry, and Service, and decide whether each country is developing or developed.

a   Click the Find tool and type **India** in the Find text box. Click Find.

Two countries are found that contain the word India in the country name, but only one of them is the one you want.

b   Right-click the India row and choose Zoom to feature(s). The map zooms to the country of India.

?   *c*   Write the correct category for the percentage of workers in agriculture in the table on your answer sheet (under step 4n).

d   Using the Select Elements tool, activate the Workforce Involved in Industry data frame.

e   Click Find in the Find dialog to search the newly active data frame. Then Zoom to India.

The Workforce Involved in Industry data frame zooms to India.

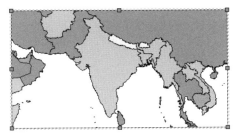

?   *f*   Interpret the legend and record India's percentage in the table on your answer sheet (under step 4n).

g   Activate the Workforce Involved in Service data frame. Click Find to find India and then zoom to it.

?   *h*   Interpret the legend and record India's percentage in the table on your answer sheet (under step 4n).

?   *i*   Based on the workforce criteria previously used to determine if a country is developed or developing, write your answer for India in the table on your answer sheet (under step 4n).

j   Repeat this process (step 5a–5i) for the other countries listed in the table.

k   Close the Find dialog.

l   Use the Select Elements tool and the Full Extent button to activate each data frame and zoom to the entire world in all three maps.

Step 6　**Create a new data frame and add data**

Now you will add economic and energy data and determine whether this data supports your initial conclusions about which countries are developed and which are developing.

Gross Domestic Product (GDP) is the total value of all goods, services, and products produced in a given country. Typically, developing countries have a low GDP. The total amount of energy consumed by a given country is also an indicator of development. If a country has a low level of energy consumption, it tends to be a developing country. Developed countries are high in both GDP and energy use.

a　Click Insert on the Main Menu toolbar and choose Data Frame to add a new data frame to the map document.

b　With the Select Elements tool, click the new data frame and drag it into the blank space in the bottom right of the layout.

c　Click and drag the blue boxes to resize the data frame to match the others.

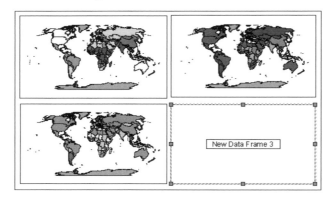

d　In the table of contents, click slowly two times on New Data Frame 3 to activate the text cursor. Name the data frame **Economic Data**.

e　Click the Add Data button.

f　Navigate to the module data folder (**C:\MapWorld9\Mod6\Data**). Open the **World6.mdb** geodatabase and choose world_economics from the list. Click Add.

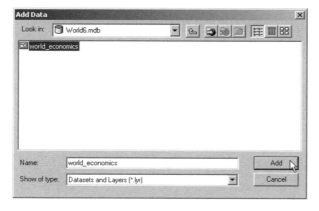

g　Add world_economics again so there are two copies of it in the data frame.

For each new layer, all countries are symbolized with one color. ArcMap randomly assigns the color, so the color on your screen may not match the color on your neighbor's screen.

Step 7   Thematically map GDP per capita and energy use

To see the pattern of GDP per capita, you need to change the legend. GDP per capita is the gross domestic product per person in a given year (1998). It is calculated by dividing the GDP of a country by its total population.

a   Double-click the first world_economics layer name in the table of contents. The Layer Properties dialog is displayed.

b   Click the Symbology tab.

c   Click Quantities in the Show list on the left side of the dialog. Graduated colors is automatically selected.

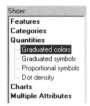

d   Click the Value drop-down list, scroll down, and select **GDP_PC8**.

The Symbology tab updates and you see a list of five colored symbols, values, and labels.

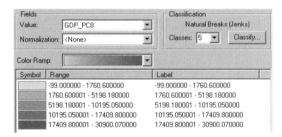

e   For the Color Ramp choose Brown Light to Dark. (Hint: To choose the color ramps by name instead of colors, right-click the drop-down list and choose Graphic View to toggle the colors off.)

By default, ArcMap has divided the GDP per capita values into five groups, or classes, which are shown in the Range column.

For some countries no data was available for GDP per capita. In this case, the GDP_PC8 field has a value of –99. Because the value of –99 represents no data, it should not be included in the range of GDP per capita values. You will assign values of –99 to a No Data class.

f   Click the Classify button on the right side of the Symbology tab. Then in the next dialog, click the Exclusion button.

g   Click the Query tab in the Data Exclusion dialog.

You will build the following query expression: **[GDP_PC8] = –99**

h   Scroll down the Fields list and find GDP_PC8. Double-click GDP_PC8 to add it to the expression box.

i   Click the = button.

j   Click the Get Unique Values button, then double-click –99 to complete the query expression.

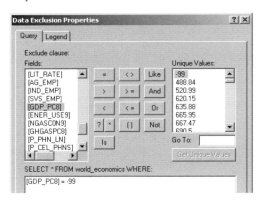

k   Click the Verify button. If you receive a message that the expression was successfully verified, click OK to close the message. If you get a different message, check your work and make sure the query expression matches the one pictured above.

l   Click the Legend tab. Check the box to Show symbol for excluded data.

m   Click the Symbol color and change the fill color to Gray 30%. Click OK on the Symbol Selector.

n   In the Label box, type **No Data**.

o   Click OK in the Data Exclusion Properties and Classification dialogs.

The ranges in the Symbology tab no longer include the value –99. Next you will change the labels for each symbol to use words instead of numbers.

p   Click on the top label (488.840000 - 2498.640000). Type **Very Low** and press Enter.

q   Change the next label down to **Low**. Then change the rest of the labels to **Moderate**, **High**, and **Very High**.

r   Click the General tab in the Layer Properties window. Change the layer name to **GDP per capita**.

s   Click OK and look at the updated legend.

Next you will map energy use. Energy use is a measure of the total energy consumption from all energy sources as measured by a trillion BTUs (British thermal units). One BTU is the amount of energy needed to heat one pound of water one degree Fahrenheit.

t   Turn off GDP per capita.

u   Double-click the remaining world_economics layer to open its Layer Properties dialog. Repeat steps 7b–7s using the information below:

- Map the value field ENER_USE9
- Use the Orange Bright color ramp
- Use the query expression **[ENER_USE9 = -99]**
- Change the layer name to **Energy Use**
- Use the same scale (Very Low – Very High) for the labels

Step 8   Analyze GDP per capita and energy use data

Now you will take this new data on GDP per capita and energy use into consideration, and you will reevaluate how you classified countries as developed or developing.

a   Turn off Energy Use and turn on GDP per capita.

b   Click the Find tool and find Bolivia. Zoom to Bolivia. Answer the following questions.

(1)   What level is the GDP per capita for Bolivia?

(2)   Based on this new information and the workforce data, should Bolivia be classified as a developing or developed country? Why?

c   Turn off GDP per capita and turn on Energy Use.

(1)   What level is the energy use for Bolivia?

(2)   Based on this new information and previous data, should Bolivia be classified as a developing or developed country?

(3)   Why does energy use increase when a country develops?

Now you will explore the data as it pertains to other countries and you will see if your earlier classification of each country has changed.

d   Explore GDP per capita and Energy Use for each of the countries listed on your answer sheet. You may consult your previous classification of these countries by looking at the table from step 4.

(1)   Complete the table on your answer sheet.

(2)   Name one country from above that you earlier classified as developed and that has GDP and energy use data that indicates it is developing.

(3)   Based on the data you collected on these six countries, do you feel that the employment criteria are good indicators of a country's economic status? Explain your answer.

Step 9   Save the map document and exit ArcMap

In this exercise, you used workforce data to determine whether countries should be classified as developed or developing. You added new data, thematically mapped the data, and reevaluated your previous classifications.

a   Click the Full Extent button so you can see the entire world in the Economic Data map.

b   Choose Save As from the File menu. Ask your teacher for instructions on where to save the map document and how to rename it. You must save this map document because you will need it for the assessment.

c   Write the new name you gave the map document and where you saved it in the space on your answer sheet.

d   Click the Restore Down button to return the ArcMap window to its original size.

e   From the File menu, click Exit.

NAME _____   DATE _____

Student answer sheet
Module 6
Human Geography III: Economic Geography

Global perspective: The Wealth of Nations

Step 3   Evaluate the legends and patterns of the maps

a-1   What do the darkest colors represent?

_____

a-2   What do the lightest colors represent?

_____

b-1   What does the description "Very High" refer to in this map?

_____

b-2   Where are the countries with a high percentage of agricultural workers generally located?

_____

b-3   Where are the countries with a low percentage of agricultural workers generally located?

_____

c-1   Where are the countries with a high percentage of service workers generally located?

_____

c-2   Where are the countries with a low percentage of service workers generally located?

_____

c-3   What relationship, if any, do you see between the agriculture and services workforce maps?

_____

d   What patterns do you see on the map?

_____

e   Using the workforce information in all three maps, in what part or parts of the world do you find the greatest number of developing countries?

_____

Step 4   Analyze data on Bolivia

l-1   What percentage of workers in Bolivia are involved in agriculture? _____

l-2   What percentage of workers in Bolivia are involved in industry? _____

l-3   What percentage of workers in Bolivia are involved in service? _____

l-4   Would you classify Bolivia as a developed or developing country? Explain.

_____

_____

_____

_____

n   In the table below, record the classification (developed or developing) you gave Bolivia in the previous question. Complete the rest of the table by following the instructions in step 5.

COUNTRY	AGRICULTURE	INDUSTRY	SERVICE	DEVELOPING	DEVELOPED
Bolivia	High	Low	High		
India					
Australia					
South Korea					
Ukraine					
Uruguay					

Step 8   Analyze GDP per capita and energy use data

b-1   What level is the GDP per capita for Bolivia? _____

b-2   Based on this new information and the workforce data, should Bolivia be classified as a developing or developed country?

_____

Why?

_____

_____

c-1   What level is the energy use for Bolivia? _____

c-2   Based on this new information and previous data, should Bolivia be classified as a developing or developed country?

_____

c-3   Why does energy use increase when a country develops?

_____

d-1   Complete the table below.

COUNTRY	GDP PER CAPITA	ENERGY USE	DEVELOPED OR DEVELOPING	IS THIS A CHANGE FROM YOUR EARLIER CLASSIFICATION?
Bolivia				
India				
Australia				
South Korea				
Ukraine				
Uruguay				

d-2   Name one country from above that you earlier classified as developed and that has GDP and energy use data that indicates it's developing.

_____

d-3   Based on the data you collected on these six countries, do you feel that the employment criteria are good indicators of a country's economic status? Explain your answer.

_____
_____
_____
_____
_____
_____
_____
_____
_____
_____

Step 9   Save the map document and exit ArcMap

c   Write the new name you gave the map document and where you saved it in the space below.

_____   _____
(Name of map document.   **(Navigation path to where map document is saved.**
For example: ABC_Global.mxd)   **For example: C:\Student\ABC)**

NAME _____   DATE _____

The Wealth of Nations
Middle school assessment

1 Open your saved version of the Global6 map document (e.g., ABC_Global6). Use the three economic activity maps (agriculture, services, and industry) to select a country you believe is a developed country and one that you believe is a developing country. Write the names of those countries at the top of the sheet provided.

2 Activate the Economic Data data frame and add a world_economics layer from the World6 geodatabase. Switch to Data View.

3 Thematically map the developing country indicators in the world_economics attribute table. Refer to the attached assessment sheet for guidelines on which data to map (page 350). (Hint: You can copy the world_economics layer and paste it into the data frame for each attribute you map. You may also want to collapse any unneeded data frames table of contents to make your work easier.)

4 Record the information you observe from the new maps on the attached sheet. You may use other developing country indicators that are not listed on the sheet. Write those in the blank lines at the end of the table.

5 Look at the data in the two columns and decide if your initial prediction is correct.

6 Write an essay that includes or answers the following information and questions:

  - Name and description of the three economic production models.

  - Does the data you found about your selected countries support what economists say about economic production as an indicator of developed or developing status?

  - Your own definitions, based on your research, of a developed country and a developing country.

The Wealth of Nations

Assessment rubric
Middle school

STANDARD	EXEMPLARY	MASTERY	INTRODUCTORY	DOES NOT MEET REQUIREMENTS
The student understands how to use maps and databases to analyze spatial distributions and patterns.	Uses GIS to analyze economic data by creating at least seven thematic maps that compare and contrast the different economic indicators for two countries. Uses additional data from outside sources.	Uses GIS to analyze economic data by creating seven thematic maps that compare and contrast the different economic indicators for two countries.	Uses GIS to analyze economic data by creating five or six thematic maps that compare and contrast the economic indicators for two countries.	Uses GIS to create four or fewer thematic maps based on economic data for one or two countries.
The student understands ways to classify economic activity.	Clearly describes the three economic production models and provides an example of each. Creates detailed and accurate original definitions of developed and developing countries. Provides ample evidence for the definition.	Clearly describes the three economic production models and then creates an accurate and original definition of developed and developing countries.	Describes the three economic production models, and attempts to create original definition of developed and developing countries.	Has difficulty describing the three economic production models, and does not attempt to create original definitions for developed and developing countries.
The student understands the spatial organization of human activities and physical systems and is able to make informed decisions.	The student understands how varying points of view about geographic context influence plans for change.	Compares own findings with the predefined economic status of the selected countries. Identifies any inconsistencies in the findings, and either accepts or rejects his/her hypothesis.	Compares and contrasts own findings with the predefined economic status of the selected countries.	Lists own ideas on the economic status of the selected country (or countries), but does not draw any comparisons with predefined economic status.

This is a four-point rubric based on the National Standards for Geographic Education. The "Mastery" level meets the target objective for grades 5–8.

NAME _____   DATE _____

The Wealth of Nations
High school assessment

1   Open your saved version of the Global6 map document (e.g., ABC_Global6). Use the three economic activity maps (agriculture, services, and industry) to select a country you believe is a developed country and one that you believe is a developing country. Write the names of those countries at the top of the sheet provided.

2   Activate the Economic Data data frame and add a world_economics layer from the World6 geodatabase. Switch to Data View.

3   Thematically map the developing country indicators in the world_economics attribute table. Refer to the attached assessment sheet for guidelines on which data to map (page 350). (Hint: You can copy the world_economics layer and paste it into the data frame for each attribute you map. You may also want to collapse any unneeded data frames table of contents to make your work easier.)

4   Record the information you observe from the new maps on the attached sheet. You may use other developing country indicators that are not listed on the sheet. Write those in the blank lines at the end of the table.

5   Look at the data in the two columns and decide if your initial prediction is correct.

6   Write an essay that includes or answers the following information and questions:

   • Name and description of the three economic production models. Give a brief explanation of the relationship between those models. For example: Will a country usually have high percentages or low percentages in all three production areas?

   • Does the data you found about your selected countries support what economists say about economic production as an indicator of developed or developing status?

   • Your own definitions, based on your research, of a developed country and a developing country.

   • Prediction of changes in your countries' economic status by the year 2020, and supporting data.

Assessment rubric

High school

The Wealth of Nations

STANDARD	EXEMPLARY	MASTERY	INTRODUCTORY	DOES NOT MEET REQUIREMENTS
The student understands how to use geographic representations and tools to analyze and explain geographic problems.	Uses GIS to analyze economic data by creating at least seven thematic maps that compare and contrast the different economic indicators for two countries. Uses additional data from outside sources.	Uses GIS to analyze economic data by creating seven thematic maps that compare and contrast the different economic indicators for two countries.	Uses GIS to analyze economic data by creating five or six thematic maps that compare and contrast the economic indicators for two countries.	Uses GIS to create four or fewer thematic maps based on economic data for one or two countries.
The student understands the classification, characteristics, and spatial distribution of economic systems.	Clearly describes the three economic production models, explains the relationship between these models, and then creates original detailed definitions of developed and developing countries. Provides ample evidence and examples for each.	Clearly describes the three economic production models, explains the relationship between these models, and then creates original detailed definitions of developed and developing countries. Provides some evidence and examples for each.	Describes the three economic production models, and attempts to create original definition of developed and developing countries.	Has difficulty describing the three economic production models, and does not attempt to create original definitions for developed and developing countries.
The student understands how to use geographic knowledge, skills, and perspectives to analyze problems and make decisions.	Compares own findings with the predefined economic status of the selected countries. Identifies any inconsistencies in the findings, and either accepts or rejects his/her hypothesis. Refines original hypothesis based on the findings. Predicts future changes for the economic status of the countries giving specific examples.	Compares own findings with the predefined economic status of the selected countries. Identifies any inconsistencies in the findings, and either accepts or rejects his/her hypothesis. Predicts any future changes for the economic status of the countries.	Compares and contrasts own findings with the predefined economic status of the selected countries. Attempts to predict future changes in economic status, but does not provide enough evidence.	Lists own ideas on the economic status of selected country (or countries), but does not draw any comparisons with predefined economic status.

This is a four-point rubric based on the National Standards for Geographic Education. The "Mastery" level meets the target objective for grades 9–12.

DEVELOPING COUNTRY INDICATORS	DEVELOPED COUNTRY _____	DEVELOPING COUNTRY _____
High birth rates		
Moderately high death rates		
High infant mortality rate		
Low life expectancy at birth		
Large percentage of population under 15		
Literacy rates low		
High urban populations		
Low gross domestic product		
Low consumption of energy		
Higher percentage per person communication facilities		
Economic production criteria—write whether they are predominantly agriculture-, industry-, or service-based.		
Other criteria: _____		
Other criteria: _____		
Other criteria: _____		

Share and Share Alike
A regional case study of North America and NAFTA

Lesson overview

Students will explore trade in North America focusing on the three trading partners in the North American Free Trade Agreement (NAFTA)—Canada, Mexico, and the United States. They will study export data for the past 10 years from each of the NAFTA countries, then use this information to identify trading trends before and after NAFTA, and to assess its effectiveness. Finally, students will make a layout containing a map and charts that support their opinions.

Estimated time   Two to three 45-minute class periods

Materials   ✔ Student handouts from this lesson to be copied:
- NAFTA objectives handout (page 356)
- GIS Investigation sheets (pages 357 to 369)
- Student answer sheet (pages 371 to 375)
- Assessment(s) (pages 376 to 379)

Standards and objectives   *National geography standards*

GEOGRAPHY STANDARD	MIDDLE SCHOOL	HIGH SCHOOL
1 How to use maps and other geographic representations, tools, and technologies to acquire, process, and report information from a spatial perspective	The student understands how to make and use maps, globes, charts, models, and databases to analyze spatial distributions and patterns.	The student understands how to use geographic representations and tools to analyze, explain, and solve geographic problems.
11 The patterns and networks of economic interdependence on Earth's surface	The student understands the basis for global interdependence.	The student understands the increasing economic interdependence of the world's countries.
13 How the forces of cooperation and conflict among people influence the division and control of Earth's surface	The student understands how cooperation and conflict among people contribute to economic and social divisions of Earth's surface.	The student understands why and how cooperation and conflict are involved in shaping the distribution of social, political, and economic spaces on Earth at different scales.

Objectives
The student is able to:
- Explain the concept of imports and exports related to trade balances of a country.
- Evaluate the effectiveness of NAFTA.
- Create a layout that displays the results of the student's research.

GIS skills and tools

 Display attributes for a specific feature

 Select a feature on the map

 Zoom the layout to view the whole page

 Zoom out a fixed amount on the map

 Move elements in a layout

 Add text to a layout

 Zoom to the full map extent

 Add data to the map

- Open the attribute table for a layer
- Relate a table to a layer attribute table
- Update a relate to reflect a change in selected features
- Clear all selected features
- Open a graph and display data for a specific map feature
- Modify a graph's appearance
- Add graphs to a layout
- Print a layout

For more on geographic inquiry and these steps, see Geographic Inquiry and GIS (pages xxiii to xxv).

Teacher notes

Lesson introduction

This lesson is designed to correspond with a unit on NAFTA. Before you start this lesson, we suggest that you review the history of NAFTA. You may want to visit the following Web sites:

- *www.customs.ustreas.gov/nafta*
  Provides links to NAFTA customs and other information from each of the three countries.
- *www.dfait-maeci.gc.ca/nafta-alena/menu-e.asp*
  Presents clear and concise information on the history of NAFTA and continuing NAFTA issues.
- *www.sice.oas.org/indexe.asp*
  Contains excellent information on NAFTA as well as on foreign trade in general.

 Note: Due to the dynamic nature of the Internet, the URLs listed above may have changed. If the URLs do not work, refer to this book's Web site for an updated link: www.esri.com/mappingourworld.

Additional resources on NAFTA are listed on this book's Web site (*www.esri.com/mappingourworld*).

Your students should have a working knowledge of basic economics in addition to background information on NAFTA. Make sure they are familiar with the terms trade, imports, exports, trade balance, and tariffs. Basic graph-reading skills will also be essential.

Begin the lesson with a discussion of the North American Free Trade Agreement (NAFTA) as spelled out in the NAFTA objectives handout included with this lesson. Ask your students to explain the objectives of NAFTA in their own words. This exercise could be done as a class activity on the board or in groups with several sets of goals being developed. Note: You may want to save the results so students can refer to them when they evaluate NAFTA in step 8.

When you are satisfied that your students have a handle on NAFTA, discuss the flow of goods into and out of a country and the concept of trade balance. Introduce the following formula for trade balance:

EXPORTS – IMPORTS = TRADE BALANCE (trade surplus or deficit)

When exports are more than imports, you have a trade surplus.

When imports are more than exports, you have a trade deficit.

Ask the students to predict the trade balance between each of the NAFTA countries.

Student activity

 Before completing this lesson with students, we recommend that you work through it yourself. Doing so will allow you to modify the activity to accommodate the specific needs of your students.

Student work on the computer component of the lesson follows the initial discussion. Ideally each student should be at an individual computer, but the lesson can be modified to accommodate a variety of instructional settings.

On the first day, distribute the GIS Investigation sheets to the students. Explain that in this activity they will create graphs and maps to explore and report on the patterns of trade over the past 10 years between the NAFTA countries. In addition, they will use the graphs to identify whether NAFTA achieved its goals as outlined in the objectives discussed in class.

Student activity
(continued)

The worksheets will provide students with detailed instructions for their investigations. As they navigate through the lesson, they will be asked questions that will help keep them focused on key concepts. Some questions will have specific answers while others will require creative thought.

Things to look for while the students are working on this activity:

- Are students using a variety of tools?
- Are they answering the questions as they work through the procedure?
- Do they need help with the lesson's vocabulary?

 Teacher Tip: Step 7 of this GIS Investigation contains instructions for students to stop and save their work. This is a good spot to stop the class for the day and to pick up the investigation the next day. Be sure your students know how to rename their project and where to save it. We recommend that you spend two class periods doing the GIS Investigation and allow one period for work on the assessment.

Conclusion

At the completion of the lesson, have the students present to the class their trading history layouts and answers to their initial predictions. Ask them to describe the trade balances between each of the countries and also whether they think NAFTA has succeeded in achieving its goals. Finally, the students should evaluate whether or not NAFTA should remain in place. Tally up the results and report to the class what their combined opinion suggests.

Assessment

Middle school: Highlights skills appropriate to grades 5 through 8

Students will create and present layouts that illustrate the history of trade between Mexico and the United States, using maps they have made, graphs, text, and graphics. Presenters will describe the trade balance between the countries, and indicate whether they think NAFTA has achieved its goals. Presentations will conclude with students offering opinions about whether NAFTA should remain in place, using information from their layout to support those opinions.

High school: Highlights skills appropriate to grades 9 through 12

Students will create and present layouts that illustrate the history of trade between Mexico and the United States, using maps they have made, charts, text, and graphics. Presenters will describe the trade balance between the countries, and indicate whether they think NAFTA has achieved its goals. Presentations will conclude with students offering opinions about whether NAFTA should remain in place, using information from their layout to support those opinions. Finally, students will write a paragraph describing how they would change the NAFTA agreement to improve or enhance future trading for all three countries.

Extensions

- Have the students research the history of NAFTA.
- Have students acquire similar data for the European Union, OPEC, and the Organization of American States, and compare and contrast the outcomes of NAFTA and the European Union.
- Have students map the movements of commodities from the NAFTA countries in which they were produced, to countries where they are bought and used.
- Go to Statistics Canada, U.S. Census Bureau, and INEGI sites to obtain total international export information and map that information.

- NAFTA affects other aspects of each nation's economy, such as employment rates. Have students research employment trends since the advent of NAFTA. Have the NAFTA countries experienced increased employment or unemployment since 1994?

- Ask the students to do research on the employment/unemployment figures in the United States, Canada, and Mexico since NAFTA was initiated in 1994, and to report if there was a substantial change in employment/unemployment figures since January 1, 1994.

- Check out the Resources by Module section of this book's companion Web site *(www.esri.com / mappingourworld)* for print, media, and Internet resources on the topics of economics, trade, and NAFTA.

NAFTA objectives handout

The North American Free Trade Agreement (NAFTA) came into effect on January 1, 1994. This agreement created the world's largest free trade area. Among the agreement's main objectives are the liberalization of trade between Canada, Mexico, and the United States, stimulation of economic growth in all three countries, and equal access to each other's markets.

NAFTA Articles 101 and 102

Article 101: Establishment of the Free Trade Area

The Parties to this Agreement, consistent with Article XXIV of the General Agreement on Tariffs and Trade, hereby establish a free trade area.

Article 102: Objectives

1   The objectives of this Agreement, as elaborated more specifically through its principles and rules, including national treatment, most-favored-nation treatment and transparency are to:

   (a) eliminate barriers to trade in, and facilitate the cross border movement of, goods and services between the territories of the Parties;

   (b) promote conditions of fair competition in the free trade area;

   (c) increase substantially investment opportunities in the territories of the Parties;

   (d) provide adequate and effective protection and enforcement of intellectual property rights in each Party's territory;

   (e) create effective procedures for the implementation and application of this Agreement, for its joint administration and the resolution of disputes; and

   (f) establish a framework for further trilateral, regional and multilateral cooperation to expand and enhance the benefits of this Agreement.

2   The Parties shall interpret and apply the provisions of this Agreement in the light of its objectives set out in paragraph 1 and in accordance with applicable rules of international law.

From the North American Free Trade Agreement between the Government of Canada, the Government of the United Mexican States, and the Government of the United States of America, published January 1, 1994, as written in the U.S. Customs Department Web site (www.customs.gov/impoexpo/nafta_newf.htm).

NAME _____   DATE _____

Share and Share Alike
A GIS investigation

Answer all questions on the student answer sheet handout

On January 1, 1994, the North American Free Trade Agreement (NAFTA) was enacted to enhance trade and increase access to the total trade market available to businesses in North America. Tariffs and quotas were eliminated to increase the competitiveness of goods produced by all North Americans. Now, a decade after NAFTA's inception, your task will be to evaluate whether these objectives have been accomplished for all three countries involved. In this lesson you will create and analyze graphs of exports and trade balances for the NAFTA trading partners. As you are doing these exercises, ask yourself the following questions:

- Have there been any changes in exports or trade balance since the inception of NAFTA?
- Is there a larger market available for business in North America?
- Is there greater competition forcing businesses to provide the best products and best prices for their goods?
- Has NAFTA been equally beneficial for all countries involved?
- Does the trade balance of a country tell the whole story of how effective NAFTA is?

Ultimately, you will have to decide if you think NAFTA is effective!

Step 1   Start ArcMap

  a   **Double-click the ArcMap icon on your computer's desktop.**

b If the ArcMap start-up dialog appears, click **An existing map** and click OK. Then go to step 2b.

Step 2   Open the Region6.mxd file

a In this exercise, a map document has been created for you. To open it, go to the File menu and choose **Open**.

b Navigate to the module folder (**C:\MapWorld9\Mod6**) and choose **Region6.mxd** (or **Region6**) from the list.

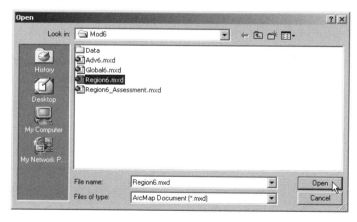

c Click Open.

When the map document opens, you see a map of North America showing the three countries participating in NAFTA.

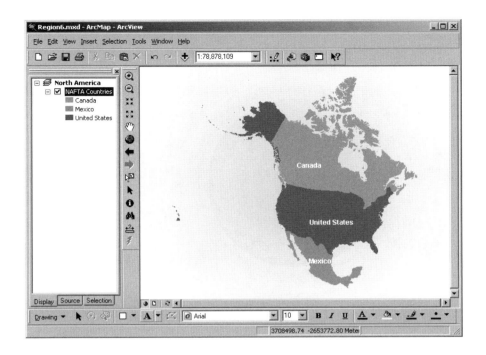

Step 3   Examine the map and attribute table

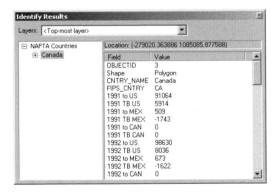

a   Click the Identify tool, then click Canada on the map.

b   Look at the attribute names in the Field column of the Identify Results window.

c   Scroll down the list and answer the following questions.

?   *(1)   Which years does the layer contain data for?*

?   *(2)   How many attributes are there for each year?*

Field names with the word "to" in them represent the value of goods and services exported from one country to another. For example, the attribute **1991 to US** represents the value of goods and services that a country exported **to** the **United States** in the year 1991. Field names with the abbreviation "TB" represent a country's trade balance.

The values are expressed in millions of U.S. dollars. That means you have to multiply the number you see by 1,000,000 to get the actual dollar value.

?   *(3)   What was the value of goods and services exported from Canada to the United States in 1991?*

d   Close the Identify Results window.

e   Right-click NAFTA Countries in the table of contents and choose Open Attribute Table.

(1)   *What is the name of the table?*

(2)   *How many rows are there for each country on the map?*

f   Click the gray box at the beginning of the United States row in the table.

Attributes of NAFTA Countries

OBJECTID*	Shape*	CNTRY_NAME	FIPS_CNTRY	1991 to US
1	Polygon	Mexico	MX	31130
2	Polygon	United States	US	0
3	Polygon	Canada	CA	91064

Step 4   **Relate another table to the layer table**

In this step you will open a second table that contains the trade data organized differently. You will relate the two tables together and then view the trading statistics in the tables and on a series of graphs.

a   Click the Source tab at the bottom of the table of contents. (Hint: If the Source tab is covered by the Attributes of NAFTA_Countries table, click the table's title bar and drag the table out of the way.)

b   Right-click NAFTA_Trading_Statistics in the table of contents and choose Open.

> *Note: The NAFTA_Trading_Statistics table is not listed on the Display tab because it is not a map layer. It is only listed on the Source tab.*

c   Make each table window smaller. Make sure that you can see all of the rows in the table as well as the Options button at the bottom. Arrange your windows so that you can see the map and both of the tables at the same time (Hint: Your ArcMap window can be small for this investigation.)

d   Examine the NAFTA_Trading_Statistics table pictured below and answer the questions.

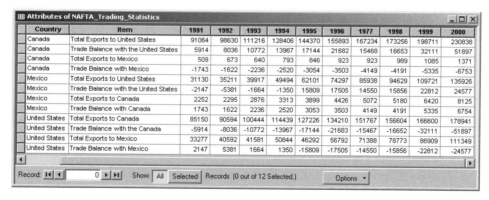

Attributes of NAFTA_Trading_Statistics

Country	Item	1991	1992	1993	1994	1995	1996	1977	1998	1999	2000
Canada	Total Exports to United States	91064	98630	111216	128406	144370	155893	167234	173256	198711	230838
Canada	Trade Balance with the United States	5914	8036	10772	13967	17144	21682	15468	16653	32111	51897
Canada	Total Exports to Mexico	509	673	640	793	846	923	923	989	1085	1371
Canada	Trade Balance with Mexico	-1743	-1622	-2236	-2520	-3054	-3503	-4149	-4191	-5335	-6753
Mexico	Total Exports to United States	31130	35211	39917	49494	62101	74297	85938	94629	109721	135926
Mexico	Trade Balance with the United States	-2147	-5381	-1664	-1350	15809	17505	14550	15856	22812	24577
Mexico	Total Exports to Canada	2252	2295	2876	3313	3899	4426	5072	5180	6420	8125
Mexico	Trade Balance with Canada	1743	1622	2236	2520	3053	3503	4149	4191	5335	6754
United States	Total Exports to Canada	85150	90594	100444	114439	127226	134210	151767	156604	166600	178941
United States	Trade Balance with the Canada	-5914	-8036	-10772	-13967	-17144	-21683	-15467	-16652	-32111	-51897
United States	Total Exports to Mexico	33277	40592	41581	50844	46292	56792	71388	78773	86909	111349
United States	Trade Balance with Mexico	2147	5381	1664	1350	-15809	-17505	-14550	-15856	-22812	-24577

Record: 0   Show: All   Selected   Records (0 out of 12 Selected.)   Options ▾

(1)   *How many rows are there for each country?*

(2)   *What information is collected under the field titled "Item" for Canada?*

(3)   *Describe in general terms the information collected in the field titled "Item."*

(4)   *How many years of data are represented in the table?*

In the next steps, you will relate the two tables by using common information to both tables (country names).

e   **Right-click NAFTA Countries in the table of contents, point to Joins and Relates, and click Relate.**

f   **In the first drop-down list in the Relate dialog, scroll down and choose CNTRY_NAME. Keep the default selections for steps 2 and 3. In step 4, replace the default name with Countries to Years. Click OK.**

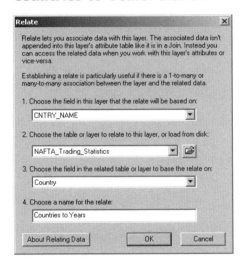

g   **Click the Options button in the Attributes of NAFTA Countries table and choose Related Tables, Countries to Years: NAFTA_Trading_Statistics.**

   What happens in the NAFTA Trading Statistics table?

   h   **Use the Select Features tool to click Canada on the map.**

What happens in the two tables and the map?

i   **Update the related rows by clicking the Options button in the Attributes of NAFTA Countries table and choosing Related Tables, Countries to Years: NAFTA_Trading_Statistics.**

Next, you'll see what happens when you make a selection in the related table.

j   **Select a row for Mexico in the Attributes of NAFTA_Trading_Statistics table.**

What happens in the two tables and the map?

k   **Now click the Options button in the related table—the Attributes of NAFTA_Trading_ Statistics table—and choose Related Tables, Countries to Years: NAFTA_countries.**

What have you observed about the way the NAFTA Trading Statistics table is tied to the NAFTA Countries attribute table and map layer?

l   **Click Selection in the Main menu in the ArcMap window and choose Clear Selected Features. All of the selections are cleared.**

Step 5   **Examine export graphs**

Now you will examine a graph of exports to Canada.

a   Click the Tools menu, point to Graphs, and choose Exports to Canada.

The graph window actually contains three small graphs, one for each country.

b   Enlarge the graph window slightly so you can read it more easily.

?   *(1)   Which country exported more goods and services to Canada—Mexico or the United States?*

?   *(2)   Why is the graph empty in the space for Canada?*

c   Drag the graph window completely off the map and table windows.

   d   Using the Select Features tool, click Mexico on the map. Mexico is outlined in blue.

?   *What happened to the graph?*

Because the graph is created from the data in the attribute table, it is tied to the map through the table. When you select a country on the map, only the data for that associated row in the table is displayed in the graph.

e   Answer the questions below using the graph and tables.

?   *(1)   How many years of data are represented on the graph?*

?   *(2)   What year does the first bar on the left represent?*

?   *(3)   Compare the numbers on the y-axis with those in the two tables. Are the numbers on the graph in thousands, millions, or billions of dollars? (Hint: Remember that the Attributes of NAFTA Countries table values are in millions of dollars.)*

?   *(4)   Looking at the graph, how would you describe the trend of Mexican exports to Canada over the 10-year period?*

?   *(5)   What was the approximate value of Mexican exports to Canada in 1991? In 2000?*

?   *(6)   Approximately how many times greater is the 2000 export figure than the 1991 export figure?*

f   Close the Exports to Canada graph.

g   Click the Tools menu and open the graph entitled Exports to U.S.

?   *(1)   How would you describe the trend of Mexican exports to the United States over the 10-year period?*

?   *(2)   Approximately how many times greater is the 2000 export figure than the 1991 export figure?*

Note: Carefully observe the y-axis values of the graphs in this investigation. The height of a column may be the same on two different graphs, but it may represent very different dollar values.

Step 6   Examine a trade balance graph

    a   Click the Tools menu and open the graph entitled Trade Balance with U.S.

    b   This graph may cover up the first graph. Drag it below the first one so you can see both graphs at the same time.

           Remember: Trade balance compares how much a country exports to a trading partner with how much it imports from the same partner.

    c   Use the Trade Balance with U.S. graph to answer the questions below.

        (1)   Did Mexico have a trade surplus or deficit with the United States for 1992?

        (2)   What was the approximate value of the trade balance for 1992? (Remember, the y-axis is in millions of dollars.)

        (3)   What was the first year that Mexico exported more to the United States than it imported from the United States?

        (4)   Describe the trend of Mexico's trade balance with the United States over the 10-year period.

    d   Click Canada on the map. Look at the Trade Balance graph again.

        Did Canada have a deficit trade balance with the United States anytime during the 10-year period?

    e   Click the map away from any country (anywhere in the yellow area) to clear the selection. Compare Mexico and Canada on the graph.

        (1)   In 1998, was Canada's trade balance with the United States greater, smaller, or about the same as Mexico's?

        (2)   In 2000, was Canada's trade balance with the United States greater, smaller, or about the same as Mexico's?

    f   Select Canada on the map, then click the Options menu in the Attributes of NAFTA Countries table and update the relate.

        Referring to the Attributes of NAFTA_Trading_Statistics table, what was the exact value of Canada's trade balance with the United States in 2000?

    g   Click the Selection menu in the ArcMap window and click Clear Selected Features.

Step 7   Change the look of a trade balance graph

    a   Close the table windows and the Exports to U.S. graph. (Keep open the Trade Balance with U.S. graph.)

    b   Click the Tools menu and open the graph entitled TB Mex.

    TB Mex is a basic graph showing the Mexican trade balance data. In the next steps you will change the properties of this graph to give it the same appearance as the other trade balance graphs.

    c   Right-click the title bar of the TB Mex graph window and choose Properties.

    d   Arrange your windows so you can see the Graph Properties dialog and both of the graphs.

You will begin by changing the graph title, adding a subtitle, and removing the graph legend.

e   In the Appearance tab, make the following changes:

- Delete the existing title and type **Trade Balance with Mexico**.
- In the Sub title box, type **All graphs cover the period 1991-2000, left to right**.
- Check the box to Label X Axis With. CNTRY_NAME should be the label field.
- Uncheck the Show Legend box.

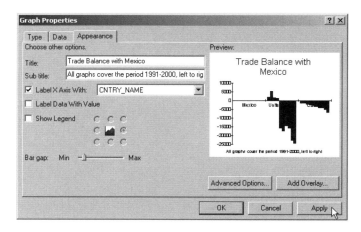

f   Click Apply. The TB Mex graph is updated.

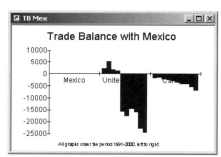

Next you will add the grid lines and change the color of the title, subtitle, and background.

g   Click the Advanced Options button on the Appearance tab of the Graph Properties dialog.

h   Click the Axis tab in the Advanced Options dialog. Choose the following settings, and then make sure your dialog looks like the one pictured below:

- Color of Axes = teal blue (the fourth color listed)
- Position = Left
- Check to Show grids
- Grid color = black

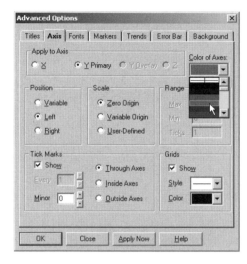

i   Click the Background tab.

j   Make sure that Graph Title is selected in the Apply To list. Then click the Text Color drop-down list and choose teal blue.

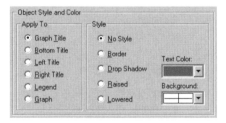

k   Click Bottom Title in the Apply To list. Choose the teal blue text color again.

l   Click Graph in the Apply To list and make the text color black.

m   Click the Palette drop-down list. Scroll up and choose 128 Pastel.

n   Locate the Background list just above the Palette list (NOT the Background Color list to the left). Select light cyan blue by scrolling down about two-thirds of the way until you see a set of cyan blues and then bright pinks. Choose the second light cyan color above the bright pink. (This is the color that is used in the Trade Balance with U.S. graph.)

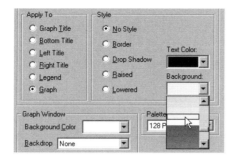

o   Click OK in both open dialog windows.

The Trade Balance with Mexico graph now looks like the other trade balance graphs. You will use this graph later in step 8.

p   Close the two graph windows.

q   Ask your teacher if you should stop here and save this ArcMap map document. Follow your teacher's instructions on how to rename the map document and where to save it. For example, you could rename it **ABC_Region6** where ABC are your initials. To help you remember, write the new name you gave the project and where you saved it on the answer sheet.

If you do not need to save the project, proceed to step 8.

Step 8   Evaluate the effectiveness of NAFTA

In this lesson, you are looking at the level of trade among the NAFTA trading partners—Canada, Mexico, and the United States. You are also looking at changes in the level of trade among the three countries over time.

Remember, it is nearly impossible for any country to produce everything it needs to support its people. Countries must import (purchase) goods from other countries. One country's imports are the other country's exports. In this map document, you have no graphs for imports because you can figure out that information by looking at the export graphs.

For example, to find the United States' imports from Canada, you switch the question in your mind and look for Canada's exports to the United States. You would find this information on the graph entitled Exports to United States.

To decide whether or not NAFTA is meeting its goals, you will need to further explore the data in the graphs and tables in the ArcMap map document. In this step, you will gather the information that you need to decide if you think NAFTA is effective. You may want to refer to the NAFTA objectives handout or the list of NAFTA goals you developed with your teacher at the beginning of the lesson. Remember these formulas as you examine the charts and answer the questions that follow:

Exports – Imports = Trade Balance

Trade Balance:
Exports > Imports = Trade Surplus
Imports > Exports = Trade Deficit

a   Go to the Tools menu and open the three export graphs. Arrange them so they are not overlapping and you can compare them.

?   *b*   Use the graphs to find the value of exports between each set of countries for the year 2000. Write the information in the table on your answer sheet.

?   *c*   Add the export values together for each pair of countries (for example, United States to Mexico plus Mexico to United States) and write that number in the Total Volume Between Partners column.

?   *d*   Rank the trading partners by the overall volume of trade between the two countries. Use 1 for the partners trading the most and 3 for the partners trading the least.

?   *(1)* *Do you think that NAFTA had a positive (+), negative (–), or neutral (n) effect on trade volume between each set of partner countries?*

?   *(2)* *Do you think that any one of these three countries benefited more than the other two by NAFTA? If so, which country? Explain your answer.*

    e   Close the three graph windows.

    f   Go to the Tools menu and open the three Trade Balance graphs. Arrange them so you can compare them.

       Remember that most countries prefer to export a higher volume and dollar value of goods than they import, but either a large trade surplus or large trade deficit may negatively affect a country's overall economic health.

    g   Look at the Trade Balance with Canada graph.

?   *(1)* *What country has a healthier trade balance with Canada—Mexico or the United States?*

?   *(2)* *On what graph do you find a set of bars that looks like a mirror image of those for the U.S. trade balance with Canada?*

    h   Compare the trade balance graphs.

?   *(1)* *What country had the most dramatic change for the better after NAFTA came into being? (Remember, NAFTA went into effect in 1994.)*

?   *(2)* *Estimate the U.S. trade deficit with Canada and Mexico for 1999 and 2000. Express the values in billions of dollars, using round numbers that you estimate from the graphs. Write the answers in the space provided.*

?   *(3)* *Did the U.S. combined trade balance get better or worse between 1999 and 2000? By how much?*

    i   Close the three graph windows.

Step 9   **Create and print a layout**

       In this step you will create a layout that presents the information about the U.S. trade balance and exports to Canada. The layout will include the NAFTA countries map and two graphs: Exports to Canada and Trade Balance with Canada. Later, in the assessment, you will use the techniques you learn to create a layout showing the results of your evaluation of NAFTA.

    a   Click the Layout View button.

    b   Maximize your ArcMap window so that it fills your screen. If your Layout toolbar is floating, dock it above the view or in your preferred location.

c   Click the Zoom Whole Page button on the Layout toolbar to make the layout fill the view.

The layout looks like a piece of paper. The layout already has a map, title, north arrow, legend, and scale bar. Before adding the graphs, you will make space for them by slightly reducing the size of the map.

d   Click the Fixed Zoom Out button on the Tools toolbar once to reduce the size of North America.

Next you will add the two graphs.

e   Click the Tools menu, point to Graphs, and then click Manage to display the Graph Manager.

f   Select Exports to Canada in the list and then click the Show on Layout button.

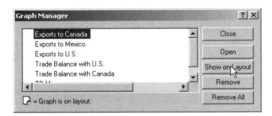

The graph is added to the middle of the layout.

g   Repeat the procedure to show the Trade Balance with Canada graph on the layout.

h   Click Close to close the Graph Manager.

You will arrange the two graphs on the left side of the layout. You will also make the map a little smaller so that it better fits the space on the right.

i   Click the Trade Balance with Canada graph using the Select Elements tool. Drag the graph to the upper left of the layout, under the title.

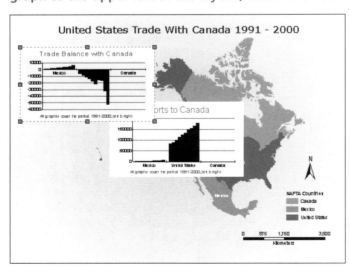

j   Click the Exports to Canada graph to activate it. Then click and drag it to a position below the first graph.

k Use the Pan tool on the Tools toolbar to move North America slightly to the right—center it between the graphs and the legend.

When you are satisfied with the look of your layout, go on to the next step. You will add your name and the date to the layout.

 l Click the New Text tool in the Draw toolbar. Click in the yellow space underneath the graphs to insert a piece of text.

m Type your name, a comma, and today's date, and then press the Enter key on your keyboard.

Amy B. Changalee, September 21, 2004

n Drag the text down to the lower left corner of the layout, just inside the map neat line. When you are finished, click anywhere in the white space outside the layout to unselect the text.

Finally, you need to focus the graphs on the United States to reflect U.S. trade with Canada.

 o Click the Select Features tool and click the United States on the map.

The graphs change to focus on U.S. data.

p From the File menu, select Print.

q Check with your teacher to be sure the correct printer is selected. Click OK to print your layout.

Step 10   Exit ArcMap

In this exercise, you used ArcMap tables, graphs, and a map to explore trade data for the three North American countries. After analyzing this data, you created a layout to present some of the information.

a Ask your teacher for instructions on where to save this map document and on how to rename it. Save it according to your teacher's instructions.

b If you are not going to save the map document, exit ArcMap by choosing Exit from the File menu. When asked if you want to save changes to the map document, click No.

NAME _____   DATE _____

Student answer sheet

Module 6
Human Geography III: Economic Geography

Regional case study: Share and Share Alike

Step 3   Examine the map and attribute table

c-1   Which years does the layer contain data for? _____   to   _____

c-2   How many attributes are there for each year? _____

c-3   What was the value of goods and services exported from Canada to the United States in 1991?
         $ _____

e-1   What is the name of the table?   _____

e-2   How many rows are there for each country on the map? _____

Step 4   Relate another table to the layer table

d-1   How many rows are there for each country?   _____

d-2   What information is collected under the field title "Item" for Canada?

         _____

         _____

         _____

         _____

d-3   Describe in general terms the information collected in the field titled "item."

         _____

         _____

d-4   How many years of data are represented in the table?   _____

g   What happens in the NAFTA_Trading Statistics table?

         _____

         _____

h   What happens in the two tables and the map?

         _____

         _____

j   What happens in the two tables and the map?

         _____

         _____

k   What have you observed about the way the NAFTA Trading Statistics table is tied to the NAFTA
        Countries attribute table and map layer?

        _____

        _____

Step 5   Examine export graphs

b-1   Which country exported more goods and services to Canada—Mexico or the United States?

        _____

b-2   Why is the graph empty in the space for Canada?

        _____

        _____

d   What happened to the graph?

        _____

        _____

e-1   How many years of data are represented on the graph?   _____

e-2   What year does the first bar on the left represent?   _____

e-3   Compare the numbers on the y-axis with those in the two tables. Are the numbers on the graph in
        thousands, millions, or billions of dollars?

        _____

e-4   Looking at the graph, how would you describe the trend of Mexican exports to Canada over the
        10-year period?

        _____

e-5   What was the approximate value of Mexican exports to Canada in 1991?

        _____

        In 2000? _____

e-6   Approximately how many times greater is the 2000 export figure than the 1991 export figure?

        _____

g-1   How would you describe the trend of Mexican exports to the United States over the 10-year period?

        _____

g-2   Approximately how many times greater is the 2000 export figure than the 1991 export figure?

        _____

Step 6   Examine a trade balance graph

c-1   Did Mexico have a trade surplus or deficit with the United States for 1992? _____

c-2   What was the approximate value of the trade balance for 1992? (Remember, the y-axis is in millions of dollars.)

_____

c-3   What was the first year that Mexico exported more to the United States than it imported from the United States?

_____

c-4   Describe the trend of Mexico's trade balance with the United States over the 10-year period.

_____

_____

d   Did Canada have a deficit trade balance with the United States anytime during the 10-year period?

_____

e-1   In 1998, was Canada's trade balance with the United States greater, smaller, or about the same as Mexico's?

_____

e-2   In 2000, was Canada's trade balance with the United States greater, smaller, or about the same as Mexico's?

_____

f   Referring to the Attributes of NAFTA_Trading_Statistics table, what was the exact value of Canada's trade balance with the United States in 2000?

_____

Step 7   Change the look of a trade balance graph

q   Write the new name you gave the map document and where you saved it.

_____   _____
(Name of map document.   **(Navigation path to where map document is saved.**
For example: ABC_Region6.mxd)   **For example: C:\Student\ABC)**

Step 8 Evaluate the effectiveness of NAFTA

b   Use the graphs to find the value of exports between each set of countries for the year 2000. Write the information in the table below.

DIRECTION OF EXPORT FLOW	VALUE OF EXPORTS (MILLION $)	TOTAL VOLUME BETWEEN PARTNERS (MILLION $)
United States to Mexico		
Mexico to United States		
United States to Canada		
Canada to United States		
Canada to Mexico		
Mexico to Canada		

c   Add the export values together for each pair of countries (for example, United States to Mexico plus Mexico to United States) and write that number in the Total Volume Between Partners column.

d   Rank the trading partners by the overall volume of trade between the two countries. Use 1 for the partners trading the most and 3 for the partners trading the least.

United States–Mexico: _____

United States–Canada: _____

Canada–Mexico: _____

d-1 Do you think that NAFTA had a positive (+), negative (–), or neutral (n) effect on trade volume between each set of partner countries?

United States–Mexico: _____

United States–Canada: _____

Canada–Mexico: _____

d-2 Do you think that any one of these three countries benefited more than the other two by NAFTA? If so, which country? Explain your answer.

_____

_____

_____

g-1 What country has a healthier trade balance with Canada—Mexico or the Untied States?

_____

g-2 On what graph do you find a set of bars that looks like a mirror image of those for the U.S. trade balance with Canada?

_____

h-1 What country had the most dramatic change for the better after NAFTA came into being? (Remember, NAFTA went into effect in 1994.)

_____

h-2   Estimate the U.S. trade deficit with Canada and Mexico for 1999 and 2000. Express the values in billions of dollars, using round numbers that you estimate from the graphs.

	1999 ($ BILLION)	2000 ($ BILLION)
U.S. trade balance with Mexico		
U.S. trade balance with Canada		
Combined deficit (total)		

h-3   Did the U.S. combined trade balance get better or worse between 1999 and 2000? _____
By how much?   _____

NAME _____   DATE _____

Share and Share Alike
Middle school assessment

1   Using the 10 years of export and trade balance data you have for the NAFTA trading partners, Canada, Mexico, and the United States, make a determination if you think that NAFTA has been effective, marginally effective, or ineffective and whether you think NAFTA should continue or not.

Remember that there are several factors that you can use in drawing your conclusions. You have looked at several of those factors during this lesson. For instance, you can compare the exports of each country as well as look at each country's trade balances. Refer back to the NAFTA Objectives handout to remind yourself of the original intent of NAFTA.

2   To display your findings, create a layout in ArcMap that graphically displays your conclusion. Use the layout in the Region6_Assessment map document that has already been started for you. You may rearrange and resize the existing layout components and add your own. Your final layout should include the following components:

(a)   Title

(b)   Map

(c)   Map legend

(d)   Orientation (north arrow or compass rose)

(e)   At least two charts

(f)   Text labels or descriptions

(g)   Author (your name)

(h)   Today's date

3   Present your layout to the class as evidence for your conclusion on NAFTA's effectiveness.

Share and Share Alike

Assessment rubric

Middle school

STANDARD	EXEMPLARY	MASTERY	INTRODUCTORY	DOES NOT MEET REQUIREMENTS
The student knows and understands how to make and use maps, globes, graphs, charts, models, and databases to analyze spatial distributions and patterns.	Creates a layout using GIS with eight or more components including several maps and charts that illustrate the effectiveness of NAFTA.	Creates a layout using GIS with eight components including a map and minimum of two charts.	Creates a layout using GIS with most of the eight specified components including a map and at least one chart.	Creates a layout using GIS with some of the eight specified components including a map.
The student knows and understands the basis for global interdependence.	Layout shows a clear understanding of the concepts of balance of trade for NAFTA trading partners through multiple charts and maps.	Layout includes charts and maps showing an understanding of the concept of balance of trade for NAFTA trading partners.	Layout includes charts and maps showing some understanding of the concept of balance of trade for NAFTA trading partners.	Layout includes chart and or map showing trade partners, but does not illustrate an understanding of balance of trade.
The student knows and understands how cooperation and conflict among people contribute to economic and social divisions of Earth's surface.	Presents a logical argument to his/her conclusions on the effectiveness of NAFTA using a variety of resources to support his/her findings including, but not limited to, data and maps.	Presents a logical argument to his/her conclusions on the effectiveness of NAFTA using data and maps to support his/her ideas.	Describes conclusions on the effectiveness of NAFTA and provides data and maps, but cannot provide ample argument or evidence.	Identifies NAFTA participants and trading partners, but does not present any personal conclusions as to its effectiveness.

This is a four-point rubric based on the National Standards for Geographic Education. The "Mastery" level meets the target objective for grades 5–8.

NAME _____   DATE _____

Share and Share Alike
High school assessment

1   Using the 10 years of export and trade balance data you have for the NAFTA trading partners, Canada, Mexico, and the United States, make a determination if you think that NAFTA has been effective, marginally effective, or ineffective and whether you think NAFTA should continue or not.

    Remember that there are several factors that you can use in drawing your conclusions. You have looked at several of those factors during this lesson. For instance, you can compare the exports of each country as well as look at each country's trade balances. Refer back to the NAFTA objectives handout to remind yourself of the original intent of NAFTA.

2   To display your findings, create a layout in ArcMap that graphically displays your conclusion. Use the layout in the Region6_Assessment map document that has already been started for you. You may rearrange and resize the existing layout components and add your own. Your final layout should include the following components:

    (a)   Title

    (b)   Map

    (c)   Map legend

    (d)   Orientation (north arrow or compass rose)

    (e)   At least two charts

    (f)   Text labels or descriptions

    (g)   Author (your name)

    (h)   Today's date

3   Present your layout to the class as evidence for your conclusion on NAFTA's effectiveness.

4   Compose a paragraph describing how you would change the NAFTA agreement to improve or enhance future trading results for all three countries. Include specific reasons why you believe your changes to NAFTA would be effective.

Share and Share Alike

Assessment rubric

High school

STANDARD	EXEMPLARY	MASTERY	INTRODUCTORY	DOES NOT MEET REQUIREMENTS
The student knows and understands how to use geographic representations and tools to analyze, explain, and solve geographic problems.	Creates a layout using GIS with eight or more components including several maps and charts that illustrate the effectiveness of NAFTA.	Creates a layout using GIS with eight components including a map and minimum of two charts that illustrates the effectiveness of NAFTA.	Creates a layout using GIS with most of the eight specified components including a map and at least one chart that illustrate the effectiveness of NAFTA.	Creates a layout using GIS with some of the eight specified components including a map that illustrates the effectiveness of NAFTA.
The student knows and understands the increasing economic interdependence of the world's countries.	Composes a brief essay that defines specific improvements to the NAFTA agreement that will benefit future trading for all three countries. Provides examples to illustrate these ideas.	Composes a paragraph that defines specific improvements to the NAFTA agreement that will benefit future trading for all three countries.	Composes a paragraph that lists ideas for improvement to the NAFTA agreement, but the improvements may not benefit all parties.	Lists ideas for improvement to the NAFTA agreement, but does not define how these improvements will enhance future trading.
The student knows and understands why and how cooperation and conflict are involved in shaping the distribution of social, political, and economic spaces on Earth at different scales.	Presents a logical argument to his/her conclusions on the effectiveness of NAFTA at a local, regional, and global level using a variety of resources to support his/her findings including, but not limited to, data and maps.	Presents a logical argument to his/her conclusions on the effectiveness of NAFTA at a local, regional, and global level using data and maps to support his/her ideas.	Describes conclusions on the effectiveness of NAFTA at a local, regional, and global level, but cannot provide ample argument or evidence.	Identifies NAFTA participants and trading partners, but does not present any personal conclusions as to its effectiveness.

This is a four-point rubric based on the National Standards for Geographic Education. The "Mastery" level meets the target objective for grades 9–12.

Live, Work, and Play
An advanced investigation

Lesson overview

Students will use the Internet to acquire the most recent data on employment in California. By exploring this data through a GIS, they will construct a thematic map for California counties that displays population, number of jobs in the finance field, and average wage data, then perform a complex query and create a map that illustrates their results. Students will be learning how to examine data to choose a place to live that satisfies a set of criteria.

Estimated time   Two to three 45-minute class periods

Materials   ✔ Student handouts from this lesson to be copied:
- Live, Work, and Play worksheet (page 384)
- GIS Investigation sheets (pages 385 to 394)

Standards and objectives   *National geography standards*

GEOGRAPHY STANDARD	MIDDLE SCHOOL	HIGH SCHOOL
1 How to use maps and other geographic representations, tools, and technologies to acquire, process, and report information from a spatial perspective	The student understands how to make and use maps, globes, graphs, charts, models, and databases to analyze spatial distributions and patterns.	The student understands how to use geographic representations and tools to analyze, explain, and solve geographic problems.
12 The processes, patterns, and functions of human settlement	The student understands the causes and consequences of urbanization.	The student understands the evolving forms of present-day urban areas.
18 How to apply geography to interpret the present and plan for the future	The student understands how varying points of view on geographic context influence plans for change.	The student understands how to use geographic knowledge, skills, and perspectives to analyze problems and make decisions.

Objectives

The student is able to:
- Use the Internet to locate data for recent economic activity.
- Use a GIS to map data acquired from the Internet.
- Thematically map various data to select an appropriate place to live.

GIS skills and tools
- Locate data on the Internet and prepare it in a comma-delimited file
- Add tables to a map document
- Join tables
- Select features by multiple attributes
- Select a county according to a number of criteria
- Create a layout
- Print a presentation-style map

For more on geographic inquiry and these steps, see Geographic Inquiry and GIS (pages xxiii to xxv).

Teacher notes

Lesson introduction

Open a class discussion with the following prompts:
- Business owners frequently use GIS to select a suitable site for a new business location.
- Individuals can also use GIS to decide where they would like to live, work, and spend their leisure time by looking for areas that have various characteristics, such as affordable housing, employment opportunities, and recreational areas.
- What criteria are important to students when they think about where they would like to live after they graduate from high school or college and enter adulthood?

Explain to your students that they will use the Internet to find data on recent economic activity. They will use a GIS to map economic data they find and will analyze the data to locate the best place to live, work, and play. This GIS Investigation will require students to work independently, with little guidance.

Student activity   **Before completing this lesson with students, we recommend that you complete it as well. Doing so will allow you to modify the activity to accommodate the specific needs of your students.**

Distribute this lesson's GIS Investigation sheets and the Live, Work, and Play Worksheet. Students should fill out the worksheet, identifying the criteria they consider the most important in choosing a place to live, then follow the guidelines to retrieve county economic data from the Internet. They will edit the data, add the data to ArcMap, and export the data to a geodatabase table. Then students will use the data to create thematic maps that display concentrations of jobs and wages, and analyze the maps to find areas with desirable characteristics (as listed on the worksheet).

 Teacher Tip: In this activity, students will download and edit data. They will export the edited data to a geodatabase table. This action results in permanent changes to the data. In order for this to work properly, each student must have their own copy of the Mod6 folder.

Conclusion   Engage students in a discussion of the patterns they discovered as they mapped different economic criteria. Were they able to find a place to live that satisfied *all* of the criteria they had? Could they add more criteria? Could they use this same method to download data for other states and broaden the search area (e.g., including Washington and Oregon)? How might site analysis be useful for businesses?

Assessment   The last step of the GIS Investigation asks students to perform their own multiple-attribute query on where to live, and create a layout of their results. In addition to the map, you may wish to ask students to submit a written report that summarizes the process they went through in selecting an optimal living site and explains their query of the data. A written report could answer any or all of the following questions:

- What steps were involved in the query process each student went through to make their site selection?
- Were they surprised at the results they got when they mapped the various criteria? Why or why not?
- Was there an alternate site that might have been as acceptable as the first site chosen? Why or why not?

 Note: Due to the independent nature of this lesson, there is no supplied assessment rubric. You are free to design assessment rubrics that meet the needs of your specific adaptation.

Extensions
- Ask students to broaden their search area to other states. Download economic data for other states from the Bureau of Economic Analysis Web site and perform the same analysis. Students should note any changes they see.
- Have students query more fields in the CA counties table to narrow their search. They can explore, for example, median rent, median house values, and the percentage of residents who are married.
- Have students download employment data for an industry that interests them and perform a query based on their individual priorities.
- Check out the Resources by Module section of this book's companion Web site (*www.esri.com/mappingourworld*) for print, media, and Internet resources on the topics of employment data and demographics.

NAME _____   DATE _____

Live, Work, and Play worksheet

Imagine: You have just graduated from college with a degree in finance. You want to move to California, but you don't know which county would be best for you. You need to find a good job in the finance industry (banking, investments, insurance, or real estate). You know that salary is important (after all, you would like to start paying off your student loans as soon as possible), but you also know there are many other things to consider when deciding where to live and work. For example, you might want to consider the cost of housing, the length of your commute, and what you'll do in your free time (do you want to be near the beach, the mountains, a city with good nightlife?).

To start your investigation of good, better, and best places to live, complete the table below. For each category, designate it as a Very Important, Average, or Not Important criterion for determining where you want to live.

LAYERS AVAILABLE	VERY IMPORTANT	AVERAGE	NOT IMPORTANT
Percentage of people ages 18–29			
Number of finance jobs available			
Average wage			

NAME _____   DATE _____

Live, Work, and Play
An advanced investigation

Note: Due to the dynamic nature of the Internet, the URLs listed in this lesson may have changed, and the graphics shown below may be out of date. If the URLs do not work, refer to this book's Web site for an updated link: www.esri.com/mappingourworld.

Answer all questions on the student answer sheet handout

Now that you have rated the importance of different criteria in determining where you want to live, you will download employment data from the Internet and use ArcMap to find places that meet your criteria. (Make sure that you have completed the Live, Work, and Play worksheet provided by your teacher before beginning this part of the activity.)

Step 1 **Locate economic data on the World Wide Web**

The World Wide Web is a great source to find geographic data. In this step, you will go to the Bureau of Economic Analysis (BEA) Web site to find economic data.

a Open your Web browser to the BEA U.S. Economic Accounts page at **www.bea.doc.gov**.

b Under Regional, click **State and Local Personal Income**.

c Under Local area annual estimates, click the link titled **Interactive tables - view or download specific data**.

d Under Step 1, choose **Detailed income and employment tables by NAICS industry, 2001-2002 (CA05 and CA25)**.

> **Step 1.** *Select a series. This will change options below.*
>
> ○ Personal income and population summary estimates (CA1-3)
> ● Detailed income and employment tables by NAICS industry, 2001-2002 (CA05 and CA25)
> ○ Detailed income and employment tables by SIC industry, 1969-2000 (CA05 and CA25)
> ○ Detailed tables of regional profiles, personal current tranfers, farm income, 1969-2002 (CA30-CA45)
> ○ Wage and salary summary estimates (CA34)
> ○ Single line of data for all counties (3183 rows returned; please limit years selected to speed process)

e Under Step 2, click **CA25N - Total full-time and part-time employment by industry**.

f   Under Step 3, click **California** for the state, then click the arrow button. `-->`

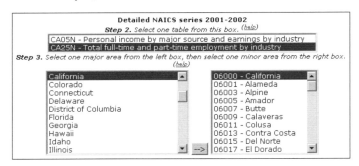

g   Click the Display button to see a report for all of California. `Display`

h   Scroll down the industry classifications and click the code **1000**. A table displays showing the number of people employed in finance and insurance jobs per county or metropolitan area (MSA) in California.

Step 2   Download data and prepare it for ArcGIS

Before you can see a map of this data in ArcMap, you need to save it to your computer and modify it so that ArcGIS can understand it. The modifications are done in a text editor program.

a   Click the Download Data button. `Download Data`
The file is in comma separated values format (.csv). Choose to save the file.

b   Save the file in your module 6 data folder (**C:\MapWorld9\Mod6\Data**) and rename it **ABC_finance.csv**, where **ABC** are your initials.

c   Open a text editor such as WordPad or Notepad. Open the ABC_finance.csv file you just saved. (Hint: Select All Files in the Files of type drop-down list to see the .csv file.) The data appears in your new document.

For example:

d Make the following changes in the first line of text:

- Replace "Finance and insurance" with **"Finance"**. (Keep the quotation marks.)
- Change "2001" to **Fin_2001** and change "2002" to **Fin_2002**. (Omit the quotation marks.)

> *Note: Fields with quotation marks will be defined as text. Fields without quotation marks will be defined as numeric. Ignore the end-of-line symbols and quotation marks at the end of each line.*

e Narrow the table window until most lines have data for a single county. (Click and drag the right edge of the window.)

f Use the Find and Replace function to find all "(D)" and replace them with **"–99"**.

Consult the graphic below to make sure you made all the necessary changes.

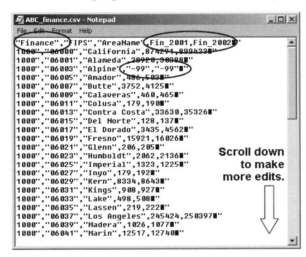

The data you saved includes statistics for counties and metropolitan areas, but you need only the county statistics.

g Delete all the data after Yuba County; it occurs approximately halfway through the file. (This is metro area data that you do not need.) The last line in your file now should read "1000", "06115", "Yuba",431,432". See example below:

```
1000","06095","Solano",4764,4942■"
1000","06097","Sonoma",11845,12078■"
1000","06099","Stanislaus",6060,6261■"
1000","06101","Sutter",1264,1207■"
1000","06103","Tehama",707,768■"
1000","06105","Trinity",101,111■"
1000","06107","Tulare",5141,5686■"
1000","06109","Tuolumne",623,588■"
1000","06111","Ventura",23494,25757■"
1000","06113","Yolo",2572,2572■"
1000","06115","Yuba",431,432■"
```

h Save the file.

i Minimize the text editor program and your Web browser. You will use them later to download additional data in step 5.

Now that your data is in a format that ArcGIS can read, you will bring the data into ArcMap and map it.

Step 3   Start ArcMap and add your data

a   Start ArcMap.

b   Open the **Adv6.mxd** (or **Adv6**) map document from your module folder (**C:\MapWorld9\Mod6\Adv6.mxd**).

When the map document opens, you see a map with two layers turned on (Population Age 18 - 29 and CA Counties).

c   Click the Add Data button and navigate to the location of **ABC_finance.csv**, select it, and click Add.

d   Open the table from the Source tab and check it to make sure the fields and rows look correct and contain data.

Finance	FIPS	AreaName	Fin_2001	Fin_2002
1000	06000	California	874291	899433
1000	06001	Alameda	28920	30388
1000	06003	Alpine	-99	-99
1000	06005	Amador	486	503
1000	06007	Butte	3752	4125
1000	06009	Calaveras	460	465
1000	06011	Colusa	179	190
1000	06013	Contra Costa	33630	35326
1000	06015	Del Norte	128	137
1000	06017	El Dorado	3435	4562
1000	06019	Fresno	15921	16026

Record: 53   Show: All | Selected   Records (0 out of 59 Selected.)   Options

e   Close the table.

Next, you will export the table to the California.mdb geodatabase.

f   Right-click ABC_finance.csv in the table of contents, point to Data, and click Export.

g   Click the Output table Browse button. Select Personal Geodatabase tables in the Save as type drop-down list. Type **finance** for the table name. Then navigate to the California geodatabase (**C:\MapWorld9\Mod6\Data\California.mdb**).

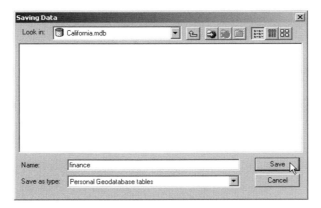

h   Click Save. Then click OK in the Export Data dialog. Click Yes to add the new table your map document.

i   Right-click the ABC_finance.csv table and click Remove. You won't be using this version of the table.

You will join the new geodatabase table to the attribute table of the CA Counties layer so that you can map the finance data.

j   Right-click the CA Counties layer, point to Joins and Relates, and click Join.

k   In the Join Data dialog, choose to Join attributes from a table. Choose FIPS as the field to base the join on. The next two boxes should be filled in automatically with finance and FIPS.

l   Click OK. Click Yes if you receive a message about indexing the table.

m   Open the CA Counties layer attribute table.

n   Scroll to the right and notice that the finance fields appear at the end of the table. (These are Finance, FIPS, AreaName, Fin_2001, and Fin_2002.) Close the table.

In the next step, you will classify the CA Counties layer so you can thematically map the finance employment data you retrieved from the Internet.

Step 4   Thematically map finance employment data

a   Click the Display tab in the table of contents. Turn off the Population Age 18 - 29 layer.

 b   Right-click CA Counties and choose Copy. Click the Paste button to paste a copy of the CA Counties layer.

c   Open the Properties dialog for the copied layer. Click the General tab and name the layer **Number of Finance Jobs**.

d   In the Symbology tab, choose Quantities, Graduated colors. For Value field, scroll to the end of the list and click finance.Fin_2002. Choose an orange or brown color ramp.

The value field (Fin_2002) represents the number of finance jobs for each county in California in 2002. By default, ArcMap has divided the number of finance jobs into five groups.

Some counties have no data available. In such cases, the Fin_2002 field has a value of –99. Because this number is a code and not a measurement of the number of jobs, counties with a value of –99 need to be excluded from the categories you just created.

e   Click the Classify button. In the next dialog, click the Exclusion button.

f   In the Query tab of the Data Exclusion Properties dialog, create the following expression to define the counties you want to exclude: **[finance.Fin_2002] = –99]**.

g   Click the Legend tab and check Show symbol for excluded data. Choose a 30% gray symbol and label the class **No Data**. Click OK.

h   Look at the histogram (graph) in the Classification dialog.

The histogram shows the number of jobs on the x-axis (bottom) and the number of counties on the y-axis (side). The gray bars show how many counties have a particular number of jobs. The blue vertical lines show where the data is broken into classes.

i   Change the classification method to Quantile and notice how the class breaks in the histogram change.

The "Quantile" method groups the counties so that each class has approximately the same number of counties. California has 58 counties in total with no data values data for two of them, leaving 56 counties with employment data for 2002. Therefore, each of the five classes has approximately 11 counties.

j   Click OK in the Classification dialog.

The Symbology tab updates the list of five colored symbols, ranges, and labels.

k   Change the labels for the five classes to the following phrases:

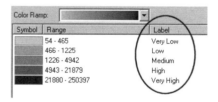

l   Click OK to update the map.

m   Save your map document according to your teacher's instructions. We recommend that you save it in your Mod6 folder and rename it **ABC_Adv6** where ABC are your initials.

n   Look at the Finance jobs map. Where are the counties with a high number of finance jobs? You will refer back to this map after you've downloaded average wage data for the California counties.

PHOTOCOPY

Step 5   Download average wage data for California counties

Now you will go back to the BEA Web site and retrieve data on average wages for California counties. Then you will add the data to ArcMap and the California geodatabase.

a   Restore your Web browser and go to the BEA economic accounts page at *www.bea.doc.gov.*

b   Under Regional, click **State and Local Personal Income**.

c   Under Local area annual estimates, click the link titled **Interactive tables - view or download specific data**.

d   Under Step 1, choose **Wage and salary summary estimates (CA34)**.

e   Under Step 2, click **Average wage per job**. Then choose **California** for the state and leave **2002** and **2001** for the years.

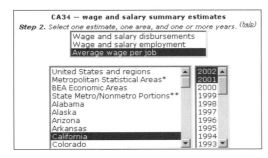

f   Click the Display button to see a report for all of California.

g   Click the Download Data button. Save the file as a comma separated values file (.csv) and rename it **ABC_wage.csv**, where **ABC** are your initials. Save it in a folder your teacher specifies.

Step 6   Prepare average wage data for ArcGIS

a   Make the following edits to the file you just saved:

   • Replace the long title "Average wage per job <sup1/</sup>" in the first line of the file with the word **"Wage"**. (Keep the quotes.)

   • Replace the double quotes and the years "2001" and "2002" with **AvWage2001** and **AvWage2002**. (Omit the quotes here.)

   • Replace any "(D)" with **"–99"**. (Note: Your data may not have any "(D)" values.)

b   Delete everything after the Yuba County data.

c   Save the file and exit the text editor program.

Step 7   Add average wage data to ArcMap

a   Repeat steps 3c–3i above to add the new **ABC_wage.csv** table to your map document and export it to the California geodatabase. Name the new table **wage**.

b   Repeat steps 3j–3l above to join the wage table to the CA Counties table. (Note: You will need to use ca_counties.FIPS as the first join field, wage as the table, and FIPS as the second join field.)

c   Click the Display tab in the table of contents. Then copy and paste the CA Counties layer.

Step 8   Thematically map California average wage data

a   Double-click the new CA Counties layer to open the Layer Properties dialog. Rename the layer **Average Wages** in the General tab.

b   Click the Symbology tab. Symbolize the layer with graduated colors using wage.AvWage2002 for the value field. Choose a color ramp that is different from the ones used for the other layers.

c   Classify the data using the Defined Interval method. Type **10000** for the Interval Size and click OK in the Classification dialog.

This places the class breaks at even intervals of ten thousand (30,000, 40,000, etc.). Because the values represent dollars, you will format the labels as currency.

d   Right-click anywhere in the white space in the symbol box and choose Format Labels. Choose the Currency format and click OK.

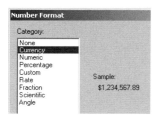

e   Click OK in the Layer Properties dialog.

 f   Save your map document.

Step 9   Query the data

With population data, number of finance jobs data, and average wage data, you will perform a query to narrow down the counties where you could live in California. For the purpose of this exercise, you will perform an example query with detailed instructions. In step 10, you will perform your own query, using the preferences you made on your worksheet.

a   Look at the different layers by turning them on and off. Try to visually identify a couple of counties that appear to meet the criteria you identified earlier.

Now you will perform a query. For this example query, assume that the number of people aged 18–29 years old is an important criterion, followed by number of finance jobs, and then average wage. Because this is countywide data, it's important to take that into consideration when you develop the query.

b   Turn off all layers except CA Counties.

c   Click Selection on the Main menu and choose Select by Attributes.

d   Choose CA Counties from the Layer list at the top of the Select by Attributes dialog. Choose "Create a new selection" for the method.

e   For Fields, scroll down and to the right and double-click **ca_counties.AGE_18_29**.

The 18- to 29-year-old population in California ranges from 203 in one county to 1,988,668 in another. Because you want to eliminate counties with a low population in this age group, you are going to query for counties with an 18- to 29-year-old population size of 10,000 or more.

f   Click the greater than or equal to button, and then type **10000** in the query box.

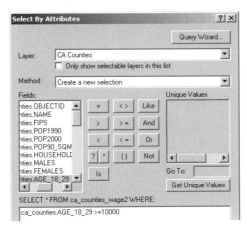

g   Verify your query and then click Apply. Move the Select By Attributes dialog so you can see all the counties in California with an 18- to 29-year-old population of 10,000 or more.

h   Click the Selection tab below the table of contents. Take note of how many counties are selected (38).

Now you will search for counties with many finance jobs as well as many people aged 18–29. This might narrow the list of places you would consider living, because some of the counties selected in your first query might have few finance jobs. You are going to query for counties with 5,000 or more finance jobs.

i   In the Select By Attributes dialog, change Method to "Select from current selection."

j   Click the Clear button.

k   Create the following new query statement: **finance.Fin_2002 >= 5000**. Verify the query statement and then click Apply.

The number of selected counties has decreased. There are now 22 selected counties that have a population aged 18–29 years of 10,000 or greater and more than 5,000 finance jobs available. Now you will add average wage data to your search. These numbers reflect annual salary in dollars. In this example, you will query for counties with an average salary of $35,000 or more.

l   Clear the current query statement and make a new one where AvWage2002 is greater than or equal to 35000. Select these features from the current selection.

The number of selected counties has decreased significantly.

m   Close the Select by Attributes dialog.

n   Click the Display tab. Open the CA Counties layer attribute table and click the Selected button at the bottom of the table.

The 13 counties that meet all the criteria you used are highlighted in blue.

o   Look at the map and ask yourself the following questions: Are the selected counties the ones you identified in step 9a? Do you know anything else about those counties that would make them an attractive place to live?

Step 10   Perform your own query and map your observations

a   Use the instructions in step 9 and the worksheet you completed at the beginning of the GIS Investigation to create your own query of CA counties. Be sure to record the criteria you used for your query. You will need to include that information in your layout.

b   After you've performed your query, prepare your map to look the way you want it to look in a layout.

c   Create a layout and title it **Live, Work and Play - Your Name**. For example, *Live, Work and Play - Joanna Smith* would be an appropriate title for a map.

Include the following items in your layout:
- Title
- Legend
- List of criteria used for query
- The Attributes of CA Counties table, displaying the results of your query
- Map scale
- North arrow
- Author of map
- Date of map creation

> *Hint: Add layout elements such as the title from the Insert menu. Add the table to the layout by opening it and then choosing Add Table to Layout from the Options button. If you experience difficulty creating the layout, refer to ArcGIS Desktop Help. Look at the topics under ArcMap, Laying out and printing maps.*

d   Print your map.

e   Save your final map document.

Human/Environment Interaction

Physical processes influence patterns of human activity, just as human activities have an effect on the environment.

Water World: A global perspective

Students will investigate and explore changes that might occur to the surface of the earth if the major ice sheets of Antarctica melted. They will begin their exploration at the South Pole by studying images and information relating to the physical geography of Antarctica. Proceeding according to the steps of the geographic inquiry method, they will consider the consequences of projected changes on human structures, both physical and political. The final assessment will call for students to create an action plan for a major city of the world that would be flooded in the event of catastrophic polar meltdown.

In the Eye of the Storm: A regional case study of Latin America and the impact of Hurricane Mitch

Students will study the destructive force of Hurricane Mitch, the deadliest storm of the twentieth century, which devastated Central America. Students will analyze information about the storm itself, compare the region before, during, and after the storm, and reflect on the impact it had on the society it ravaged. The students will end their study by developing a hurricane relief/rebuilding plan for the region, and then compare their theories with the plans that were actually made.

Data Disaster: An advanced investigation

Students will act as data detectives in this scenario-based investigation. Due to a recent storm, the main computer was flooded at the International Wildlife Conservancy (a fictitious organization). They lost all their metadata for their ecozones. The data and maps have survived, but they are unsure what each zone represents. They know that the zones represent areas that are critically challenged and are in danger. The students must analyze the data that is available to them and create new metadata for the data.

Water World
A global perspective

Lesson overview

Students will investigate and explore changes that might occur to the surface of the earth if the major ice sheets of Antarctica melted. They will begin their exploration at the South Pole by studying images and information relating to the physical geography of Antarctica. Proceeding according to the steps of the geographic inquiry method, they will consider the consequences of projected changes on human structures, both physical and political. The final assessment will call for students to create an action plan for a major city of the world that would be flooded in the event of catastrophic polar meltdown.

Estimated time   Two to three 45-minute class periods

Materials   ✔ Student handouts from this lesson to be copied:
- GIS Investigation sheets (pages 403 to 411)
- Student answer sheets (pages 412 to 415)
- Assessment(s) (pages 416 to 420)

Standards and objectives

National geography standards

	GEOGRAPHY STANDARD	MIDDLE SCHOOL	HIGH SCHOOL
1	How to use maps and other geographic representations, tools, and technologies to acquire, process, and report information from a spatial perspective	The student understands the relative advantages and disadvantages of using maps, globes, aerial and other photographs, satellite-produced images, and models to solve geographic problems.	The student understands how to use geographic representations and tools to analyze, explain, and solve geographic problems.
4	The physical and human characteristics of places	The student understands how different physical processes shape places.	The student understands the changing physical and human characteristics of places.
11	The patterns and networks of economic interdependence on Earth's surface	The student understands the basis for global interdependence and the reasons for the spatial patterns of economic activities.	The student understands the increasing economic interdependence of the world's countries.
18	How to apply geography to interpret the present and plan for the future	The student understands how the interaction of physical and human systems may shape present and future conditions on Earth.	The student understands how to use geographic knowledge, skills, and perspectives to analyze problems and make decisions.

Standards and objectives (continued)

Objectives

The student is able to:

- Manipulate map projections using GIS technology.
- Compare locations on a map to photos and satellite imagery using GIS.
- Analyze the impact on major human systems (such as transportation networks) if various parts of the Antarctic ice sheet melted causing a significant rise in sea level.
- Predict how such a catastrophe might change the nature of cities and societies around the world, and propose ways to minimize danger and hardship.

GIS skills and tools

 Add layers to the map

 View a hyperlink to a photograph or image

     Zoom in and out of the map

 Identify a feature to learn more about it

- Change the map projection for a data frame
- Activate a second data frame

For more on geographic inquiry and these steps, see Geographic Inquiry and GIS (pages xxiii to xxv).

Teacher notes

Lesson introduction

Begin the lesson with a discussion of Antarctica. Use these questions as a guide.

- What is the climate of Antarctica like?
- What does the place look like?
- Are there any human settlements there?

After a brief discussion introducing the subject, share with your students some of the work that scientists have been doing in the region. You may want to have them explore some of the Internet sites associated with this lesson on the book's Web site. These resources provide information on the latest research into snow and ice melt in Antarctica and its impact on mean sea level. In the GIS investigation, students will look at visual representations of the rising sea level and analyze current data about cities, transportation networks, and other important human structures. They will be challenged in the closing assessment to save a major city from the rising floodwaters by using the data they gather in the course of their investigation.

Student activity

 Before completing this lesson with students, we recommend that you complete it as well. Doing so will allow you to modify the activity to accommodate the specific needs of your students.

After the initial discussion, have your students work on the computer component of the lesson. Ideally, each student should be at an individual computer, but the lesson can be modified to accommodate a variety of instructional settings.

Distribute the GIS Investigation sheets to the students. This investigation has two main parts: observing Antarctica, and analyzing the earth at various sea levels. The investigation sheets will provide students with detailed instructions for their investigations. As they work through the steps, they will explore how the changes in sea level could affect important human structures. They will begin to make hypotheses that they will attempt to prove in their final assessment.

The investigation sheets include questions to help students focus on key concepts. Some questions will have specific answers; others require creative thought.

Things to look for while the students are working on this activity:

- Are the students using a variety of tools?
- Are the students answering the questions as they work through the procedure?
- Are the students beginning to ask their own questions of the data they are observing?

Conclusion

Before beginning the assessment, briefly discuss the initial observations your students have made. Review their findings on the areas most affected by the changing sea level. How similar or different were these observations?

Assessment

Middle school: Highlights skills appropriate to grades 5 through 8

The middle school assessment will provide students a list of cities that will be greatly affected by the 50-meter rise in sea level. Students will work in teams of three. They will select one city from the list (or you can assign each group a city) and create an action plan for relocating the city and its resources. They will focus on basic modes of transportation such as major roads and railways. Each group may also address how the change in sea level will affect industry and commerce.

Within each group, students should assign themselves to the following roles:

* Team leader—organizes the group and assists in creating the final product
* Cartographer—focuses on manipulating the GIS and printing of maps (if necessary) for the final product
* Data expert—focuses on research and determines which data is best to use

Each student group will work together as a team to create the final product. Team presentations can be made at your discretion. Some suggestions include oral presentation, a formal written report with printed maps, or a "science fair"– type poster presentation.

High School: Highlights skills appropriate to grades 9 through 12

The high school assessment will provide students a list of 10 major cities that will be greatly affected by the 50-meter rise in sea level. Students will work in teams of three. They will select one city from the list and create an action plan to relocate the city, shift the roles that city plays in the national and international scheme of things to another city, adapt the city to its new environment, or choose another option. They must take into account transportation, utilities, economics, and politics in their solutions.

Within each group, students should assign themselves to the following roles:

* Team leader—organizes the group and assists in creating the final product
* Cartographer—focuses on manipulating the GIS and printing of maps (if necessary) for the final product
* Data expert—focuses on research and determines which data is best to use

Each student group will work together as a team to create the final product. Team presentations are at your discretion. Some suggestions include oral presentation, a formal written report with printed maps, or a "science fair"–type poster presentation.

Extensions

* Have students research other details for the affected cities and countries by obtaining additional data from outside sources such as the Internet.
* Have students create action plans for an entire country by having small groups research specific details. For example, one group could focus on political boundary issues, another on transportation issues, another on export and trade. Each group would need to present its findings to the whole class (or country) and then the group must decide as a whole team how the different plans will work together.
* Identify a city along the Mississippi River that frequently floods. Research and analyze its flood disaster plans. How could those plans be improved?
* Check out the Resources by Module section of this book's companion Web site (*www.esri.com/mappingourworld*) for print, media, and Internet resources on the topic of Antarctica.

NAME _____   DATE _____

Water World
A global perspective

ACQUIRE

ASK

EXPLORE

ACT

ANALYZE

Answer all questions on the student answer sheet handout

Imagine that the year is 2100. Scientists have determined that the rapidly warming climate of the earth will cause the ice sheets of Antarctica to break apart and melt at a much faster rate than was predicted a hundred years earlier. You and your GIS investigation team are presented with the challenge of studying the impact this change will have on the planet.

In part 1 of this GIS investigation, you will explore and compare maps of Antarctica offering different views of the continent. You will also investigate specific Antarctic sites to learn more about the continent. In part 2, you will use world maps to investigate changes in ocean levels associated with melting of the Antarctic ice sheets.

Part 1: A South Pole point of view

Step 1 Start ArcMap

a Double-click the ArcMap icon on your computer's desktop.

b If the ArcMap start-up dialog appears, click **An existing map** and click OK. Then go to step 2b.

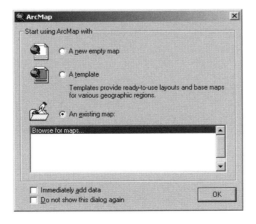

Step 2 **Open the Global7.mxd file**

a In this exercise, a map document has been created for you. To open it, go to the File menu and choose **Open**.

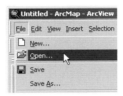

b Navigate to the module folder (**C:\MapWorld9\Mod7**) and choose **Global7.mxd** (or **Global7**) from the list.

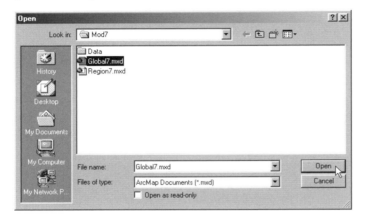

c Click Open.

When the map document opens, you see a map with three layers turned on (Latitude & Longitude, Continents, Ocean). The check mark next to the layer name tells you the layer is turned on and visible in the map.

Step 3 **Look at Antarctica**

a Use the legend to locate the continent of Antarctica.

Take a look at where the melted water is coming from. Scientists believe that the first area to melt will be the Western Ice Shelf of Antarctica. The western part of Antarctica is considerably smaller than the eastern portion. It lies on the west of the Transantarctic Mountain Range, which is basically all the land to the west of the prime meridian. The prime meridian is the line that runs north-south on the map.

The map is now viewed in a geographic map projection. Because this is a flat projection of the spherical earth, some parts of the map are skewed (out of shape). This affects either size or shape of features (such as landforms), distance, direction, or all four.

? *Do you think this map gives you a realistic representation of Antarctica? Explain.*

b   At the top of the table of contents, right-click South Pole and click Properties.

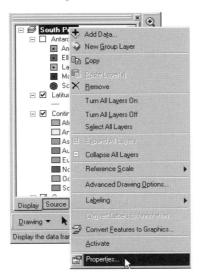

Data Frame Properties opens.

c   Click the Coordinate System tab.

The current coordinate system is GCS_WGS_1984 (Geographic Coordinate System, World Geodetic System of 1984).

d   Under Select a coordinate system, click the plus signs next to the following folders to expand them: Predefined, Projected Coordinate Systems, and World. Scroll down the list of world projections and click **Mercator (world)**.

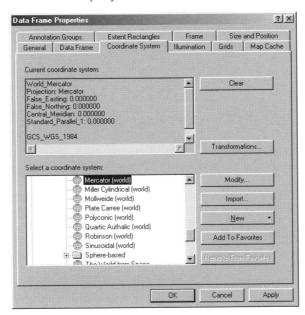

e   Click OK, then click the Full Extent button to see all the continents.

The map looks rectangular.

?   *Does this projection give you a better view of the region around the South Pole? Why or why not?*

f   Repeat steps b through e to change the map projection to each one listed below:

- Aitoff (world)
- Bonne (world)
- Cylindrical equal area (world)
- Equidistant conic (world)
- Mollweide (world)
- Polyconic (world)
- Robinson (world)
- Sinusoidal (world)
- Van der Grinten I (world)

?   *Do any of these projections work well for viewing Antarctica?*

Step 4   View the South Pole

As you reviewed the various projections, you may have thought that none of them would give you a good perspective of the South Pole, or you may have wanted to flip the map upside down or change its center. The recommended projection to view either pole is the polar orthographic projection. This is a picture of the earth as though you were looking at it from space. In order to see the South Pole, you will choose a projection that centers the map on the South Pole.

a   Open Data Frame Properties. Make sure that the Coordinate Systems tab is selected.

b Expand the following folders: Predefined, Projected Coordinate Systems, and Polar. Scroll down the list of polar projections and click **South Pole Orthographic**.

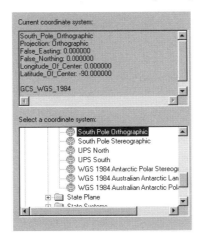

 c Click OK, then click the Full Extent button.

Step 5   Picture Antarctica

a In the table of contents, click the check box next to Antarctic Sites to display the point layer.

b In the table of contents, right-click Antarctic Sites and click Zoom To Layer to get a closer view of Antarctica.

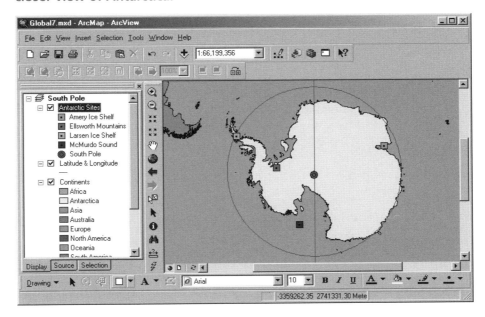

Each point on the map is hyperlinked to an image. Like an Internet link, if you click on a hyperlink in ArcMap, you are taken to additional information.

 c Click the Hyperlink tool. When you move the cursor over the map, it turns into a lightning bolt. Notice that when the Hyperlink tool is selected, a blue dot displays at the center of each Antarctic site, indicating that it has an active hyperlink.

d   Move the tip of the lightning bolt over the South Pole until the cursor changes to a hand, then click. A photograph of the South Pole opens in the default Windows image viewer.

e   If the photograph appears cut off you may need to resize the image or the image viewer.

f   Read the caption at the bottom of the photograph to learn more about the picture.

g   Minimize the image viewer.

h   Find out more about what Antarctica looks like by clicking on the other points on the map. Read the caption at the bottom of each image. Hint: You can stretch or maximize the images and the image viewer to see the images and captions at a larger size.

i   In the table of contents, click the check box next to Antarctic Sites to turn it off.

j   Ask your teacher if you should stop here and save this ArcMap document. Follow your teacher's instructions on how to rename the map document and where to save it. Exit ArcMap by choosing Exit from the File menu.

If you do not need to save the map document and exit ArcMap, proceed to part 2.

?   *On the answer sheet, write the new name you gave the map document and where you saved it.*

Part 2: Just add water

Antarctica has two major ice sheets: the western and the eastern. The western sheet is smaller than the eastern and covers Antarctica from the Transantarctic Mountains, which run from the South Pole westward. The eastern sheet is on the opposite side of the mountain range and includes the majority of the continent. Both of these enormous sheets of ice are moving from the continental center toward the ocean. For example, as the western ice sheet moves into the ocean it forms the Ross and Ronne Ice Shelves that float on top of the ocean. It is at this point where the ice begins to break apart and melt.

Now you will examine what might happen to the water levels of the oceans if part of these ice sheets were to melt.

Step 1   Activate the Water World data frame

a   Make sure the Global7.mxd (or Global7) map document is open. If you renamed this document in part 1, open it using the name you recorded. Otherwise, continue with step 1b.

b   Click the minus sign next to the South Pole data frame to collapse it in the table of contents.

c   Click the plus sign next to the Water World data frame to expand its contents.

d   In the table of contents, right-click Water World and click Activate.

The Water World map displays on your screen. You see a world map in Robinson projection. This projection is commonly used for world maps.

You see the Country Outlines from the year 2004. You also see a layer named "20,000 Years Ago." This layer shows an elevation map of the earth as scientists believe it looked 20,000 years ago. At that time, sea level was 400 feet lower than it is today.

e Stretch the ArcMap window by dragging the lower right corner with your mouse.

f What significant differences do you see between today's country outlines and the elevation map of 20,000 years ago? List at least three.

Step 2 Analyze global sea levels if Antarctic ice sheets melted

If the western ice sheet melted, scientists predict that the oceans would rise about 5 meters. If the eastern ice sheet melted, sea level would rise about 50 meters. If all the ice at the South Pole melted, including all the ice shelves and glaciers, sea level would increase by 73 meters.

One by one, you will turn on the layers: Today, Plus 5 Meters, Plus 50 Meters, and Antarctic Total Thaw and make observations in the table on your answer sheet. Remember: ArcMap draws the layers starting with the bottom of the table of contents and moves upward. Therefore, a layer that's turned on at the top of the table of contents will "draw over" a layer below it.

a Turn the layers on and off and compare each change in sea level. Record your general observations of each layer in the table on the answer sheet.

b Turn off all layers except Plus 50 Meters.

Step 3 View changes in water levels

a Click the Add Data button.

b Navigate to the module 7 layers files folder (**C:\MapWorld9\Mod7\Data \LayerFiles**). Click **Lakes.lyr** and hold the Ctrl key down and click **Rivers.lyr**. Click Add.

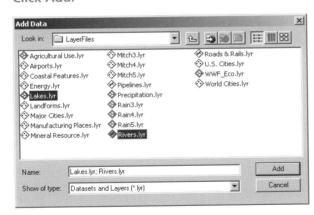

Both layers appear in the table of contents.

c Use the Zoom In tool to drag a box around South America.

(1) *What kinds of changes do you see in the rivers and lakes? Provide a specific example.*

(2) *With a sea level increase of 50 meters, what kinds of consequences do you foresee for the major river ecosystems of South America? Provide a specific example.*

(3) *There are several locations around the globe that are on the interior of land-masses and are below sea level. One of them is in South America. Hypothesize how these low-lying areas were formed.*

d Click the Full Extent button to see the entire world on the map.

e Turn off the Rivers and Lakes layers.

Step 4 View changes in political boundaries

The oceans of the world form the coastlines of many nations. In this step, you will focus on coastal boundaries and how the 50-meter rise would affect political boundaries.

a Click the Add Data Button.

b Navigate to the module 7 layer files folder (**C:\MapWorld9\Mod7\Data\LayerFiles**) and double-click **Major Cities.lyr**.

c Click the Zoom In tool. Zoom to focus on the Middle East.

d Turn on the Major Cities layer.

The dark blue dots that represent city locations appear on the map. Note how some of them are now in the ocean or on the edge of the ocean.

e Turn on Country Outlines so you can view current country boundaries.

f Take note of significant changes in the amount of land now remaining in the countries of that region. Predict possible consequences to the societies that live in those areas (political disputes, trade and economic issues, transportation problems, and so on) and record them in the table on your answer sheet. Use the Zoom, Pan, and Identify tools to help you complete the table.

Note: Refer to the ArcMap Toolbar Quick Reference for a brief explanation of the Zoom and Pan tools.

g Click the Full Extent button. Repeat the process of zooming and identifying potential consequences of the rising sea level for the other major regions of the world.

 Record your results in the table on the answer sheet.

 h Click the Full Extent button.

Step 5 Look for additional data to explore

You have now only begun to scratch the surface of understanding a world with a significant rise in sea level.

Based on your previous observations, list other possible layers of data you would like to analyze to study the impact of this phenomenon. In your final assessment, you will have the opportunity to explore many other datasets; this list will help to guide you in further explorations.

Step 6 Exit ArcMap

In this exercise, you used ArcMap to investigate the continent of Antarctica. You also explored and analyzed the potential effect of thawing the Antarctic ice sheets on the global environment.

a Ask your teacher for instructions on where to save this ArcMap map document and on how to rename the map document.

b If you are not going to save the map document, exit ArcMap by choosing Exit from the File menu. When asked if you want to save changes to Global7.mxd (or Global7), click No.

NAME _____ DATE _____

Student answer sheet

Module 7
Human/Environment Interaction

Global perspective: Water World

Part 1: A South Pole point of view

Step 3   Look at Antarctica

a   Do you think this map gives you a realistic representation of Antarctica? Explain your answer.

_____

_____

_____

_____

_____

e   Does this projection give you a better view of the region around the South Pole? Why or why not?

_____

f   Do any of these projections work well for viewing Antarctica? _____

Step 5   Picture Antarctica

j   Write the new name you gave the map document and where you saved it.

_____   _____
(Name of map document.   **(Navigation path to where map document is saved.**
For example: ABC_Global7.mxd)   **For example: C:\Student\ABC)**

Part 2: Just add water

Step 1   Activate the Water World data frame

f   What significant differences do you see between today's country outlines and the elevation map of 20,000 years ago? List at least three.

_____

_____

_____

_____

_____

Step 2   Analyze global sea levels if Antarctic ice sheets melted

a   Record your general observations of each layer in the table below.

SEA LEVEL	OBSERVATIONS
Today	
Plus 5 meters	
Plus 50 meters	
Total Thaw (plus 73 meters)	

Step 3   View changes in water levels

c-1   What kinds of changes do you see in the rivers and lakes? Provide a specific example.

_____

_____

_____

_____

_____

c-2   With a sea level increase of 50 meters, what kinds of consequences do you foresee for the major river ecosystems of South America? Provide a specific example.

_____

_____

_____

_____

_____

c-3   There are several locations around the globe that are on the interior of landmasses and are below sea level. One of them is in South America. Hypothesize how these low-lying areas were formed.

_____

_____

_____

_____

_____

_____

_____

_____

_____

_____

_____

Step 4   View changes in political boundaries

f, g   Record your results in the table below.

REGION	COUNTRIES/AREAS AFFECTED	POSSIBLE CONSEQUENCES
Middle East		
Asia		
Europe		
Africa		
Oceania		
North America		
Latin America		

Step 5   Look for additional data to explore

Based on your previous observations, list other possible layers of data you would like to analyze to study the impact of this phenomenon. In your final assessment, you will have the opportunity to explore many other datasets; this list will help to guide you in further explorations.

_____

_____

_____

_____

_____

_____

_____

_____

_____

MODULE 7 • HUMAN/ENVIRONMENT INTERACTION

NAME _____ DATE _____

Water World
Middle school assessment

You and your teammates have been selected to be part of an elite team of GIS experts who will determine the fate of one of the following six world cities. Over the next 50 years, the rise in sea level up to 50 meters will affect these cities.

List of world cities affected by 50-meter rise in sea level:

San Francisco, USA	Miami, USA
London, England	Calcutta, India
Tokyo, Japan	Houston, USA

As part of your task, select a city and develop an action plan for relocating the city and its resources. The plan must take into account the following factors:

Major roads	Ocean ports
Railroads	Utilities
Airports	Relocation of people

Your Project Director (teacher) will provide you with a detailed listing of the available data you can use. You will add this data to the Water World data frame in the Global7 map document you used in the GIS Investigation.

Your action plan must include each of the following:

- Time line—It describes the various phases of your plan. For example, one five-year phase might include relocating people, while another includes relocating specific businesses.
- Map—It displays proposed changes and could be a series of maps generated in ArcMap or on paper.
- Data—It will support your suggested changes. This data will come from the GIS Investigation and can be displayed through maps, charts, or tables.
- Written report—This is a written explanation of your plan and can include the time line.

Your Project Director (teacher) will provide you with instructions on how you will present your action plan to the class.

Water World

Assessment rubric

Middle school

STANDARD	EXEMPLARY	MASTERY	INTRODUCTORY	DOES NOT MEET REQUIREMENTS
The student knows and understands the relative advantages and disadvantages of using maps, globes, aerial and other photographs, satellite-produced images, and models to solve geographic problems.	Uses GIS to create a map in an appropriate projection (Ortho-graphic) to display information on the region of the South Pole. Uses GIS to create a digital map illustrating the effects of a 50-meter rise in sea level for the selected city.	Uses GIS to create a map in an appropriate projection (Ortho-graphic) to display information on the region of the South Pole. Uses a GIS to create a paper map illustrating the effects of a 50-meter rise in sea level for the selected city.	Uses GIS to create a map in an appropriate projection (Ortho-graphic) to display information on the region of the South Pole. Cre-ates a map of the selected city, but it does not adequately illus-trate the effects of the 50-meter rise in sea level.	Does not select an appropriate projection for view data on the South Pole, and cannot make dis-tinctions between various projec-tions. Does not include a map of the selected city.
The student knows and understands how differ-ent physical processes shape places.	Creates a clear and concise hypothesis on the impact of a 50-meter rise in sea level on a city. Provides data and maps that support ideas.	Creates a clear and concise hypothesis on the impact of a 50-meter rise in sea level on a city. Provides a map that supports ideas, but does not highlight sup-porting data.	Identifies some factors that will affect a city with a significant rise in sea level, but does not create a formal hypothesis. Provides some data or a basic map.	Identifies some places that will be affected by a rise in sea level, but does not identify what changes could take place.
The student knows and understands the basis for global interdependence and the reasons for the spatial patterns of eco-nomic activities.	Analyzes the effect on trade and transportation routes for a flooded city. Develops an action plan that relocates the city and provides details on plan imple-mentation. The plan uses data and maps to support ideas and includes a time line.	Analyzes the impact on trade and transportation routes for a flooded city. Develops an action plan on how to deal with this change that includes a time line.	Attempts to analyze appropriate data and creates an outline for a plan, but does not provide detail on implementation. Time line is incomplete.	Does not select appropriate datasets (such as transportation) to analyze the issue of spatial pat-terns of economic activities, and therefore has difficulty creating a plan. The plan lacks a time line.
The student knows and understands how the interaction of physical and human systems may shape present and future condi-tions on Earth.	Develops a coherent argument that supports the relocation plan. Uses a variety of data to support the findings and creates addi-tional data and maps that illus-trate changes to the human infrastructure.	Develops a coherent argument that supports the relocation plan. Uses a variety of data to support the findings.	Creates a relocation plan for the selected city, but does not offer a variety of data to support ideas.	Relocates a selected city, but does not formalize a plan on how the change will take place. Uses little or no data to support ideas.

This is a four-point rubric based on the National Standards for Geographic Education. The "Mastery" level meets the target objective for grades 5–8.

NAME _____   DATE _____

Water World
High school assessment

You and your teammates have been selected to be part of an elite team of GIS experts who will determine the fate of one of the following 10 world cities. Over the next 50 years, the rise in sea level up to 50 meters will affect these cities.

List of World Cities affected by 50-meter rise in sea level:

San Francisco, USA	Houston, USA
London, England	Odessa, Ukraine
Tokyo, Japan	Rome, Italy
Miami, USA	Sydney, Australia
Calcutta, India	Buenos Aires, Argentina

As part of your task, select a city and develop an action plan that relocates the city, relocates the roles of the city to another city, adapts the city to its new environment, or chooses another option. The plan must take into account the following factors:

Major roads	Utilities
Railroads	Relocation of people
Airports	Economics and trade relations
Ocean ports and shipping lanes	Agriculture and manufacturing

Your Project Director (teacher) will provide you with a detailed listing of the available data you can use. You will add this data to the Water World data frame in the Global7 map document you used in the GIS Investigation. You may need to consult an atlas or the Internet to research some factors.

Your action plan must include each of the following:

* Time line—It describes the various phases of your plan. For example, one five-year phase might include relocating people, while another includes relocating specific businesses.
* Map—It displays proposed changes and could be a series of maps generated in ArcMap or on paper.
* Data—It will support your suggested changes. This data will come from the GIS Investigation and can be displayed through maps, graphs, or tables.
* Written report—This is a written explanation of your plan and can include the time line.

Your Project Director (teacher) will provide you with instructions on how you will present your action plan to the class.

Water World

Assessment rubric

High school

STANDARD	EXEMPLARY	MASTERY	INTRODUCTORY	DOES NOT MEET REQUIREMENTS
The student knows and understands how to use geographic representations and tools to analyze, explain, and solve geographic problems.	Uses GIS to create a map in an appropriate projection (Orthographic) to display information on the region of the South Pole. Uses GIS to create a digital map illustrating the effects of a 50-meter rise in sea level for the selected city.	Uses GIS to create a map in an appropriate projection (Orthographic) to display information on the region of the South Pole. Uses a GIS to create a paper map illustrating the effects of a 50-meter rise in sea level for the selected city.	Uses GIS to create a map in an appropriate projection (Orthographic) to display information on the region of the South Pole. Creates a map of the selected city, but it does not adequately illustrate the effects of the 50-meter rise in sea level.	Has difficulty projecting maps using GIS, and does not show an understanding of variations in projections. Does not include a map of the selected city.
The student knows and understands the changing physical and human characteristics of places.	Clearly defines how the characteristics of a city could change, should a 50-meter rise in sea level occur. Includes data from the GIS Investigation to create an original map to illustrate the changes.	Clearly defines how the characteristics of a city could change, should a 50-meter rise in sea level occur. Includes data from the GIS Investigation.	Defines how some characteristics of a city change, should a 50-meter rise in sea level occur. Provides little evidence from the data.	Attempts to define how some characteristics of a city change with a 50-meter rise in sea level. Does not provide any evidence from the data.
The student knows and understands the increasing economic interdependence of the world's countries.	Provides specific examples of how the loss of the selected city would affect the global marketplace. Provides an economic action plan to prevent this loss from occurring.	Identifies how the global economic infrastructure will be altered with a 50-meter rise in sea level, using the perspective of the selected city.	Identifies how the selected city will be affected economically, but does not make a connection to the global marketplace.	Attempts to identify how the selected city will be affected economically, but does not make a connection to the global marketplace.
The student knows and understands how to use geographic knowledge, skills, and perspectives to analyze problems and make decisions.	Develops a detailed action plan to sustain the selected city in the event of a 50-meter rise in sea level. Creates original maps with a GIS that uses a variety of data sources to support the ideas in the plan.	Develops a detailed action plan to sustain the selected city in the event of a 50-meter rise in sea level. Uses a variety of data sources to support the ideas in their plan.	Develops a plan to sustain the selected city in the event of a 50-meter rise in sea level. Provides little or no data to support ideas.	Creates an outline of a plan to sustain the selected city in the event of a 50-meter rise in sea level. Provides little or no data to support ideas.

This is a four-point rubric based on the National Standards for Geographic Education. The "Mastery" level meets the target objective for grades 9–12.

Student assessment handouts data sheet

The layers below can be found in the World7.mdb geodatabase in your module 7 Data folder (C:\MapWorld9\Mod7\Data\World7.mdb).

FEATURE CLASS NAME	DESCRIPTION OF DATA
roads_rail	Lines that represent main and smaller roads and railroads for the world.
w_cities	Points that represent cities that are major population centers of the world and includes whether a city is a shipping port.
us_cities	Points that represent a detailed listing of U.S. cities.
airports	Points representing the world's airports.
energy	Points that represent major power plants for the world and includes type of energy source (atomic, thermal, and so on).
pipelines	Lines that represent major oil and gas pipelines throughout the world.
manufact_plc	Points that list major manufacturing places around the world.
mineral_res	Points of mineral resource mining sites around the world.
rivers	Lines that represent the major rivers of the world.
lakes	Polygons that represent the major lakes of the world.

In the Eye of the Storm
A regional case study of Latin America and the impact of Hurricane Mitch

Lesson overview

Students will study Hurricane Mitch, the deadliest storm of the twentieth century, and the havoc it wreaked on several Central American countries. They will analyze information about the storm itself, compare the region before, during, and after the storm, and consider the consequences of such a disaster for the society it ravaged. The lesson will conclude with the development of a hurricane relief and rebuilding plan for the region, and comparison of student ideas with the plans that were actually made.

Estimated time　Three 45-minute class periods

Materials　✔ Student handouts from this lesson to be photocopied:
- GIS Investigation sheets (pages 425 to 436)
- Student answer sheets (pages 437 to 442)
- Assessment(s) (pages 443 to 448)

Standards and objectives

National geography standards

GEOGRAPHY STANDARD	MIDDLE SCHOOL	HIGH SCHOOL
1　How to use maps and other geographic representations, tools, and technologies to acquire, process, and report information from a spatial perspective	The student understands the characteristics, functions, and applications of maps, globes, aerial and other photographs, satellite-produced images, and models to solve geographic problems.	The student understands how to use technologies to represent and interpret Earth's physical and human systems.
7　The physical processes that shape the patterns of Earth's surface	The student understands how to predict the consequences of physical processes on Earth's surface.	The student understands the spatial variation in the consequences of physical processes across Earth's surface.
15　How physical systems affect human systems	The student understands how natural hazards affect human activities.	The student understands strategies of response to constraints and stresses placed on human systems by the physical environment, and how humans perceive and react to natural hazards.
18　How to apply geography to interpret the present and plan for the future	The student understands how to apply the geographic point of view to solve social and environmental problems by making geographically informed decisions.	The student understands how to use geographic knowledge, skills, and perspectives to analyze problems and make decisions.

Standards and objectives (continued)

Objectives

The student is able to:

- Analyze human and physical characteristics of the region of Central America.
- Follow the development and impact of Hurricane Mitch on Central American countries.
- Compare satellite imagery of Hurricane Mitch to ground tracking observations.
- Develop a disaster relief plan for the areas most heavily affected by Hurricane Mitch.

GIS skills and tools

   Add layer files and images to a map

       Zoom in or out on the map

   Pan the map

   Identify the attributes of a feature

   Measure distance on a map

- Zoom to the geographic extent of a layer
- Label features including overlapping labels
- Turn labels on and off
- Change a layer's symbol
- Activate and expand a second data frame
- Open an attribute table
- Sort a field in a table in ascending order
- Show only selected records in a table
- Show all records in a table
- Rearrange layers in the table of contents
- Hide or show a layer's legend in the table of contents

For more on geographic inquiry and these steps, see Geographic Inquiry and GIS (pages xxiii to xxv).

Teacher notes

Lesson introduction

First, discuss the weather hazards of your own region. Use the following questions as a guide:

- What weather hazards are specific to our hometown or region?
- When do they typically occur? (Year-round, or in a particular season?)
- What are some characteristics of these phenomena?
- How do you prepare for one? (You may want to review school procedures, or have the students share plans they have from home.)

After the discussion, tell your class that in this GIS Investigation, they will be studying the impact of Hurricane Mitch on a large part of Central America. They will explore characteristics of the storm, including how it developed, and how it changed the physical and human characteristics of the region. In the final assessment, they will be part of a special team developing an action plan for dealing with the devastation the storm caused.

Student activity

 Before completing this lesson with students, we recommend that you complete it yourself. Doing so will allow you to modify the activity to suit the specific needs of your students.

After the initial discussion, have the students work on the computer component of the lesson. Ideally, each student should be at an individual computer, but the lesson can be modified to accommodate a variety of instructional settings.

Distribute the GIS Investigation sheets to the students and the "Central America Prior to Hurricane Mitch" handout. This project has two parts: a general investigation of the region, then tracking of the storm and consideration of its impact. The GIS Investigation will provide them with detailed instructions for their ArcMap exploration.

 Teacher Tip: This GIS Investigation is divided into two parts, each appropriate for a 45-minute class period. At the end of part 1, students are instructed to ask you how to rename the map document and where to save it. Because this data will be valuable to the students during the assessment, be sure they record the new name and location on their answer sheet.

In part 2, the students will add image data on Hurricane Mitch and track its path over Central America. Because these image files are large, you may not want students to have individual copies of the module 7 folder. You can have students access the same copy of the data but save individual map documents (e.g., ABC_Region7 where ABC are a student's initials). If students save their map documents periodically during the activity, they will be able to use them to complete the assessment.

In addition to instructions, the handout includes questions to help students focus on key concepts. Some questions will have specific answers, while others require creative thought.

Things to look for while the students are working on this activity:

- Are the students using a variety of tools?
- Are the students answering the questions as they work through the procedure?
- Are the students beginning to ask their own questions of the data they are observing?

Conclusion

Once students have completed the GIS Investigation, have them share their findings, either in small groups or as a class. They should have a basic understanding of the region and the effect of the storm in terms of rainfall amounts, wind speeds, and so forth. Ask your students which parts of the region suffered the most damage and why. To find out more about the storm and its impact on Central America, refer to this book's companion Web site *(www.esri.com/mappingourworld)* for print, media, and Internet resources on Hurricane Mitch. The USGS has an enormous amount of data on the impact of Hurricane Mitch and disaster recovery efforts in the affected countries *(http://mitchnts1.cr.usgs.gov)*. Encourage students to research this topic while completing their assessment. Allow class time for each team to meet and plan how they will complete the assessment.

Assessment

The middle school and high school assessments vary in the specific tasks the students undertake, but the basic design of the assessments is the same.

Assign students to small groups of three to four students per team. Each team will have the following (some roles may be combined):

Within each group, students should assign themselves to the following roles:

- Team leader—organizes the group and takes the lead in creating the final product/presentation.
- Cartographer—creates the maps using GIS.
- Data expert—gathers data sources and determines which data is best to use.
- Multimedia specialist—responsible for creating the multimedia presentation of your choice.

Assign each team to a particular job focus: flood hazards or volcano and landslide hazards. The choices are detailed in the assessment handouts. There is a different assessment handout created for each team type. Also assign each team a particular country to focus on. That way, they will have more depth to their research. The goal for each team is to assess the possible damage and then to develop a disaster relief plan based on their findings. You may modify this assessment into a longer research project or shorten it to a two-night homework assignment.

Extensions

- Have students create a disaster relief plan for your city using a local natural hazard as the springboard. Research disasters in your city or region's past to find out how they have changed the local area.
- Watch excerpts from films such as *The Perfect Storm* or *Twister* and discuss how these fictional storms relate to those in real life.
- Conduct a book study on *Isaac's Storm,* by Erik Larson. Compare the Galveston storm of 1900 to Hurricane Mitch.
- Check out the Resources by Module section of this book's companion Web site *(www.esri.com/mappingourworld)* for print, media, and Internet resources on the topics of Hurricane Mitch, Central America, and tropical storms.

NAME _____   DATE _____

In the Eye of the Storm
A GIS investigation

ACQUIRE

ASK

EXPLORE

ACT

ANALYZE

Answer all questions on the student answer sheet handout

Part 1: The calm before the storm

October 21, 1998

A tropical storm is brewing in the Atlantic Ocean. It began as a tropical wave a few weeks earlier, off the coast of western Africa. Today it is causing some rain and thunderstorms over the Caribbean. Later, the barometric pressure of the system will continue to drop and it will soon be identified as a tropical depression—the beginning of a hurricane. By the time Hurricane Mitch left the Central America region, 9,086 people were dead and 9,191 were declared missing.

Central America consists of the small chain of countries that link the North and South American continents. Explore these countries using ArcView and gather data to complete your chart titled Central America Prior to Hurricane Mitch. The data you are collecting will help you to gain an understanding of the complexity of this region's delicate infrastructure. The steps that follow will guide you in obtaining the information using GIS.

Step 1   Start ArcMap

a   Double-click the ArcMap icon on your computer's desktop.

b   If the ArcMap start-up dialog appears, click **An existing map** and click OK. Then go to step 2b.

MODULE 7 • HUMAN/ENVIRONMENT INTERACTION

Step 2   Open the Region7.mxd file

a   In this exercise, a map document has been created for you. To open it, go to the File menu and choose **Open**.

b   Navigate to the module folder (**C:\MapWorld9\Mod7**) and choose **Region7.mxd** (or **Region7**) from the list.

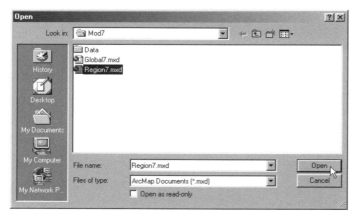

c   Click Open.

The map document opens and you see a world map. Scroll down in the table of contents. A check mark next to the following layers tells you they are displayed: Central America, Continents, and Ocean.

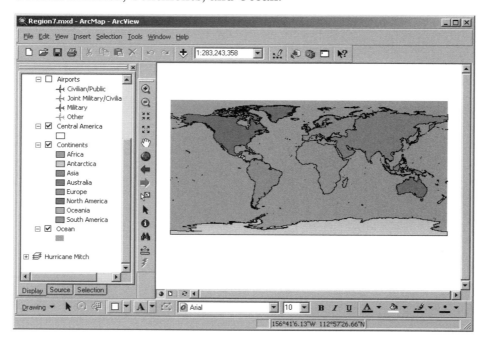

Step 3   Focus on the capital cities of Central America

a   **Right-click on the Central America layer name and click Zoom to Layer. The map immediately centers around the countries of Central America.**

Before you look at the path and effect of Hurricane Mitch, you will collect data about Central America prior to Mitch. Record this information in the table titled "Central America Prior to Hurricane Mitch" on your student answer sheet. The information for the country of Belize has been completed for you as an example of the data you will need to find. First, you will record the country capitals.

b   **Scroll to the top of the table of contents and turn on the Capitals layer by clicking the box next to the layer name.**

You can find the names of the capitals by using the Identify tool or by labeling features. Labeling features is a quick way to get information about a group of features.

c   **In the table of contents, right-click Capitals and click Properties. Click the Labels tab.**

d   **At the top of the Labels tab, click the small white box next to "Label features in this layer." Notice that Name is already chosen as the field to use for labeling.**

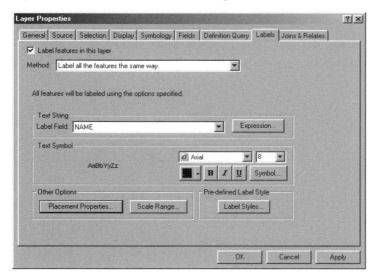

e   Near the bottom of the dialog under Other Options, click the Placement Properties button. Click the Conflict Detection tab. At the bottom of the tab, click the small white box next to "Place overlapping labels."

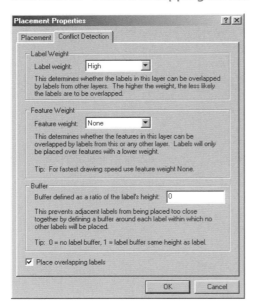

f   Click OK on the Placement Properties dialog and the Layer Properties dialog. The name of each capital city displays on the map.

❓   *g*   Use this information to record the capital cities for each of the countries in Central America. Record them in the Populated Places column.

⚡ *Note: If you don't know the name of a country, use MapTips to identify it.*

h   When you have finished recording the names of each capital, go to the table of contents, right-click Capitals, and click Label Features. The labels disappear from the map.

Step 4   Focus on Central America prior to Hurricane Mitch

There are four layers in the table of contents that are not turned on. Each of these layers provides important data about the transportation network in Central America. This table is a summary of the layers.

Populated Places	Points showing major cities and populated areas of Central America
Roads	Lines representing Central American roads and trails
Railroads	Lines representing Central American railroads
Airports	Points showing airport location and type (civilian, military/civilian, military, other)

a   Turn these layers on one at a time to obtain important data for your handout.

⚡ *Remember: Layers that are at the top of the table of contents will cover up layers that are listed lower. Turn on each layer individually to see it clearly. When using the Identify tool, make sure the appropriate layer is selected in the Layers list in the Identify Results window.*

b Record this data in the Populated Places and Transportation Network columns on your answer sheet.

Now you will add more data needed to complete your handout.

c Click the Add Data button.

d Navigate to the module 7 layer files folder (**C:\MapWorld9\Mod7\Data\LayerFiles**).

e Hold down the Ctrl key and click the following four layer files:

- **Agricultural Use.lyr**
- **Coastal Features.lyr**
- **Landforms.lyr**
- **Precipitation.lyr**

f Click Add. The point layers (Coastal Features and Landforms) appear at the top of the table of contents; the polygon layers (Agricultural Use and Precipitation) appear lower down, below the Airports layer. All four layers are turned off.

The table below summarizes the data in each layer:

Agricultural Use	Polygons showing agricultural use of land in Central America
Coastal Features	Points representing coastal features of Central America
Landforms	Points representing mountains, mountain chains, and volcanoes in Central America
Precipitation	Polygons showing annual precipitation in millimeters (mm) for Central America

g Scroll down and turn on Precipitation.

The precipitation data appears in the map, but you cannot determine the average precipitation for each individual country. You will move the Central America layer and change its legend so you can view each country's average precipitation data.

h Click and drag the Central America layer above Agricultural Use in the table of contents. It covers up Precipitation.

i Click on the yellow symbol for the Central America layer to open the Symbol Selector.

j Click the Hollow symbol.

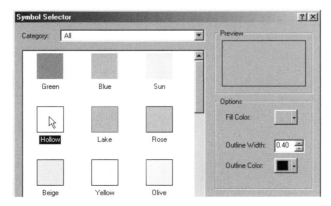

k In the Options panel on the right, increase the Outline Width to 1.5.

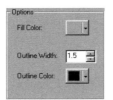

l Click OK.

The map now displays Central American countries as outlines.

? *m* Analyze the precipitation for each country and record the precipitation data in the Average Precipitation column on your answer sheet.

n Turn off Precipitation.

o Turn on each of the remaining layers individually. Use the available data to complete the remaining columns in your table.

p After you complete the data table, scroll to the top of the table of contents and click the minus sign to the left of Snapshot of Central America to collapse the data frame.

? *(1) Which country has the most area devoted to agriculture?*

? *(2) Which country has the most area covered by mountains?*

? *(3) Which country has the most extensive transportation network?*

Step 5 Save the map document and exit ArcMap

In the first part of the GIS Investigation, you analyzed and recorded information about Central America from before Hurricane Mitch.

? *a* Ask your teacher for instructions on where to save this ArcMap map document and on how to rename the map document. If you are not going to save the map document, proceed to step 5b now. Otherwise, record the map document's new name and where you saved it on your answer sheet.

b From the File menu, click Exit.

Part 2: The storm

October 24–26, 1998

In a span of less than two days, Tropical Storm Mitch develops into a category 5 hurricane with winds in excess of 155 mph. Category 5 is the deadliest rating on the Saffir-Simpson Hurricane Potential Damage Scale. Barometric pressure drops to 905 millibars, the lowest pressure ever observed in the Atlantic basin.

Step 1   Open the map document

a   Double-click the ArcMap icon on your computer's desktop.

b   Refer to your answer sheet and determine where Region7.mxd (or Region7) is saved and how you renamed it.

c   Navigate to the location of the saved map document and open it. If you didn't rename or save the map document in part 1, navigate to the module 7 folder (**C:\MapWorld9\Mod7**) and select **Region7.mxd** (or **Region7**).

Step 2   Track Hurricane Mitch

a   In the table of contents, right-click the Hurricane Mitch data frame and click Activate. Click the plus sign to expand its contents.

The Hurricane Mitch data frame displays with the following layers turned on: Latitude & Longitude, Central America, Continents, and Ocean.

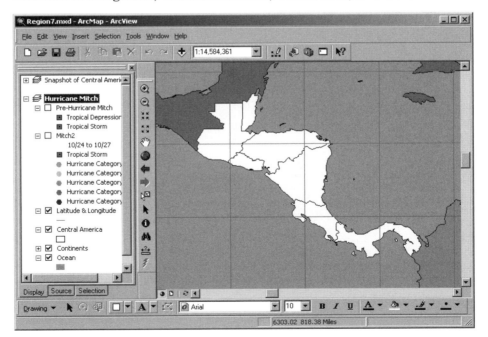

b   Turn on Pre-Hurricane Mitch and Mitch2.

Both of these layers, and similarly named ones to follow, will show the location of the center or eye of the storm. At locations where it was declared a hurricane, the legend reflects the category number. Placing your mouse pointer over a location displays the MapTip for that location. You will now explore the data that these layers represent.

c   **Click the Identify tool. Click the most southeastern dark square that is a tropical storm. An Identify Results window displays with information about Tropical Storm Mitch, including: latitude and longitude, time (Zulu), wind velocity (miles per hour), pressure reading (millibars), and status in reference to the Saffir-Simpson scale. The time is written in this format: Month/Day/Hour (of 24).**

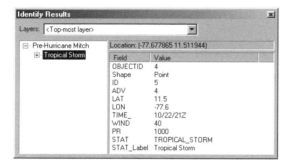

d   **Answer the following questions:**

   (1)   At what time was Tropical Storm Mitch at this location?

   (2)   What does the "z" mean in the time?

   (3)   What was Mitch's wind speed at this location?

e   **Click the location of Hurricane Mitch category 1 on the map. The Identify Results window updates.**

f   **Answer the following questions:**

   (1)   What are the latitude and longitude coordinates for Hurricane Mitch at this location?

   (2)   At what time was Hurricane Mitch at this location?

   (3)   What was Mitch's wind speed at this location?

g   **Click the last mapped location for Hurricane Mitch category 5 before it struck land. The Identify Results window updates.**

h   **Answer the following questions:**

   (1)   At what time was Hurricane Mitch at this location?

   (2)   What was Mitch's wind speed at this location?

Now you will determine how much time it took Mitch to develop from a tropical storm to a category 5 hurricane.

i   **Close the Identify Results window.**

j   **In the table of contents, right-click Mitch2 and click Open Attribute Table. The Attributes of Mitch2 table opens.**

k   Scroll right in the table until you see a field named TIME_ and a field named STAT. Click the TIME_ column heading to select it. The heading depresses like a button and the column turns light blue.

l   Right-click the TIME_ column heading and click Sort Ascending.

m   Click the small gray box to the left of the first record to select it. The first record turns light blue to show it is selected. Notice that this record represents Hurricane Mitch when it was a tropical storm.

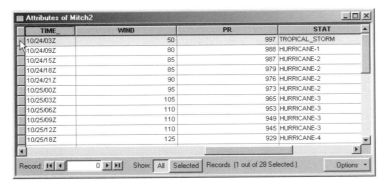

TIME_	WIND	PR	STAT
10/24/03Z	50	997	TROPICAL_STORM
10/24/09Z	80	988	HURRICANE-1
10/24/15Z	85	987	HURRICANE-2
10/24/18Z	85	979	HURRICANE-2
10/24/21Z	90	976	HURRICANE-2
10/25/00Z	95	973	HURRICANE-2
10/25/03Z	105	965	HURRICANE-3
10/25/06Z	110	953	HURRICANE-3
10/25/09Z	110	949	HURRICANE-3
10/25/12Z	110	945	HURRICANE-3
10/25/18Z	125	929	HURRICANE-4

Record: 0   Show: All Selected   Records (1 out of 28 Selected.)   Options ▼

n   Scroll down and locate the record representing the first time Hurricane Mitch became a category 5 hurricane. Hold the Ctrl key down and click the small gray box for the Hurricane-5 record. Both records are highlighted blue in the table and on the map.

Remember: You must click the small gray box to the left of a row in the table in order to select the entire row.

o   Click the Selected button at the bottom of the attribute table. Now you see only the two selected records and they are easier to compare.

p   Use the information in the attribute table to determine how long it took for Tropical Storm Mitch to become a category 5 hurricane. On the answer sheet, write down the times for each event and determine the time difference. Remember, the time is written in this format: Month/Day/Hour (of 24).

q   Click the All button at the bottom of the attribute table to see all the records again.

r   Examine the attribute table further and identify the maximum wind speed. Record this on your answer sheet.

s   Close the Attributes of Mitch2 table.

Step 3   Measure the size of the storm

The National Oceanic and Atmospheric Administration (NOAA), in partnership with the National Aeronautics and Space Administration (NASA), used special storm-tracking satellites to take several high-resolution photographs of Mitch from space. You will view these images and measure the massive size of this storm.

a   Click the Add Data button and navigate to the module 7 images folder (**C:\MapWorld9\Mod7\Data\Images**).

b Click mitch2sat.tif and click Add.

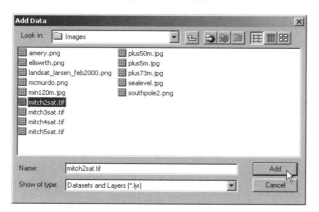

The satellite image now sits underneath your basemap and you cannot see it. In order to see both the satellite image and the storm track, you will rearrange the layers in the table of contents.

c Click and drag the mitch2sat.tif layer so it's just below Mitch2 in the table of contents.

Now you can see the highlighted points of Tropical Storm and Hurricane Category 5. Hurricane Category 5 is almost directly over the eye of the storm. The eye is the center of the cloud mass that looks like a doughnut hole.

You will use the Measure tool to measure several parts of the storm.

d Turn off Mitch2 so you have a better view of the eye of Hurricane Mitch.

e Use the Zoom In and Pan tools so the satellite image fills the map display. Do not zoom in too close or it will be difficult to view the image.

f Click the Measure tool. Your cursor turns into a right-angle ruler with cross hairs.

g Click the left edge of the eye once and move the cursor directly across the diameter of the eye. Double-click when your cursor is at the right edge of the eye.

> *Note: If you accidentally clicked the wrong spot, you can double-click to end the line and simply start over.*

A segment length appears on the bottom left of the ArcMap window.

 Segment: 24.901145 Total: 24.901145 Miles

? *What is the diameter of the eye of Hurricane Mitch?*

? *h* Use the Measure tool to measure the total diameter of the storm at its widest point and the distance of the eye to the coastline of Honduras. Record your answers for Mitch2sat.tif in the table on the answer sheet.

> *Note: You will need to use the Zoom and Pan tools to focus on the area of the storm you need to measure.*

i Turn off mitch2sat.tif.

Now that you have recorded data for mitch2sat.tif, you will follow the same procedures for adding new layers and measuring the storm at other locations.

j   Click the Add Data button. Navigate to the module 7 images folder (**C:\MapWorld9\Mod7\Data\Images**).

k   Hold down the Ctrl key and add the following images: mitch3sat.tif, mitch4sat.tif, and mitch5sat.tif.

l   Turn off all three images. Click the minus sign in front of each image to collapse its legend.

 m   Click the Add Data button again. This time, navigate to the module 7 layer files folder (**C:\MapWorld9\Mod7\Data\LayerFiles**).

n   Hold down the Ctrl key and add the following layer files: Mitch3.lyr, Mitch4.lyr, and Mitch5.lyr.

With six new layers added, it's important to organize your table of contents so you can view the layers easily.

o   In the table of contents, click mitch3sat.tif and drag it directly below Mitch3. Do this for the other satellite images.

p   Turn each set of layers on and off individually. Use the Zoom and Pan tools to see different parts of each layer. Use the Measure tool to collect data on the size of Hurricane Mitch at these different locations.

> *Note: If you don't remember how to use the Measure tool, refer to steps 3f–3g.*

? *q*   Record your measurements and observations in the table on the answer sheet.

Step 4   Analyze rainfall from Hurricane Mitch

Once Hurricane Mitch made landfall, the winds weakened to the point where it was downgraded to a tropical storm. Nonetheless, Mitch still had not shown its worst side. In the days that followed, Mitch poured more than 30 inches of rain in the region. You will take a closer look at the precipitation that fell in the region on October 30 and 31, 1998.

a   Turn off all layers except mitch3sat.tif, Central America, Continents, and Ocean.

 b   Click the Add Data button. Navigate to the module 7 layer files folder (**C:\MapWorld9 \Mod7\Data\LayerFiles**). Add the following layer files to your map: Rain3.lyr, Rain4.lyr, and Rain5.lyr.

c   Turn on Rain3. The rain pattern appears overlaid on top of mitch3sat.tif. Answer the following questions:

?   *(1)*   *What pattern do you notice in the amount of rainfall within the storm?*

?   *(2)*   *Is this a pattern you expected to find? Why or why not?*

    d   In the table of contents, click Rain4 and drag it above the corresponding satellite image, mitch4sat.tif. Do this for the other rain layers.

    e   While turning layers on and off, look at each set of rain and satellite layers. Answer the following questions:

?   *(1)*   *At the Mitch4 location, what was the highest range of rainfall measured?*

?   *(2)*   *Which country received the majority of this heavy rain?*

?   *(3)*   *Describe the difference between the rainfall patterns on October 30 and October 31, 1998.*

?   *(4)*   *What kind of damage do you expect to find with this type of storm? What aspects of the region will be most affected? Elaborate on your answer using your table, Central America Prior to Hurricane Mitch, as a resource.*

Step 5   **Save the map document and exit ArcMap**

In this GIS investigation, you used ArcMap to analyze a large region of Central America and to track Hurricane Mitch as it approached and made landfall. You more than likely have many questions as to the extent of the damage Mitch caused. The assessment will have you take on the role of emergency management personnel. Your job will be to identify those areas where danger is still high as a result of the storm and to develop emergency action plans for affected Central American countries. You will use the data from this investigation to help you in the assessment.

?   *a*   Ask your teacher for instructions on where to save this ArcMap map document and on how to rename the map document. If you already renamed the map document in part 1, save it under that name.

    b   Record the new name of the map document and its location on your answer sheet.

    c   Choose Exit from the File menu.

NAME _____   DATE _____

Student answer sheet
Module 7
Human/Environment Interaction

Regional case study: In the Eye of the Storm

Part 1: The calm before the storm

Step 3   Focus on the capital cities of Central America

g   Record the capital cities for each of the countries in Central America in the Populated Places column in the table on the next two pages.

Step 4   Focus on Central America prior to Hurricane Mitch

b   Record this data in the Populated Places and Transportation Network columns in the table on the next two pages.

m   Analyze the precipitation for each country and record the precipitation data in the Average Precipitation column in the table on the next two pages.

p-1   Which country has the most area devoted to agriculture? _____

p-2   Which country has the most area covered by mountains? _____

p-3   Which country has the most extensive transportation network? _____

Step 5   Save the map document and exit ArcMap

a   Write the new name you gave the map document and where you saved it.

_____   _____
(Name of map document.   **(Navigation path to where map document is saved.**
For example: ABC_Region7.mxd)   **For example: C:\Student\ABC)**

Central America Prior to Hurricane Mitch

COUNTRY	POPULATED PLACES	TRANSPORTATION NETWORK	MAJOR EXPORTS
Belize	Capital: Belmopan Distribution: Throughout the country, but concentrated around the capital	Roads: Sparse road network Railways: none Airports: 1 civilian	Bananas
Guatemala	Capital: Distribution:	Roads: Railways: Airports:	Coffee
Honduras	Capital: Distribution:	Roads: Railways: Airports:	Coffee
El Salvador	Capital: Distribution:	Roads: Railways: Airports:	Coffee
Nicaragua	Capital: Distribution:	Roads: Railways: Airports:	Coffee
Costa Rica	Capital: Distribution:	Roads: Railways: Airports:	Garments
Panama	Capital: Distribution:	Roads: Railways: Airports:	Bananas

MODULE 7 • HUMAN/ENVIRONMENT INTERACTION

PHOTOCOPY

Central America Prior to Hurricane Mitch (continued)

COUNTRY	AGRICULTURAL LAND USE	AVERAGE PRECIPITATION	PHYSICAL LANDMARKS
Belize	Primarily forest with some irrigated land and little cropland	Primarily 1,401–2,800 mm	Maya Mountains
Guatemala			
Honduras			
El Salvador			
Nicaragua			
Costa Rica			
Panama			

Part 2: The storm

Step 2   Track Hurricane Mitch

d-1   At what time was Tropical Storm Mitch at this location? _____

d-2   What does the "z" mean in the time? _____

d-3   What was Mitch's wind speed at this location? _____

f-1   What are the latitude and longitude coordinates for Hurricane Mitch at this location?

_____

f-2   At what time was Hurricane Mitch at this location? _____

f-3   What was Mitch's wind speed at this location? _____

h-1   At what time was Hurricane Mitch at this location? _____

h-2   What was Mitch's wind speed at this location? _____

p   Write down the times for each event and determine the difference. The time is written in this format: Month/Day/Hour (of 24).

Hurricane - 5: _____

Tropical_Storm: _____

Time Difference: _____

r   Examine the attribute table further and identify the maximum wind speed. _____

Step 3   Measure the size of the storm

g   What is the diameter of the eye of Hurricane Mitch?   _____

h–q   Record your measurements and observations in the table below.

	DIAMETER OF EYE	DIAMETER OF STORM	DISTANCE OF EYE TO COASTLINE OF HONDURAS	HOW HAS THE STORM CHANGED FROM THE PREVIOUS IMAGE?
Mitch2sat.tif				
Mitch3sat.tif				
Mitch4sat.tif				
Mitch5sat.tif				

Step 4   Analyze rainfall from Hurricane Mitch

c-1   What pattern do you notice in the amount of rainfall within the storm?

_____

c-2   Is this a pattern you expected to find? Why or why not?

_____

e-1   At the Mitch4 location, what was the highest range of rainfall measured?

_____

e-2   Which country received the majority of this heavy rain?

_____

e-3   Describe the difference between the rainfall patterns on October 30 and October 31, 1998.

_____

_____

_____

e-4   What kind of damage do you expect to find with this type of storm? What aspects of the region will be most affected? Elaborate on your answer using your table, Central America Prior to Hurricane Mitch as a resource.

_____

_____

_____

Step 5   Save the map document and exit ArcMap

a   Write the new name you gave the map document and where you saved it.

_____   _____
(Name of map document.   **(Navigation path to where map document is saved.**
For example: ABC_Region7.mxd)   **For example: C:\Student\ABC)**

NAME _____   DATE _____

Volcano and Landslide Hazard Team
Middle school assessment

Your assigned country is: _____

The focus of this team is potential hazards of landslides and debris flows from volcanoes in the region devastated by the intense rainfall of Hurricane Mitch. Begin by studying the volcanoes in your assigned country and the amounts of rainfall they received from the storm. Compare this rainfall to the typical precipitation received in that area. Use the ArcMap map document and add data from the module 7 data folder (CentralAmerica.mdb\ca_utility) to help you with your presentation. This additional data provides information on utility lines in Central America.

In your report/presentation, you need to do the following:
- Predict which towns are in the most danger from flooding.
- Create an emergency action plan for these towns, including:
  - An evacuation plan in case of major landslides/debris flows
  - Ways to reroute power
  - A plan to provide medical and humanitarian aid to the affected areas
  - Alternative routes to transportation networks that will be affected
- Identify agricultural areas that will be damaged (if any).

NAME _____ DATE _____

Flood Hazard Team
Middle school assessment

Your assigned country is: _____

The focus of this team is damage from rising floodwaters caused by Hurricane Mitch. It will be helpful to look at the rainfall layers from Hurricane Mitch, typical precipitation patterns for the region, and data from the module 7 data folder (CentralAmerica.mdb\ca_drain and ca_utility). This additional data provides information on drainage features and utility lines in Central America.

In your report/presentation, you need to do the following:

- Predict which towns are in the most danger from flooding.
- Create an emergency action plan for these towns, including:
  - An evacuation plan in case of major landslides/debris flows
  - Ways to reroute power
  - A plan to provide medical and humanitarian aid to the affected areas
  - Alternative routes to transportation networks that will be affected
- Identify agricultural areas that will be damaged (if any).

In the Eye of the Storm

Assessment rubric

Middle school

STANDARD	EXEMPLARY	MASTERY	INTRODUCTORY	DOES NOT MEET REQUIREMENTS
The student knows and understands the relative advantages and disadvantages of using maps, globes, aerial and other photographs, satellite-produced images, and models to solve geographic problems.	Uses GIS to gather a variety of data about Hurricane Mitch including new data from outside sources. Analyzes this information to determine cities at greatest risk for particular hazards.	Uses GIS to gather a variety of data about Hurricane Mitch and analyzes this information to determine cities at greatest risk for particular hazards.	Uses GIS to gather data about Hurricane Mitch and attempts to determine cities at risk for particular hazards. Because their selection of data is limited, some predictions may not be accurate.	Only uses data provided from the original GIS project and does not make accurate predictions.
The student knows and understands how to predict the consequences of physical processes on Earth's surface.	Accurately predicts the impact of Hurricane Mitch on the physical environment of their assigned Central American country. Creates a map of the affected areas using a variety of data.	Accurately predicts the impact of Hurricane Mitch on the physical environment of their assigned Central American country. Provides ample data to support their predictions.	Attempts to predict the impact of Hurricane Mitch on the physical environment of their assigned Central American country. Provides some data to support their predictions.	Is not able to identify major effects of Hurricane Mitch on the physical environment of their assigned Central American country.
The student knows and understands how natural hazards affect human activities.	Accurately predicts the impact of Hurricane Mitch on human activities in their assigned Central American country. Creates a map of the affected areas using a variety of data.	Accurately predicts the impact of Hurricane Mitch on human activities in their assigned Central American country. Provides ample data to support their predictions.	Attempts to predict the impact of Hurricane Mitch on human activities in their assigned Central American country. Provides some data to support their predictions.	Is not able to identify the effects of Hurricane Mitch on human activities in their assigned Central American country.
The student knows and understands how to apply the geographic point of view to solve social and environmental problems by making geographically informed decisions.	Creates an emergency action plan for their country that takes into account infrastructure and environmental changes. Uses a variety of data to support the ideas in their plan. Creates a map with evacuation routes and other important factors.	Creates an emergency action plan for their country that takes into account infrastructure and environmental changes. Uses a variety of data to support the ideas in their plan, including an evacuation route.	Creates an emergency action plan, but does not take into account both infrastructure and environment changes—focuses only on one. Uses data to support the ideas in their plan, but is missing some important factors.	Creates an outline for an emergency action plan, but does not take into account both infrastructure and environmental changes—focuses only on one. Provides little or no data to support their plan.

This is a four-point rubric based on the National Standards for Geographic Education. The "Mastery" level meets the target objective for grades 5–8.

NAME _____   DATE _____

Volcano and Landslide Hazard Team
High school assessment

Your assigned country is: _____

The focus of this team is potential hazards of landslides and debris flows from volcanoes in the region devastated by the intense rainfall of Hurricane Mitch. Begin by studying the volcanoes in your assigned country and the amounts of rainfall they received from the storm. Compare this rainfall to the typical precipitation received in that area. Use the ArcMap map document and add data from the module 7 data folder (CentralAmerica.mdb\ca_utility) to help you with your presentation. This additional data provides information on utility lines in Central America.

In your report/presentation, you need to do the following:

- Predict which towns are in the most danger from debris flow and landslide threat.
- Create an emergency action plan for these towns, including:
  - An evacuation plan in case of major landslides/debris flows
  - Ways to reroute power
  - A plan to provide medical and humanitarian aid to the affected areas
  - Alternative routes to transportation networks that will be affected
- Identify agricultural areas that will be damaged (if any) and discuss how this would hinder the economy of the country.

NAME _____   DATE _____

Flood Hazard Team
High school assessment

Your assigned country is: _____

The focus of this team is damage from rising floodwaters caused by Hurricane Mitch. It will be helpful to look at the rainfall layers from Hurricane Mitch, typical precipitation patterns for the region, and data from the module 7 data folder (CentralAmerica.mdb\ca_drain and ca_utility). This additional data provides information on drainage features and utility lines in Central America.

In your report/presentation, you need to do the following:

- Predict which towns are in the most danger from flooding.
- Create an emergency action plan for these towns, including:
  - An evacuation plan in case of major landslides/debris flows
  - Ways to reroute power
  - A plan to provide medical and humanitarian aid to the affected areas
  - Alternative routes to transportation networks that will be affected
- Identify agricultural areas that will be damaged (if any) and discuss how this would hinder the economy of the country.

In the Eye of the Storm

Assessment rubric

High school

STANDARD	EXEMPLARY	MASTERY	INTRODUCTORY	DOES NOT MEET REQUIREMENTS
The student knows and understands how to use technologies to represent and interpret Earth's physical and human systems.	Uses GIS to analyze a variety of data from satellite imagery to social infrastructure themes to determine what areas are at greatest risk from Hurricane Mitch. In addition, the students import data from outside sources.	Uses GIS to analyze a variety of data from satellite imagery to social infrastructure themes to determine what areas are at greatest risk from Hurricane Mitch.	Uses GIS to analyze data to determine what areas are at greatest risk from Hurricane Mitch. Because their selection of data is limited, some predictions may not be accurate.	Only uses data provided from the original GIS project and does not make accurate predictions.
The student knows and understands the spatial variation in the consequences of physical processes across Earth's surface.	Analyzes and makes accurate predictions on the impact of Hurricane Mitch on the physical environment of their assigned country. Uses GIS to create a map that details how the storm affects the region.	Analyzes and makes accurate predictions on the impact of Hurricane Mitch on the physical environment of their assigned country. Provides details on how the storm affects the region.	Reviews data and attempts to make predictions about the storm impact on the physical environment of their assigned country.	Makes inaccurate predictions on the storm impact on the physical environment of their assigned country.
The student knows and understands strategies to respond to constraints placed on human systems by the physical environment and how humans perceive and react to natural hazards.	Analyzes and makes accurate predictions on the impact of Hurricane Mitch on human systems of their assigned country. Uses GIS to create a map that details how the storm affects the region.	Analyzes and makes accurate predictions on the impact of Hurricane Mitch on human systems of their assigned country. Provides details on how the storm affects the region.	Reviews data and attempts to make predictions about the storm impact on human systems of their assigned country.	Makes inaccurate predictions on the storm impact on human systems of their assigned country.
The student knows and understands how to use geographic knowledge, skills, and perspectives to analyze problems and make decisions.	Creates an emergency action plan for their country that takes into account infrastructure and environmental changes. Uses a variety of data to support the ideas in their plan. Creates a map with evacuation routes and other important factors.	Creates an emergency action plan for their country that takes into account infrastructure and environmental changes. Uses a variety of data to support the ideas in their plan, including an evacuation route.	Creates an emergency action plan, but does not take into account both infrastructure and environment changes—focuses only on one. Uses data to support the ideas in their plan, but is missing some important factors.	Creates an outline for an emergency action plan, but does not take into account both infrastructure and environmental changes—focuses only on one. Provides little or no data to support their plan.

This is a four-point rubric based on the National Standards for Geographic Education. The "Mastery" level meets the target objective for grades 9–12.

Data Disaster

An advanced investigation

Lesson overview

Students will act as data detectives in this scenario-based investigation. Due to a recent storm, the main computer was flooded at the International Wildlife Conservancy (a fictitious organization). All documentation for their ecozone maps was lost. The data and maps have survived, but staff members are unsure what each field in the database represents. What they do know is that the maps show areas in critical situations, and even in some cases in outright danger. The students must analyze the data that is available to them and create metadata for the data.

Estimated time　Two 45-minute class periods

Materials　✔ Student handouts from this lesson to be copied:
- GIS Investigation sheets (pages 453 to 462)
- Student answer sheet (pages 463 to 466)

Standards and objectives　*National geography standards*

	GEOGRAPHY STANDARD	MIDDLE SCHOOL	HIGH SCHOOL
1	How to use maps and other geographic representations, tools, and technologies to acquire, process, and report information from a spatial perspective	The student understands how to make and use maps, globes, graphs, charts, models, and databases to analyze spatial distributions and patterns.	The student understands how to use maps and other graphic representations to depict geographic problems.
5	That people create regions to interpret Earth's complexity	The student understands the connections among regions.	The student understands how multiple criteria can be used to define a region.
8	The characteristics and spatial distribution of ecosystems on Earth's surface	The student understands the local and global patterns of ecosystems.	The student understands the distribution and characteristics of ecosystems.

Objectives

The student is able to:
- Analyze and interpret various kinds of ecoregion data.
- Understand the importance of metadata in using spatial data.
- Create metadata for a dataset to make it more useful.

GIS skills and tools
- Create a new map document
- Change map symbols using the Symbol Selector
- Copy a layer and symbolize it based on different attributes
- Determine the best color scheme for depicting spatial information
- Start ArcCatalog and create a new folder connection
- View metadata for a feature class and change the metadata stylesheet
- Update metadata using the metadata editor

For more on geographic inquiry and these steps, see Geographic Inquiry and GIS (pages xxiii to xxv).

Teacher notes

Lesson introduction

Begin the lesson with a discussion of what metadata is. Metadata is data about data—it is information that describes the content and quality of GIS data. Metadata includes information about who created the data, when it was created, what resources were used, how it was made, and also defines various fields in an attribute table. Without metadata, tables can end up as meaningless sets of numbers and words.

Student activity

 Before completing this lesson with students, we recommend that you complete it as well. Doing so will allow you to modify the activity to accommodate the specific needs of your students. The lesson is designed for students working individually at the computer, but it can be modified to accommodate a variety of instructional settings.

After the initial discussion, distribute and explain the Data Disaster GIS Investigation. In the GIS Investigation, students will act as data detectives. They will create an ArcMap map document from the ground up, beginning with a new data frame. The will also create metadata for an ecoregions layer using ArcCatalog.

 Teacher Tip: Make sure you advise students on how to name and where to save their map document. Be sure to have instructions ready for the students if you would like them to save their work.

 Teacher Tip: In step 6, students are asked to update metadata for the WWF_Eco layer. This action results in permanent changes to the WWF_Eco feature class in the World7.mdb geodatabase (C:\MapWorld9\Mod7\Data\ World7.mdb). In order for this to work properly, each student must have their own copy of the Mod7 folder.

Steps 5–7 can be done as part of the assessment. They include instructions on how to look at a sample metadata document and how to write some metadata.

Things to look for while the students are working on this activity:

- Are students selecting appropriate colors for the data they are trying to map?
- Are they documenting their hypotheses about the data as they create the various maps?
- When creating metadata, are they referring back to the sample metadata and their answer sheet for guidance?

Conclusion

Each student will have completed the holes in the data definition tables provided in the student answer sheet handout. The GIS Investigation reviews sample metadata documentation. You should be present to answer any questions students might have. Show them how to access the metadata for the data included with this book through ArcCatalog. Visit *www.esri.com / mappingourworld* to access the Federal Geographic Data Committee (FGDC) Web site.

Assessment

Middle school: Highlights skills appropriate to grades 5 through 8

In the middle school assessment, students will create a metadataset based on their findings in the GIS Investigation. Items that they should include are:

- Identification—identifies what the data represents.
- Attribute information—defines each of the fields in the attribute table.
- Metadata reference—identifies who created the metadata and contact information.

High school: Highlights skills appropriate to grades 9 through 12

In the high school assessment, students will create metadata based on their findings in the GIS Investigation. Items that they should include are:

- Identification—identifies what the data represents.
- Attribute information—defines each of the fields in the attribute table.
- Metadata reference—identifies who created the metadata and contact information.
- Attribute information—defines each of the values for the THREAT and FINAL attributes.

Extensions

- Have students create their own attribute tables and corresponding metadata using school information or by conducting a local survey, and so forth.
- Students can create their own data disasters and try to stump the class in trying to define other mystery datasets.
- View metadata documents for other data included in this book. Have students determine basic information about the data.
- Check out the Resources by Module section of this book's companion Web site *(www.esri.com / mappingourworld)* for print, media, and Internet resources on the topic of metadata.

NAME _____   DATE _____

Data Disaster
An advanced investigation

ACQUIRE

ASK

EXPLORE

ACT

ANALYZE

Answer all questions on the student answer sheet handout

International Wildlife Conservatory (a fictitious organization) headquarters have been flooded. All of the metadata for their ecosystems data has been destroyed. This metadata contained detailed information describing the ecosystems database. It included definitions of the fields of the database, the names of types of data, and definitions for the data.

The IWC has created a list of endangered spaces known as Earth 200. According to the IWC, Earth 200 represents a list of the world's most unique and diverse environments. If any of these areas were lost or destroyed, the impact on the planet's ecosystem would be felt at a global level.

The dataset that you must investigate contains important information about these regions and identifies which are most threatened. The problem is that without the metadata, no one knows which areas are under the greatest threat. Your job is to review and map the data, rebuild the metadata, and use it to determine which regions are at a critical stage.

Step 1   Start ArcMap

   a   Double-click the ArcMap icon on your computer's desktop.

   b   If the ArcMap start-up dialog appears, click **A new empty map** and click OK. Then go to step 1d.

        In this exercise, you will create your own map document from scratch.

   c   Click the New Map File button. A new blank map document opens.

   d   Click the Add Data button.

   e   Navigate to the World7.mdb geodatabase in the module 7 Data folder (**C:\MapWorld9 \Mod7\Data\World.mdb**) and add the following feature classes:

        world30—ocean background with 30-degree grid

        continents—continents of the world

        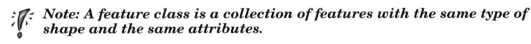 *Note: A feature class is a collection of features with the same type of shape and the same attributes.*

   f   Stretch your ArcMap window to make it larger.

ArcMap randomly assigns a symbol color to each added layer. You will change these colors using the symbol selector and then change the layer names.

g   In the table of contents, click the symbol below world30 to open the Symbol Selector.

h   In the Options panel on the right, change the Fill Color to a light shade of blue and change the Outline Color to Gray 30%.

i   In the table of contents, click the world30 layer name once, then click it again slowly. Change the layer name to **Ocean**.

j   Move continents to the top of the table of contents.

k   Click the symbol below continents to open the Symbol Selector. Change the symbol for continents to a Hollow symbol.

l   Change the layer name to **Continents**.

m   At the top of the table of contents, change the name of the data frame from Layers to **Data Disaster**.

Step 2   **Add the mystery data**

a   Click the Add Data button and navigate to the module 7 layer files folder (**C:\MapWorld9\Mod7\Data\LayerFiles**). Add WWF_Eco.lyr to the map.

The layer file loads with a complex legend with many colors.

Remember: A layer file contains the complete definition of a layer including its name, data source, symbology, and other properties. By saving a layer outside a map document as a layer file, you can reuse it in other maps.

Next you will determine the data source for the layer file so you can use this information later when you rebuild the metadata.

b   In the table of contents, right-click WWF_Eco and choose Properties. Click the Source tab.

(1)   *What is the name of the feature class that the layer is based on?*

(2)   *What is the name of the geodatabase that contains the feature class?*

c   Close the Layer Properties dialog.

d   Widen the table of contents so you can read the legend or pause the cursor over a long legend label to view the complete text.

What do you think this map is showing?

Step 3 **Evaluate the attribute table and map two layers**

The layer file's complex legend is created from the attribute table that is associated with the WWF_Eco features. Take a closer look at the information in the table to begin to solve the mystery of the data.

a In the table of contents, right-click WWF_Eco and select Open Attribute Table. The Attributes of WWF_Eco table displays.

b Use the scroll bar at the bottom of the attribute table to see all the fields in the table.

? *(1) There are nine fields in the table. Three of these fields, OBJECTID, Shape_Length, and Shape _Area won't help you identify the mystery data. You will concentrate on the other six fields. Record the names of the other six fields and the types of data they contain (numeric or text) in the table on the answer sheet.*

? *(2) Write a hypothesis on what you think the data in this table represents.*

c Compare the data in the table to the WWF_Eco layer legend.

? *Which field does this layer map?*

In order to evaluate all of the data, you will map each of the six fields you recorded on the answer sheet. You will also name each legend heading to match the field being mapped and compare each layer to the others.

d Close the attribute table.

e In the table of contents, click the LEGEND label below the WWF_Eco layer name, then click it again slowly. Type **MHT_NAME** in the text box and press Enter.

Now you will remember that the data mapped in this layer matches the field MHT_NAME in the attribute table.

f Rename the WWF_Eco layer to WWF_Eco1.

In order to map the other five fields, you will make copies of this layer and change the legend and layer names appropriately.

g Turn off the WWF_Eco1 layer. Right-click WWF_Eco1 and click Copy.

h Click the Paste button. A duplicate of the WWF_Eco1 layer is now added to the top of the table of contents.

i Right-click WWF_Eco1 and click Properties. Click the General tab.

j Change the layer name to WWF_Eco2.

k Click the Symbology tab.

l In the Value Field drop-down list, choose ECOREGION. Notice that all the fields from the attribute table, except OBJECTID and SHAPE, can be mapped.

m Click the Add All Values button. Click Yes to generate the list of more than 500 unique values. Notice that ECOREGION is the heading for the legend.

n Keep the default color scheme or change it by using the Color Scheme drop-down list.

o Click OK to close the Layer Properties and apply the new layer name and legend.

p   Turn on WWF_Eco2.Look at the map and observe the similarities and differences between WWF_Eco2 (ECOREGION) and WWF_Eco 1 (MHT_NAME).

?   *Record the similarities and differences between the two layers.*

q   Click the minus sign to the left of WWF_Eco2 to hide its legend.

?   *r*   Ask your teacher where to save this map document and how to rename it. Save the map document. Record the new name and location on your answer sheet.

Step 4   Map the data and analyze it

In order to get a good look at all the data, you will need to map each of the three remaining fields the same way you mapped the ECOREGION field.

a   In the table of contents, right-click WWF_Eco1 (MHT_NAME) and click Copy.

b   Click the Paste button three times.

c   Rename the layers WWF_Eco3, WWF_Eco4, and WWF_Eco5.

d   Use these layers to map the three remaining fields: BDI, Threat, and Final. Even though BDI, Threat, and Final are numeric fields, map them using unique values.

e   Save the Map document, then examine these layers.

    (1)   *After mapping out all the data fields, what additional conclusions can you make about the data?*

    (2)   *Use this new information to revise your hypothesis about what the data represents.*

★★★ NEWS FLASH ★★★

A cleanup volunteer has just discovered a document that appears to have some information about the data you are researching. The document has been heavily damaged by floodwater and is difficult to interpret. The table on your answer sheet has the information that could be deciphered.

?   *f*   Read the table on the answer sheet and use the mapped data to complete the missing pieces. The definitions in the table belong to the six fields you mapped. Fill in the table on the answer sheet with the appropriate field name from the attribute table.

With this information, you are very close to being able to create a simple metadata document. You know what the data looks like because you've mapped it, and you have made educated guesses as to the definitions for each field. An important piece of information still missing is how to interpret the legend for the numeric fields.

Each of the numeric fields represents a scale. The question is what order does the scale follow? Is 1 the lowest or the highest value in the scale? The ranking definitions without their numeric values are found on the answer sheet for each of the numeric fields. The title for each table is its field name.

g Complete the tables on your answer sheet by filling in the numeric values for each description. Keep in mind that the definitions may not be listed in numeric order and don't forget about zero. The first table has been filled in for you.

Congratulations! All of the details that you have discovered will go in the attribute information section of a metadata document. That section defines each of the fields in the database or attribute table, identifies what type of data it is, and provides the user with details on how types of data are ranked or identified.

Step 5　View metadata documentation

Now that you have had the opportunity to fill in the blanks in the definitions of the data, you will create a formal metadata document. Metadata means data about data. Metadata for each GIS data source is viewed and created in ArcCatalog. You will start ArcCatalog from ArcMap.

Before you create your own metadata, you will look at a sample metadata document.

 a Save your ArcMap map document.

b Click the ArcCatalog button on the Standard toolbar. ArcCatalog opens in a separate window.

c Minimize the ArcMap window.

On the left side of the ArcCatalog window you see the Catalog tree. The Catalog tree is where you expand and contract folders to navigate to data stored on your computer.

On the right side of the ArcCatalog window is the preview pane. This pane has three tabs that show you different views of the item that is selected in the Catalog tree.

In ArcCatalog, you can create direct connections to folders to speed up the process of navigating to data. You will create a connection to the Mod7 folder so you can get to the module data more quickly.

d Click the Connect To Folder button on the Standard toolbar. In the Connect to Folder dialog, navigate to the Mod7 folder (**C:\MapWorld9\Mod7**) and click on it, then click OK. The new folder connection displays in the Catalog tree.

e Expand the Mod7 and Data folders. Expand the World7.mdb geodatabase.

f   Click on the continents feature class, then click the Metadata tab.

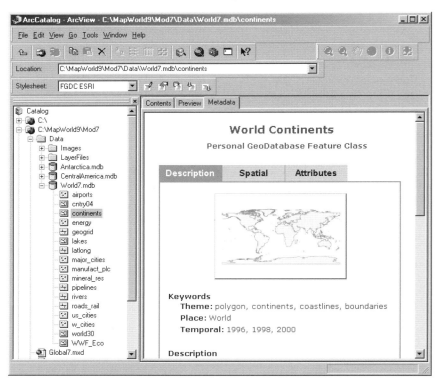

You see the metadata for World Continents presented in FGDC ESRI style. This style reflects the metadata standard developed by the Federal Geographic Data Committee with some additional information from ESRI. You will change the style to a question-and-answer format.

g   Make sure the Metadata toolbar is visible. (In the graphic above, the Metadata toolbar is docked above the Catalog tree.)

> *Note: If you don't see the Metadata toolbar, right-click in the gray area near the top of the ArcCatalog window and click Metadata on the menu that appears.*

h   Click the Stylesheet down arrow and choose FGDC FAQ from the list.

i   Stretch the ArcCatalog window so you can view more of the metadata.

j   Explore the links in the World Continents metadata and answer the questions below. (Hint: After you click a link, you can go back to the original display by scrolling to the top of the page or by right-clicking on the page and choosing Back on the menu that appears.

    (1)   How would you describe this dataset?

    (2)   Who made this data?

    (3)   When was this data published?

    (4)   Which section explains whether or not you can reuse the data for a project or publish a work that uses the data?

Now that you have viewed sample metadata, you will write a description (abstract) of the WWF_Eco data that you will add to a metadata document in step 6.

k Look at the hypothesis you wrote on your answer sheet in steps 3b-2 and 4e-2. Use the World Continents metadata abstract as a sample and answer the following question:

? *How would you describe the data in the WWF_Eco layer that you investigated?*

Step 6　Create metadata documentation

Now you will create your own metadata for WWF_Eco. For simplicity, you will create metadata in only a few key areas.

In step 2, you added the WWF_Eco.lyr layer file to the map and determined that the data source for the layer file is the WWF_Eco feature class located in the World7.mdb geodatabase. You want to create metadata for the feature class rather than the layer file that references it.

a In the Catalog tree, look through the list of feature classes in the World7 geodatabase and click on WWF_Eco. Its metadata displays in the metadata tab on the right.

b Scroll down through the metadata page. Some information is filled in automatically by ArcCatalog, but other information is missing.

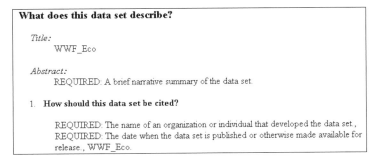

You see the word "REQUIRED," followed by a description of the required information, in many places. This indicates missing information that is required to comply with FGDC metadata standards. You will fill in some, but not all, of the missing information.

 c Click the Edit Metadata button on the Metadata toolbar. This opens the metadata editing window.

The information is divided into seven main sections. The seven section titles, beginning with Identification, appear across the top of the window. Each section also contains multiple tabs.

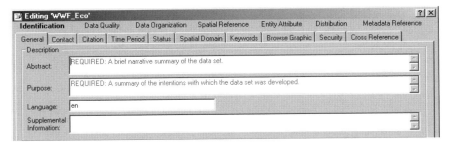

d In the Identification section, make sure the General tab is selected.

You will begin by filling in general description information.

e   **In the Description area, in the Abstract text box, highlight the red text and replace it with the description of the WWF_Eco data that you wrote on your answer sheet in step 5k.**

f   **In the Supplemental Information text box, type a statement explaining that the original metadata was destroyed in a flood and that this metadata is reconstructed information.**

Next you will define each of the fields in the attribute table.

g   **Click on the Entity Attribute section, then click on the Attribute tab.**

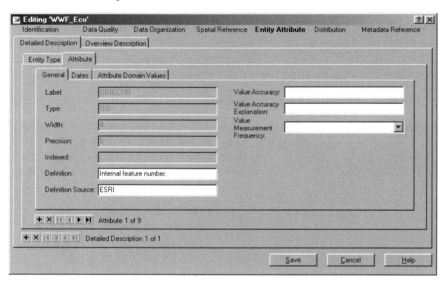

You see the first attribute, OBJECTID, in the WWF_Eco attribute table. Notice that a definition and definition source are already entered for this attribute. Also notice that this is attribute 1 of 9. You will advance through the other eight attributes and add definitions for five of them.

h   **Click the Move Next button near the bottom of the Attribute tab to display the next attribute in the WWF_Eco attribute table.**

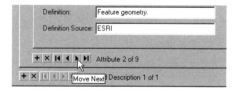

You see the definition for the SHAPE attribute.

i   **Click the Move Next button again to advance to the third attribute, ECOREGION. In the Definition text box, type the definition from step 4f on your answer sheet.**

j   **Use the Move Next button to advance to the next four attributes (MHT_NAME, BDI, THREAT, and FINAL). Fill in the definitions based on your answers in step 4f.**

k   Use the Move Previous button to navigate back to the fifth attribute, BDI.

Now you will define each of the values in the BDI attribute field.

l   Under the Attribute tab, click the Attribute Domain Values tab.

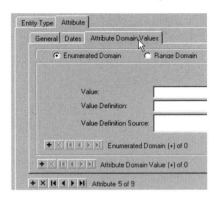

Notice that Enumerated Domain is already selected. An enumerated domain is used to describe a list of values. Each value should be listed in the metadata along with a definition of the value and, if known, the source of that definition. For the BDI attribute, you will enter each attribute value and its definition.

m   In the Value text box, Type **1**.

n   In the Value Definition box, type the description of the value listed in the BDI table in step 4g on your answer sheet.

o   Click the Add button to add this value to the metadata document.

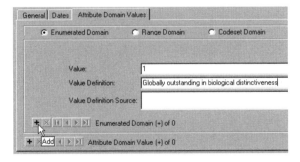

p   Continue adding values and their definitions for the BDI attribute as listed in the BDI table on your answer sheet. Don't forget to add the "0" value and its definition.

Now you will add information about who created the metadata.

q   Click the Metadata Reference section at the top of the dialog, then click the Details button next to Contact.

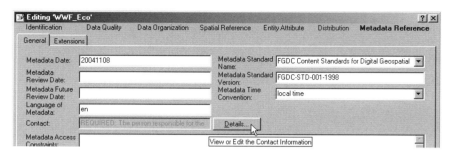

r   Select Person as the Primary Contact. Fill in your name for the Person, your school for the Organization, and your grade for the Position. Click OK.

s   Click Save on the bottom of the editing window to exit and save your changes to the metadata document.

t   Look for your updates in the metadata document.

u   Ask your teacher if there are other items that you should add to the metadata document.

Two additional items could include:
* Values and definitions for the THREAT attribute
* Values and definitions for the FINAL attribute

v   Click the File menu and click Print to print your metadata document.

Step 7   Exit ArcCatalog and ArcMap

Congratulations on a job well done! Your efforts will help the IWC get their data records back on track.

a   In ArcCatalog, click the File menu and click Exit.

b   In ArcMap, click the File menu and click Exit.

NAME _____   DATE _____

Student answer sheet
Module 7
Human/Environment Interaction

Advanced investigation: Data Disaster

Step 2   Add the mystery data

b-1   What is the name of the feature class that the layer is based on?

_____

b-2   What is the name of the geodatabase that contains the feature class?

_____

d   What do you think this map is showing?

_____

_____

Step 3   Evaluate the attribute table and map two layers

b-1   Write each field name and the type of data it contains (numeric or text) in the table below.

FIELD NUMBER	FIELD NAME	FIELD TYPE
1		
2		
3		
4		
5		
6		

b-2   Write a hypothesis on what you think the data represents.

_____

_____

_____

_____

c   Which field does this layer map?

_____

p   Record the similarities and differences between these two layers.

_____
_____
_____
_____

r   Write the new name you gave the map document and where you saved it.

_____   _____
(Name of map document.   **(Navigation path to where map document is saved.**
For example: ABC_Adv7.mxd)   **For example: C:\Student\ABC)**

Step 4   Map the data and analyze it

e-1   After mapping out all the data fields, what additional conclusions can you make about the data?

_____
_____
_____
_____

e-2   Use this new information to revise your hypothesis about what the data represents.

_____
_____
_____
_____

f   Read the table below and use the mapped data to complete the missing pieces. The definitions in the table belong to the six fields you identified in step 3b-1. Fill in the table with the appropriate field name from the attribute table.

FIELD NAME	DEFINITION
	The final assessment of the ecoregion as the estimated threat to the ecoregion over the next 20 years.
	Descriptive name for the _____ that are relatively large areas of land or water that share a large majority of their species, dynamics, and environmental conditions.
	The biological distinctiveness index. It is based on the species richness, endemism, rareness, and so forth.
	Represent global terrestrial and freshwater areas defined as ecoregions.
	The major habitat for the area.
	Degree of threat to the ecoregion. Some examples include logging, conversion to agriculture/ urbanization, and so on.

g   Complete the following tables by filling in the numeric values for each description. Keep in mind that the definitions may not be listed in numeric order and don't forget about zero. The first table has been filled in for you.

BDI	
1	Globally outstanding in biological distinctiveness
2	Regionally outstanding in biological distinctiveness
3	Bioregionally outstanding in biological distinctiveness
4	Locally important in biological distinctiveness
0	Not assessed or unknown

THREAT	
	Relatively stable in degree of threat
	Relatively intact in degree of threat
	Vulnerable in degree of threat
	Not assessed or unknown
	Endangered in degree of threat
	Critical in degree of threat

FINAL	
	Not assessed or unknown
	Critical in estimated threat over the next 20 years
	Endangered in estimated threat over the next 20 years
	Relatively intact in estimated threat over the next 20 years
	Vulnerable in estimated threat over the next 20 years
	Relatively stable in estimated threat over the next 20 years

Step 5   View metadata documentation

j-1   How would you describe this dataset?

_____

j-2   Who made this data?

_____

j-3   When was this data published?

_____

j-4   Which section explains whether or not you can reuse the data for a project or publish a work that uses the data?

_____

k   How would you describe the data in the WWF_Eco layer that you investigated?

_____

_____

_____

_____

Module 1   ArcMap: The Basics

Part 1: Introducing the software

Step 4   Work with layers

g   Which layers are not visible on the map but are turned on in the table of contents? Rivers, Lakes

i   What happened on your map? The rivers show up on the map.

j-1   What happened on your map? The lakes show up on the map.

j-2   What would happen if you dragged the Lakes under Land Areas?
    The lakes will disappear again; or, the lakes will be covered up by the Land Areas layer.

Step 5   Change the active data frame

b   What is the name of the layer that is turned on in the World Population data frame? World Countries

Step 7   Identify a country and record country data

g-1   What is the fourth listing in this column? CNTRY_NAME

g-2   What is the fifth listing in this column? TOT_POP

g-3   What is the final listing in this column? Shape_Area

h-1   What do you guess the field entitled "SQMI" stands for? Square miles

h-2   What is the number to the right of the field "SQMI"? 3,648,399.8

Step 8   Compare the Identify Results data with the table data

a   Which row in this table has the attributes for the United States? The third row (OBJECTID 154)

b-1   Where are these field names displayed in the table? Across the top of the table; or, as column headings in the table

b-2   How many square miles of land are in the United States? 3,648,399.8

b-3   Give a brief explanation of the relationship between the Identify Results window and the table.
    The Identify Results window has the same items of information that are in the table. The Identify Results window has information about only one country while the table has all the countries. (Students may not pick up on the second part just yet.)

Step 9   Explore city data on the world map

i   Use the Identify tool to find the name and country of any two cities you choose.
    Answers will vary. Possible answers include:

CITY NAME	COUNTRY WHERE THE CITY IS LOCATED
Zaragoza	Spain
Hamburg	Germany

Step 10   Explore Europe with an attribute table

a-1   What is the name of the table you opened? Attributes of World Countries

a-2   What country is listed in the first row of the table? Falkland Is.

MODULE 1
ArcMap

MODULE 2
global

MODULE 2
regional

MODULE 3
global

MODULE 3
regional

MODULE 3
advanced

MODULE 4
global

MODULE 4
regional

MODULE 4
advanced

MODULE 5
global

MODULE 5
regional

MODULE 5
advanced

MODULE 6
global

MODULE 6
regional

MODULE 7
global

MODULE 7
regional

MODULE 7
advanced

MODULE 1
ArcMap

MODULE 2
global

MODULE 2
regional

MODULE 3
global

MODULE 3
regional

MODULE 3
advanced

MODULE 4
global

MODULE 4
regional

MODULE 4
advanced

MODULE 5
global

MODULE 5
regional

MODULE 5
advanced

MODULE 6
global

MODULE 6
regional

MODULE 7
global

MODULE 7
regional

MODULE 7
advanced

b   What country is listed in the last row of the table? Vanuatu

g   What happens to the map when you click on these rows in the table? The three additional countries are outlined in blue.

h-1   What happens to Poland and the other countries that were highlighted? The blue outlines disappear.

h-2   Did you see the United States become outlined in blue on the map? If not, why not?
        Most students will not be able to see the United States outlined in blue because it isn't in the part of the world they are looking at. Some students may see the U.S. if they didn't zoom in enough or if they have a large screen.)

j   Why can you see the United States now when you couldn't see it in the previous step?
        The instructions had the students zoom in too far for them to be able to see the United States. Once they zoomed to the full extent of the map, they were able to see the United States.

Step 11   Practice identifying features
b   What do you see on your map? South America only

d-1   What country is it? Brazil

d-2   What is this country's total population? 172,860,000

e-1   What city is it? Manaus

e-2   What population class is this city in? 1,000,000–5,000,000

h-1   What are the names of these two large cities? São Paulo, Rio de Janeiro

h-2   What population class are these cities in? 5,000,000 and greater

Step 12   Practice zooming out
c-1   What does your map look like? It's small or reduced in size.

c-2   Which button could you use to return your map to full size? Full Extent (the button that looks like a globe)

Step 13   Practice finding a feature
k-1   How many tourists arrive in Sudan each year? 30

k-2   How many people live in Sudan? 35,080,000

k-3   Does this seem like a low or high number of tourists for this population?
        It is a very low number for that population. Students should wonder why this is such a low number. This is a good example of how geographic analysis can lead to further questions.

Step 14   Zoom to a feature and create a bookmark
d   Is Qatar a large country or a small one? A small one

k-1   How many people live in Qatar? 744,000

k-2   How many cell phones do they have? 322,152

k-3   How many people are there for every cell phone in Qatar? 2.3 people per cell phone

m   What large country is directly west of Qatar? Saudi Arabia

Step 15   Continue to explore the World Population map

c-1   What boot-shaped country do you see on the map? Italy

c-2   What is the population of that country? 57,634,000

c-3   How many cell phones does that country have? 54,118,327

c-4   How many people are there for every cell phone in this country? 1.06 people per cell phone

d-1   What is the population of Japan? 126,550,000

d-2   How many cell phones does Japan have? 80,612,353

d-3   How many people are there for every cell phone in Japan? 1.57 people per cell phone

g   What happened to Qatar? It is no longer outlined in blue (the blue outline disappeared).

Step 17   Label and print a map

b   Where do you think these labels come from?
The labels come from the country name attribute field (CNTRY_NAME) in the World Countries layer.

Part 2. The geographic inquiry model

Step 4   Ask a geographic question and develop a hypothesis

a   What makes this a geographic question?
Answers will vary. Students should recognize that the question involves the distribution of something (phone lines) in different places (countries).

b   Write a hypothesis that answers the geographic question:
Answers will vary. Possible answers include: The number of phone lines increases proportionately with the number of people in the world's most populous countries; the number of phone lines does not increase proportionately with the number of people.

Step 5   Add a layer to your map

a   What other attribute of countries do you need in order to investigate your hypothesis? Number of phone lines

e   What is the name of the layer that has been added to your table of contents? World Phone Lines

Step 6   Explore the World Phone Lines map

b-1   What color in the legend indicates countries with the fewest phone lines? Yellow or light orange

b-2   What color indicates countries with the most phone lines? Light brown or dark brown

b-3   What color indicates countries with no data available for this layer? Gray

c   What other layer in your map has a graduated color legend? World Countries

d-1   Which two countries had the most phone lines in 2002? China and the United States

d-2   On which continent are most of the countries with the fewest phone lines? Africa

d-3   Which two countries have the largest populations? China and India

MODULE 1
ArcMap

MODULE 2
global

MODULE 2
regional

MODULE 3
global

MODULE 3
regional

MODULE 3
advanced

MODULE 4
global

MODULE 4
regional

MODULE 4
advanced

MODULE 5
global

MODULE 5
regional

MODULE 5
advanced

MODULE 6
global

MODULE 6
regional

MODULE 7
global

MODULE 7
regional

MODULE 7
advanced

d-4   Name three countries that are in the same population class (color) as the United States.
Answers will vary. Examples include Brazil, Nigeria, Indonesia, Japan, Pakistan, Bangladesh, Russia.

d-5   Which of the three countries, if any, are in the same phone line class (color) as the United States?
None of the countries

h   What two fields might help in answering the geographic question?
POP_2000
LINES_2002 (or LINES_1997)

Step 7   Research and record phone line and population data for China

g, h   Use your Find and Identify tools to locate China. Record the population and phone lines in the appropriate columns.

COUNTRY NAME	COUNTRY POPULATION	PHONE LINES 2002	NUMBER OF PEOPLE PER PHONE LINE
China	1,261,832,000	212,929,439	5.93

l   Record the number of people per phone line for China in the last column of the table.

Step 9   Research and record population, phone line, and phone line density data for all the countries

c-1   What is the number of people per phone line (PHONE_DENS) for China? 5.93

c-2   Does this number agree with the value you calculated in step 7? Yes

d   Use the Find and Identify tools to locate the countries in the table below. Record the population, phone lines, and number of people per phone line for each country.

COUNTRY NAME	COUNTRY POPULATION	PHONE LINES 2002	NUMBER OF PEOPLE PER PHONE LINE
China	1,261,832,000	212,929,439	5.93
India	1,014,004,000	40,560,159	25.00
United States	275,563,000	178,013,706	1.55
Indonesia	224,784,000	8,317,008	27.03
Brazil	172,860,000	38,547,781	4.48
Russia	146,001,000	35,332,242	4.13
Pakistan	141,554,000	3,538,850	40.00
Japan	126,550,000	70,614,903	1.79

Step 11   Analyze results of your research

a   In the table below, the column on the left ranks the countries by population from highest to lowest. In the column on the right, rank the countries from the lowest number of people per phone line to the highest number of people per phone line, using the data you recorded in step 9. Then draw lines connecting the same country in each column.

RANKED BY POPULATION (HIGHEST TO LOWEST)	RANKED BY NUMBER OF PEOPLE PER PHONE LINE (LOWEST NUMBER OF PEOPLE PER PHONE LINE TO HIGHEST)
China	United States
India	Japan
United States	Russia
Indonesia	Brazil
Brazil	China
Russia	India
Pakistan	Indonesia
Japan	Pakistan

b-1   Which country has the fewest people per phone line? United States

How many people have to share a phone line in this country? 1.55

b-2   How does the country in question b-1 rank in population size with the other seven countries in your table?
It's the third largest in population.

b-3   Which country has the most people per phone line? Pakistan

How many people have to share a phone line in this country? 40.00

b-4   How does the country in question b-3 rank in population size with the other seven countries in your table?
It's the seventh largest.

b-5   What is the population of Japan? 126,550,000

How many people have to share a phone line in Japan? 1.7900

b-6   What country has the most phone lines? China

How does the number of people per phone line in this country compare with the seven other countries in your table? It is fifth.

b-7   Russia and Pakistan have about the same number of people. Why do you suppose these two countries have such a different number of people who have to share a phone line? What factors do you think contribute to this disparity?
Answers will vary. Examples: The countries have different cultures, a different economic base, and different economics of citizens.

b-8   What do you think the answer to the geographic question is?
The number of phone lines does not necessarily increase at the same rate as an increase in population.

b-9   How does your initial hypothesis (step 4b) compare with your answer to the geographic question?
Answers will vary.

Step 12   Develop a plan of action

a   Use the information in your table to describe the current phone line situation in your chosen country.

Answers will vary. Possible answers include:

China has the highest number of phone lines of any country in the world, but its extremely large population (over one billion people) means that nearly 6 people must share a phone line. Approximately 4 Brazilians must share a phone line. Brazil ranks fourth among countries that have shared phone lines, after the United States, Japan, and Russia. Brazil is ahead of the fifth-ranked country, China, where nearly 6 people must share a phone line.

In Indonesia, 27.03 people must share a phone line. Indonesia falls far behind other countries of similar size such as the United States (1.55 people per phone line) and Brazil (4.48 people per phone line).

The United States has more than 178 million phone lines—second-most in the world. This country leads the world in access to phone lines because only 1.55 people have to share a phone line.

b   Do you think that increasing the number of phone lines operating in your chosen country would improve the quality of life there? Why or why not?

Answers will vary.

c   List three concerns you have about increasing the number of phone lines in your chosen country.

Answers will vary. Students may list issues such as (1) difficulty of building phone lines in rural areas or among the many islands of Indonesia, (2) cost of building additional phone lines, or (3) a preference for expanding more modern cell-phone infrastructure.

d   List two new geographic questions that you would like to investigate to help you develop a sound plan.

Answers will vary but should include a geographic component.

Module 2   Global perspective: The Earth Moves

Step 3   Look at earthquake location data

b-1   Do earthquakes occur in the places you predicted? List the regions you predicted correctly for earthquake locations.

Answers will vary depending on where students predicted earthquakes.

b-2   What patterns do you see in the map?

Answers will vary, but they should indicate that earthquakes largely occur on the western coast of North and South America, along the eastern coast of Asia, and along the islands of the Pacific Rim. The pattern follows the Ring of Fire. They may also note a pattern of earthquakes down the center of the Atlantic Ocean and a string in the southern Atlantic, from South America eastward through to the Indian Ocean. Another pattern that is evident is a string that runs east and west through the south-central part of Asia and the southern part of Europe.

Step 4   Sort and analyze earthquake magnitudes

g   How do the 15 selected locations compare to your original paper map? List three ways.

Answers will vary based on their original predictions. If they selected any spots around the Ring of Fire, then their predictions were fairly close to reality.

Step 5   Look at volcano data

a-1   How do the volcano locations compare with your original predictions? List the regions of volcanic activity you predicted correctly.

Answers will vary based on their original predictions. If they selected any spots around the Ring of Fire, then their predictions were fairly close to reality.

a-2   What patterns do you see in the volcano points and how do they compare with the earthquake patterns?

The earthquake and volcano points line up so that they are similar in their patterns with the exception of the volcanoes along the eastern side of Africa.

Step 6   Select all active volcanoes

h-1   Does this data provide any patterns that were not evident before? Identify those patterns.

The majority of the world volcanoes on the map are active, particularly in the islands of the Pacific Rim.

h-2   Create a hypothesis as to why volcanoes and earthquakes happen where they do.

Student answers will vary. Their hypothesis should begin to allude to the idea of plate tectonics and the fact that movement at plate boundaries causes disruptions on the earth's surface.

Step 7   Identify the active volcanoes on different continents

e   Use the Identify tool to find the name, elevation, activity level, and country location of three volcanoes. Write that information in the space below.

Student answers will vary. Here is a correct example: On Take, 3,063 m, Solfatara Stage, Japan; Banahao, 2,177 m, Active, Philippines; and Ibu, 1,340, Active, Indonesia.

Step 8   Add the plate boundaries layer

i   Compare the actual plate boundaries to the ones you drew on your paper map. Record all similarities and differences.

Answers will vary based on students' original hypotheses on plate boundary locations. If they drew the boundaries to follow the patterns of earthquakes and volcanoes, then they are on target.

MODULE 1
ArcMap

MODULE 2
global

MODULE 2
regional

MODULE 3
global

MODULE 3
regional

MODULE 3
advanced

MODULE 4
global

MODULE 4
regional

MODULE 4
advanced

MODULE 5
global

MODULE 5
regional

MODULE 5
advanced

MODULE 6
global

MODULE 6
regional

MODULE 7
global

MODULE 7
regional

MODULE 7
advanced

Step 9   Add a layer file and an image

g, h   Use MapTips to find out the names of all major landforms formed at plate boundaries. Write them below and label them on your paper map. Next to the name of each landform, write how you think the landform was created.

NAME OF LANDFORM	FORMED BY
Mid-Atlantic Ridge	The separation of South American and African plates
Aleutian Trench Aleutian Islands	The boundary between the North American and Pacific plates
Rocky Mountain Range Cascade Mountain Range Sierra Nevada Mountain Range Baja California	The boundary of the eastern edge of the Pacific plate and the western edge of the North American plate
Sierra Madre Occidental Sierra Madre del Sur	The boundary of the North American, Pacific, and Coca plates
Andes Mountain Range Peru–Chile Trench	The boundary of Nazca and the South American plate
Alps Mountain Range Atlas Mountains	The boundary of the African and Eurasian plates
East Africa Rift Valley	The boundary of the African, Arabian, and Somali plates
Himalayas Tibetan Plateau	The boundary of the Indo-Australian, Indo-Chinese, Eurasian, and Amur plates
Mariana Trench	The boundary of the Philippine and the Pacific plates

Note: The above list covers landforms identified in MajorLandforms.lyr in the ArcMap map document. Students can find additional landforms by consulting an atlas or physical map of the world.

Step 10   Identify major cities at high and low risk for seismic activity

b   Find the names of specific cities that are high-risk or low-risk for a seismic event. Write those names in the table.

HIGH RISK		LOW RISK	
1	Tokyo, Japan	1	Kazan, Russia
2	Reykjavik, Iceland	2	Tombouctou, Mali
3	San Francisco, United States	3	Kansas City, United States
4	Quito, Ecuador	4	Minsk, Belarus
5	Managua, Nicaragua	5	Godhavn, Greenland

Module 2   Regional case study: Life on the Edge

Step 2   Open the region2.mxd file and look at cities data

g   Use the Identify tool to locate one city within each country listed in the table below and record that city's population.

Student answers will vary. This is an example of a table with correct information:

CITY NAME	COUNTRY NAME	CITY POPULATION
Kunming	China	1,280,000
Delhi	India	7,200,000
Tokyo	Japan	23,620,000

Step 3   Look at population density

c   Use the Identify tool to locate two world cities in East Asia in areas where the population density is greater than 200 people per square kilometer. Record below.

Student answers will vary. This is an example of a table with correct information:

WORLD CITIES THAT HAVE A POPULATION DENSITY GREATER THAN 200 PEOPLE PER SQUARE KILOMETER
Nanjing, China
Calcutta, India

Step 4   Look at earthquake magnitudes

a-1   Where did the largest earthquakes occur?

Student answers will vary. They should include a statement that large earthquakes tend to fall along the islands of the Pacific Rim, including some that occur under the ocean.

a-2   Did large earthquakes occur near densely populated areas? Where?

No. For the most part, the largest earthquakes occurred underwater.

Step 5   Measure the distance between active volcanoes and nearby major cities

g-1   Are there many active volcanoes located close to highly populated areas? What is the closest distance you found? Record the name of the volcano and the city, and their distance apart.

Yes, there are many active volcanoes near highly populated areas. There are several possible answers for close distances. For example, Manado, Indonesia, is only 4 miles from Mahawu, an active volcano.

g-2   What patterns do you see in the volcano points and how do they compare with the earthquake patterns?

The patterns of earthquake and volcanic activity are virtually the same, especially along the islands of the Pacific Rim (the western edge of the Ring of Fire).

MODULE 1
ArcMap

MODULE 2
global

MODULE 2
regional

MODULE 3
global

MODULE 3
regional

MODULE 3
advanced

MODULE 4
global

MODULE 4
regional

MODULE 4
advanced

MODULE 5
global

MODULE 5
regional

MODULE 5
advanced

MODULE 6
global

MODULE 6
regional

MODULE 7
global

MODULE 7
regional

MODULE 7
advanced

Module 3   Global perspective: Running Hot and Cold

Step 3   Observe annual world temperature patterns

b   Write three observations about the pattern of temperatures displayed on the map.

Student answers will vary. Possible observations include the following:

The warmest temperatures are clustered halfway between the North and South poles.

Temperatures get steadily colder as you go from the equator toward the North Pole.

There are many cities with cold temperatures in the Northern Hemisphere, but none in the Southern Hemisphere.

Step 4   Label the latitude zones

s   Use the Identify tool to get information on cities and complete the table below.

ZONE	TYPICAL TEMPERATURE RANGE	EXAMPLE CITY (IT REFLECTS TYPICAL TEMPERATURES OF THAT ZONE)	ANOMALIES (CITIES THAT DO NOT FIT THE PATTERN OF THEIR ZONE)
Tropical	65°–85° F	Any of the 28 cities colored red or orange and that are between the Tropic of Capricorn and the Tropic of Cancer.	Quito (51°–60°) and La Paz (41°–50°)
North Temperate Zone	31°–64° F	Any of the cities colored purple, blue, or green and that are between the Tropic of Cancer and the Arctic Circle.	Students should look for cities that seem to differ from the others around them or from other cities at the same latitude. Examples include Lhasa or Ankara.
South Temperate Zone	54°–64° F	Any of the cities colored green or orange and that are between the Tropic of Capricorn and the Antarctic Circle. (Buenos Aires, Cape Town, Johannesburg, Sydney, Melbourne, Wellington, Auckland.)	None

s-1   Why do you think there aren't any cities in the North or South polar zones?

The temperatures are probably too cold to support major cities.

s-2   How is the North Temperate Zone different from the South Temperate Zone?

Student answers will vary. Possible observations include the following:

There is more land area in the North Temperate Zone.

There are cities with average temperatures below 51° in the North Temperate Zone, but none in the South Temperate Zone.

Step 5   Observe climate distribution

b-1   Complete the table.

LATITUDE ZONES	CHARACTERISTIC CLIMATE(S)
Tropical zones	Tropical Wet, Tropical Wet and Dry, and Dry Some areas of Arid, Semiarid, Humid Subtropical, Highlands
Temperate zones	Humid Subtropical, Humid Continental, Marine, Mediterranean, Subarctic Some areas of Arid, Semiarid, Highlands, Tundra
Polar zones	Subarctic, Tundra, Ice Cap

b-2   Which zone has the greatest number of climates? North Temperate Zone

c   Give an example of a city in each of the following climate zones.
Student answers will vary. This list represents sample answers:

Arid Khartoum
Tropical Wet Kisangani
Tropical Wet and Dry Bamako
Humid Subtropical Atlanta
Mediterranean Rome
Marine Paris
Humid Continental Warsaw
Subarctic Irkutsk
Highland Lhasa

Step 6   Observe monthly temperature patterns in the Northern Hemisphere

j-1   What does the graph show now?
Average monthly temperatures in Miami

j-2   What city is highlighted on the map? Miami

k-1   What does the graph show now? Average monthly temperatures in Miami and Boston

k-2   What city or cities are highlighted on the map? Miami and Boston

l   Use the Monthly Temperature graph to complete the table below.

CITIES	COLDEST MONTH	LOWEST TEMPERATURE (°F)	HOTTEST MONTH	HIGHEST TEMPERATURE (°F)	TEMPERATURE RANGE OVER 12 MONTHS
Boston	January	28°	July	73°	45°
Miami	January, February	68°	August	83°	15°

m-1   What is the name of the city? Quebec

m-2   How does its monthly temperature pattern differ from Boston's?
The overall pattern is the same, but winter temperatures are colder and summer temperatures are slightly cooler. The annual temperature range is slightly greater: −55°. Summer is slightly shorter and winter slightly longer in the more northern city.

n-1   What is the name of the city? Kingston

n-2   How does its monthly temperature pattern differ from Miami's?
Kingston has a smaller temperature range (5°) than Miami. Both cities are warm year-round, but Miami shows more seasonal variation. They have identical high temperatures, but Kingston's lows are not as cool as those in Miami.

o   List the name of each of the cities displayed in the graph and complete the information in the table below.

CITY	LATITUDE	COLDEST MONTH	LOWEST TEMPERATURE (°F)	HOTTEST MONTH	HIGHEST TEMPERATURE (°F)	TEMPERATURE RANGE OVER 12 MONTHS
Quebec	46.9°	January	15°	July	70°	55°
Boston	42.5°	January	28°	July	73°	45°
Miami	25.9°	January, February	68°	August	83°	15°
Kingston	18.8°	January, February	78°	July, August	83°	5°

p   Based on the information displayed in the graph, the map, and the table on your answer sheet, state a hypothesis about how the monthly temperature patterns change as latitude increases.
Student answers will vary. Answers should include the following points:
As latitude increases, the range of temperatures over the year increases.
The lower latitudes have less seasonal variation and tend to be warm year-round.
Temperatures get steadily colder as latitude increases.
January and February are the coldest months and July and August are the hottest months. (Note: this is a correct observation based on the four cities students are observing in this problem. A later step will focus on the difference between the Northern and Southern hemispheres.)

Step 7   Test your hypothesis

g-1   Complete the table below.
Student answers will vary.

CITY	LATITUDE
Stockholm	59°
Berlin	53°
Warsaw	53°
Prague (Praha)	50°
Vienna	48°
Budapest	48°
Athens (Athinai)	38°
Rome (Roma)	42°

g-2   Do the cities you selected confirm or dispute your hypothesis? Explain.
Student answers will vary depending on their hypothesis. These cities show a similar pattern to that observed in North America.

Step 8   Analyze temperature patterns in the Southern Hemisphere

e   Complete the table below.

CITY	LATITUDE	COLDEST MONTH	LOWEST TEMPERATURE (°F)	HOTTEST MONTH	HIGHEST TEMPERATURE (°F)	TEMPERATURE RANGE OVER 12 MONTHS
Darwin	−13°	July	78°	January, February	84°	6°
Brisbane	−27°	July	59°	January, February	77°	18°
Sydney	−34°	July	53°	January, February	72°	19°
Melbourne	−38°	July	48°	February	68°	20°

f   Compare the monthly temperature patterns in the Southern Hemisphere to those in the Northern Hemisphere.

Patterns in the Southern Hemisphere mirror those in the Northern Hemisphere. Winter temperatures are not as cold in the Southern Hemisphere because none of the cities has a latitude greater than −38°. The major difference is that the warmest and coldest months are reversed.

Formulate a hypothesis about the relationship between monthly temperature patterns and increases in latitude.

Student answers will vary.

Step 9   Test your hypothesis on how latitude affects monthly temperature patterns in the Southern Hemisphere

e-1   Complete the table below.

CITY	LATITUDE
Cape Town	−34°
Johannesburg	−26°
Gabarone	−25°
Luanda	−9°

e-2   Based on your observations, do the cities you selected confirm or dispute your hypothesis about how latitude affects monthly temperature patterns in the Southern Hemisphere? Explain.

Student answers will vary. The patterns are the same as those seen in Australia.

Step 10   Investigate the ocean's influence on temperature

b-1   In which Canadian city would you experience the coldest winter temperatures? Winnipeg

b-2   In which Canadian city would you experience the warmest winter temperatures? Vancouver

b-3   Looking at the map, why do you think the warmest city has temperatures that are so much warmer than the others in the winter?

Vancouver is the only Canadian city (on this map) located on the coast. The proximity to the ocean has a steadying effect on the air temperature in Vancouver throughout the year. Therefore, the fluctuation between summer and winter temperatures is not as large as with inland cities at the same latitude.

MODULE 1
ArcMap

MODULE 2
global

MODULE 2
regional

**MODULE 3
global**

MODULE 3
regional

MODULE 3
advanced

MODULE 4
global

MODULE 4
regional

MODULE 4
advanced

MODULE 5
global

MODULE 5
regional

MODULE 5
advanced

MODULE 6
global

MODULE 6
regional

MODULE 7
global

MODULE 7
regional

MODULE 7
advanced

k-1   Complete the table below.

CITY	LATITUDE
London	51°
Amsterdam	52°
Berlin	52°
Warsaw	52°
Kiev	50°

k-2   What do these cities have in common as to their location on the earth?

All the cities are in the Northern Hemisphere, on the continent of Europe, and at approximately 50° north latitude.

k-3   Which cities have the mildest temperatures? London and Amsterdam

k-4   What happens to the winter temperatures as you move from London to Kiev?

Winter temperatures get steadily colder as you move east and inland.

k-5   Why do you think some cities have milder temperatures than the others?

Students should note that these cities are the warmest near the ocean (London on an island, Amsterdam on the coast).

l   Based on your observations of Canada and Western Europe, state a hypothesis about the influence of proximity to the ocean (or distance from it) on patterns of temperature.

Student answers will vary. They should note that those cities closest to the ocean have milder temperatures than cities at the same or similar latitudes located on the interior of continental landmasses.

Step 11   Investigate the impact of elevation on temperature patterns

e-1   Complete the table below.

CITY	LATITUDE
Kisangani	1° (0.7)
Libreville	0°
Quito	1° (−.38)
Singapore	1° (1.11)

e-2   What do these cities have in common as to their location on the earth? All are located very close to the equator.

e-3   What temperature pattern do these four cities have in common?

All five cities show very little range in monthly temperatures throughout the year (6° or less).

e-4   How is Quito different from the other three?

Its temperatures are significantly cooler than the other three cities.

e-5   Since all these cities are located on or very near the equator, what other factor could explain the difference in their temperature patterns?

Student answers will vary. They should not predict that Quito's close proximity to the ocean causes its cooler temperatures, because they just learned that proximity to the ocean causes milder temperatures.

h   Analyze the selected records and complete the table below.

CITY	ELEVATION (FEET)
Kisangani	1,361
Libreville	32
Quito	9,226
Singapore	104

j   Based on your observation of temperatures along the equator and the information in the table above, state a
        hypothesis about the influence of elevation on patterns of temperature.

        Student answers will vary. Students should note that cities at significantly higher elevations have cooler temperatures
        than other cities at a similar latitude.

Step 12   Revisit your initial ideas

g   Rank the 13 cities from coldest to hottest according to their average January temperatures.

        1. Irkutsk   8. Quito
        2. Minneapolis   9. Wellington
        3. Helsinki   10. Miami
        4. Lhasa   11. Khartoum
        5. Vancouver   12. Buenos Aires
        6. London   13. Singapore
        7. Tunis

j   Rank the 13 cities from hottest to coldest according to their average July temperatures.

        1. Khartoum   8. Vancouver
        2. Miami   9. Helsinki
        3. Singapore   10. Lhasa
        4. Tunis   11. Quito
        5. Minneapolis   12. Buenos Aires
        6. London   13. Wellington
        7. Irkutsk

l   Put a check mark (✔) next to those answers that you predicted correctly.

MODULE 1
ArcMap

MODULE 2
global

MODULE 2
regional

MODULE 3
global

MODULE 3
regional

MODULE 3
advanced

MODULE 4
global

MODULE 4
regional

MODULE 4
advanced

MODULE 5
global

MODULE 5
regional

MODULE 5
advanced

MODULE 6
global

MODULE 6
regional

MODULE 7
global

MODULE 7
regional

MODULE 7
advanced

Module 3   Regional case study: Seasonal Differences

Step 3   Observe patterns of rainfall

a-1   Which month gets the most rainfall in Bombay?   July

a-2   Which months appear to get little or no rainfall in Bombay?   December–April

a-3   Approximately how much rainfall does Bombay get each year (in inches)?   83

a-4   Write a sentence summarizing the overall pattern of rainfall in Bombay in an average year.
Bombay gets more than 80 inches of rain per year in a concentrated period from June to September.

d-1   How did this change the map?   The new city is selected. It turns blue on the map.

d-2   How did this change the graphs?   Both graphs now reflect data for the new city.

e   Analyze the graphs and fill in the Mangalore section of the table below.

CITY	MONTHS WITH RAINFALL	HIGHEST MONTHLY RAINFALL (INCHES)	TOTAL ANNUAL RAINFALL (INCHES)
Mangalore	April–November	40	135
Bombay			
Ahmadabad			

f-1   Complete the rest of the table in step e above.

CITY	MONTHS WITH RAINFALL	HIGHEST MONTHLY RAINFALL (INCHES)	TOTAL ANNUAL RAINFALL (INCHES)
Mangalore	April–November	40	135
Bombay	May–October	26	84
Ahmadabad	June–September	13	32

f-2   As you move northward along the subcontinent's west coast, how does the pattern of rainfall change?
The rainy season gets shorter. It starts later in the year and ends earlier. The monthly and yearly rainfall totals decline.

f-3   Although the monthly rainfall amounts differ, what similarities do you see among the overall rainfall patterns of these three cities?
In all three cities, the rainy seasons and dry seasons are at the same time of year. In each city, July has the highest rainfall total of any month, and the period from December through March is dry.

Step 4   Compare coastal and inland cities

a   Complete the table below.

CITY	MONTHS WITH RAINFALL	HIGHEST MONTHLY RAINFALL (INCHES)	TOTAL ANNUAL RAINFALL (INCHES)
Bangalore	April–December	7	36

b   How does the rainfall pattern of Bangalore compare with that of Mangalore?
Similarities: The two cities have a rainy season between April and December.
Differences: Mangalore gets approximately four times as much rain as Bangalore.

d   What is the distance between the two cities? Approximately 170 miles

e   How can this data help you explain the differences between patterns of rainfall in inland Bangalore and coastal
        Mangalore?

        Mangalore is on the coast while Bangalore is on the interior (Deccan) plateau. A narrow coastal mountain range
        (the Western Ghats) separates the two cities. The significant difference in total and monthly rainfall results from
        the orographic effect produced by the Western Ghats. Moist monsoon winds are forced to rise to go over these
        mountains as they come ashore. Condensing in the cooler upper atmosphere, most of the monsoon's moisture
        falls on the windward side of the mountains, leaving the inland side much drier.

Step 5   Compare eastern and western South Asian cities

a-1   Analyze the graphs and complete the table below.

CITY	MONTHS WITH RAINFALL	HIGHEST MONTHLY RAINFALL (INCHES)	TOTAL ANNUAL RAINFALL (INCHES)
Kabul	December–May	3	11
Herat	December–April	2	10

a-2   Describe the pattern of rainfall in these two cities.

        Both of these cities are extremely dry. What little rainfall they do receive falls in the early months of the year.

a-3   How do you think Afghanistan's rainfall pattern will affect the way of life in that country?

        There is not enough rainfall to support agriculture. They will have to rely on activities such as nomadic herding or
        extractive industry if any appropriate resources exist.

b-1   Analyze the graphs and complete the table below.

CITY	MONTHS WITH RAINFALL	HIGHEST MONTHLY RAINFALL (INCHES)	TOTAL ANNUAL RAINFALL (INCHES)
Calcutta	February–November	13	64
Dhaka	February–November	16	79

b-2   Describe the pattern of rainfall in these two cities.

        These two cities have significant annual rainfall total with a distinct rainy season that lasts longer than the rainy
        season on the southwest coast. The dry season lasts from November to February. The majority of the rain falls
        between May and October.

c   What is happening to the patterns of rainfall as you move from west to east across South Asia?

        The amount of annual rainfall increases as you move eastward and the length of the rainy season gets longer.

Step 6   Observe yearly precipitation

e-1   Which regions within South Asia get the least rainfall? The northwest (Afghanistan and Pakistan)

e-2   Which regions within South Asia get the most rainfall? The southwest coast and the northeast

e-3   In step 5c you were comparing Calcutta, Herat, New Delhi, and Dhaka. Does the map of yearly rainfall that is on
        your screen now reflect the observation you made at that time? Explain.

        Student answers will vary, but essentially, students should observe that precipitation does increase as you move
        from west to east across South Asia.

f   What relationships do you see between South Asia's patterns of yearly rainfall and its physical features?

The region's heaviest rainfall is on the windward side of the Western Ghats and the Himalayas. Orographic lift is responsible for these areas of heavy rainfall. Cities on the Deccan Plateau, on the subcontinent's interior, get significantly less rainfall because they lie in the rain shadow of the mountains.

Step 7   Explore the monsoon's impact on agriculture and population density

b-1   Which regions or countries of South Asia are suitable for agriculture and which are not? Explain.

Student answers will vary. The western section of South Asia (Afghanistan, Pakistan, and western India) does not get enough rainfall to support agriculture. Additionally, much of Afghanistan and Pakistan is in the mountains, making agriculture unlikely there. Most of the remainder of the subcontinent is suitable for farming because it gets sufficient rainfall and is either a plain or plateau.

b-2   In which regions of South Asia do you expect to see the lowest population density? Explain.

Student answers will vary. Students should expect the dry mountainous west to have the lowest population density because the region cannot produce enough food to support a large population.

b-3   In which regions of South Asia do you expect to see the highest population density? Explain.

Student answers will vary. Students should recognize the importance of rivers to agriculture (alluvial flood plain, fertile deltas, and a steady source of water) and predict a high population density there.

e-1   Does the Agriculture layer reflect the predictions you made in step 7b? Explain.

Answers will vary depending on their answer in step 7b. However, the data does illustrate lack of farming in the dry mountainous regions.

e-2   Why are grazing, herding, and oasis agriculture the major activities in Afghanistan?

Mountainous terrain and scarce rainfall make these the only viable economic activities for most people.

e-3   What do you know about rice cultivation that would help explain its distribution on the agriculture map?

Students familiar with rice cultivation will note that this is a crop that is often grown in flooded fields (wet rice cultivation) and requires a lot of water. Therefore, it makes sense that rice is cultivated in areas with the highest rainfall.

e-4   Is there any aspect of the agriculture map that surprised you? Explain.

Student answers will vary. Some students may be surprised about the agricultural activity in Pakistan since the area is so dry.

i-1   Does the Population Density layer reflect the predictions you made in step 7b? Explain.

Student answers will vary depending on their prediction.

i-2   Why is Afghanistan's population density so low?

Its harsh conditions make this an area that cannot support a large population.

i-3   Since most of Pakistan gets little to no rainfall, how do you explain the areas of high population density in that country?

The Indus River provides a rich alluvial flood plain and a year-round supply of water for irrigation.

i-4   What is the relationship between population density and patterns of precipitation in South Asia?

Overall population density is highest where rainfall amounts are conducive to agriculture. The notable variation to this pattern is the high population density along the rivers—particularly in the west. The rich soil and dependable source of water on the Indo-Gangetic Plain enable agriculture to support dense populations in spite of insufficient rainfall in some areas or at some times of year.

i-5   What is the relationship between population density and physical features in South Asia?

Population density is lowest in mountainous areas of Afghanistan and Pakistan and highest on the Indo-Gangetic Plain.

Module 3   Advanced investigation: Sibling Rivalry

Step 4   Analyze characteristics of El Niño and La Niña

e, f   Record your temperature observations in the table. Record your observations of precipitation characteristics for the same time period.

YEAR / SO	TEMPERATURE CHARACTERISTICS	PRECIPITATION CHARACTERISTICS
1997 / El Niño	Ranges above normal from 1–5 degrees Celsius	500 mm above normal over much of the Pacific
1999 / La Niña	Ranges below normal from 1–2 degrees Celsius	200 mm below normal over Pacific west of South America

h   Synthesize the information you've recorded and develop a definition of El Niño and La Niña. Write these definitions on your answer sheet.

Student answers will vary.

Step 5   Are El Niño and La Niña equal and opposite?

a   Complete the table on your answer sheet by filling in your observations for El Niño and La Niña's effect on the different regions of the world.

WORLD REGION	TEMPERATURE		PRECIPITATION		OTHER		OTHER	
	DEC. 1997 EL NIÑO	MAR. 1999 LA NIÑA	DEC. 1997 EL NIÑO	MAR. 1999 LA NIÑA	DEC. 1997 EL NIÑO	MAR. 1999 LA NIÑA	DEC. 1997 EL NIÑO	MAR. 1999 LA NIÑA
North America	Ranges above normal from 1–2° C.	1–1.5° above normal in north and 1–1.5° C below normal in south.	Parts of southeast 10–50 mm above normal with scattered areas 10–20 mm below normal.	Predominantly normal with small areas in north 10–20 mm below and 10–50 mm above in south.				
South America	Northern half of continent ranges above normal from 1–5° C. Southern parts are 0–1.5° C below normal.	Predominantly normal.	10–500 mm below normal in north with small areas 10–100 mm above normal.	10–40 mm below normal in east and 10–200 mm above normal in central part of the region.				
Europe	Central Europe 0.5–1° C above normal.	Normal.	South and west 10–50 mm above normal.	Normal.				

Step 5   Are El Niño and La Niña equal and opposite? (continued)

a   (continued)

WORLD REGION	TEMPERATURE		PRECIPITATION		OTHER		OTHER	
	DEC. 1997 EL NIÑO	MAR. 1999 LA NIÑA	DEC. 1997 EL NIÑO	MAR. 1999 LA NIÑA	DEC. 1997 EL NIÑO	MAR. 1999 LA NIÑA	DEC. 1997 EL NIÑO	MAR. 1999 LA NIÑA
Africa	.5–2° C above normal.	Central Africa .5–1.5° C above normal.	Mixed central and south 10–200 mm below normal. Small areas 10–100 mm above normal.	South 10–50 mm above normal. Small areas 10–250 mm below normal.				
Asia	.5–1.5° C below normal in north and .5–2° C above normal in south.	.5–2° C above normal in the South. Small scattered areas 1° C below normal.	Southeast 10–200 mm above normal. Small areas 10–20 mm below normal.	Small areas 10–30 mm above normal. Much of south 10–40 mm below normal.				
Oceania	.5–3° C above normal in most areas. Some spots slightly below normal.	Predominantly normal except western Australia .5–2° C above normal. Small sections of east 1° C below normal.	10–400 mm below normal with isolated areas 200 mm above normal.	Western areas 10–250 mm below normal. Eastern areas 10–200 mm above normal.				
Pacific Ocean	Ranges above normal from .5–5° C.	Ranges from 1° C above normal to 2° below normal	400 mm above normal over much of the Pacific.	200 mm below normal over Pacific west of South America.				

b   Based on the data you recorded in the previous question, is La Niña equal and opposite to El Niño? Explain your answer.

No. El Niño is a much longer-lasting event and produces a larger difference in precipitation and temperature than La Niña. It is true that they are opposite. Where El Niño is associated with an increase in water temperature (off the coast of South America), La Niña is associated with a decrease in water temperature. However, the increase in water temperature during El Niño is greater than the decrease in water temperature during La Niña. Therefore, they are not equal effects.

c-1   Was one year better than the other for you and your community?

Student answers will vary depending on which region of the world they are living in or addressing.

c-2   If in one year you received greater than normal rainfall, did your town have problems with flooding?

Student answers will vary depending on which region of the world they are living in or addressing.

c-3   If weather was unseasonably warm and mild, did outdoor activities such as amusement parks have greater turnout?

Student answers will vary depending on which region of the world they are living in or addressing.

c-4   How did these things affect the local economy?

Student answers will vary depending on which region of the world they are living in or addressing.

Module 4   Global perspective: The March of Time

Step 3   Look at cities in 100 C.E.

b-1   Where are they located on the earth's surface?

Many of the cities are approximately 30 degrees north latitude. All of the cities are in the northern or eastern hemispheres.

b-2   Where are they located in relation to each other?

Five of the cities are located in the lands surrounding the Mediterranean Sea. All but three of the cities are in Asia. None of the cities is located in North or South America or Australia.

b-3   Where are they located in relation to physical features?   All of the cities are located near rivers or near the coast.

c   What are possible explanations for the patterns you see on this map?

Answers will vary. Possible answers include the influence of climate, the extent of the Roman Empire, trade, suitability for agriculture, and so on.

Step 4   Find historic cities and identify modern cities and countries

e, g   Use the Find and Identify tools to complete the information in the table below.

HISTORIC CITY NAME	MODERN CITY NAME	MODERN COUNTRY
Carthage	Tunis	Tunisia
Antioch	Antioch	Turkey
Peshawar	Peshawar	Pakistan

Step 5   Find the largest city of 100 C.E. and label it

a   What's your estimate of how many people lived in the world's largest city in 100 C.E.?   Answers will vary.

g-1   What was the largest city in 100 C.E.?   Rome

g-2   What was the population of the world's largest city in 100 C.E.?   450,000

Step 6   Look at cities in 1000 C.E. and label the most populous city

c-1   What notable changes can you see from 100 C.E. to 1000 C.E.?

The Mediterranean Sea is no longer the site of half the world's largest cities.

c-2   What similarities can you see between 100 C.E. and 1000 C.E.?

Cities still cluster around 30 degrees north latitude.
All of the cities are under 1,000,000 population.
All but two of the cities are in Asia.
None of the top 10 cities is located in the Americas or Australia.

f-1   What was the largest city in 1000 C.E.?   Cordova

f-2   What was the population of the world's largest city in 1000 C.E.?   450,000

Step 7   Observe and compare other historic periods

b   Explore the map to complete the table below.

YEAR C.E.	LARGEST CITY	POPULATION OF LARGEST CITY	MAJOR DIFFERENCES FROM PREVIOUS TIME PERIOD
100	Rome	450,000	Not applicable
1000	Cordova	450,000	Mediterranean Sea is no longer a center of urban development.
1500	Beijing	672,000	Four of the 10 largest cities are now in China.
1800	Beijing	1,100,000	A city exceeds 1,000,000 for the first time. Europe now has three of the largest cities. Japan now has three of the largest cities.
1900	London	6,480,000	Nine of the 10 largest cities are now in Europe (six) and North America (three). Five of the European cities are in Western Europe. Only one of the cities is in Asia. All 10 cities are over 1,000,000. For the first time, major cities move into more northern middle latitudes. The size of the largest city is six times what it was 100 years earlier.
1950	New York	12,463,000	First time a South American city is on the Top Ten list. First city in Southern Hemisphere is listed. Europe drops back to three of the top 10 cities with only two in Western Europe.
2000	Tokyo	23,620,000	Half of cities are over 15,000,000. Western Europe has only one of the top 10 cities.

Step 8   State a hypothesis

b   In the table below, state in the left column which periods in history are associated with the greatest changes. In the right column, state possible explanations for the changes that you see.

TIME PERIODS OF SIGNIFICANT CHANGE	EXPLANATION FOR CHANGE
1800 through 1900	The Industrial Revolution caused rapid growth of European and North American cities. It also led to the immigration of millions of people to the United States from Europe.
1950 through 2000	The size of the world's largest cities mushroomed because of rapid economic growth in the developed world and the loss of agricultural jobs in the developing world (sending people to cities in hopes of finding a job).

Step 9   Investigate cities in the present time

a-1   How many of your original guesses are among the cities in Top Ten Cities, 2000 C.E.? Answers will vary.

a-2   Which cities did you successfully guess? Answers will vary.

f, g   In the table below, write the name, population, and rank for the other cities on your list. For cities that are not in the top 30, leave the population column blank and write >30 in the rank column.

CITY NAME	POPULATION	RANK
Los Angeles	9,763,600	14
Hong Kong	5,395,997	27
Bombay	9,950,000	12
Detroit		>30
Chicago	7,717,100	18

i   In general, how far are these other cities from the top 10? Answers will vary.

Module 4   Regional case study: Growing Pains

Step 3   Compare birth rate and death rate data

b-1   Which world region or regions have the highest birth rates? Sub-Saharan Africa, Southwest Asia

b-2   Which world region or regions have the lowest birth rates?
North America (Canada and the United States), Australia, Europe (with Asian Russia)

c-1   Which world region or regions have the highest death rates? Sub-Saharan Africa

c-2   Which world region or regions have the lowest death rates?
Mexico, Central America, Australia, China, and the Middle East

d-1   If the overall rate of growth is based on the formula BR – DR = NI, which world regions do you think are growing the fastest?
Most sub-Saharan African countries

d-2   Which world regions do you think are growing the slowest?
Many European countries

l   Choose two European countries and two African countries and record their birth and death rates in the table below.
Answers will vary. The answers provided below are examples of correct responses:

COUNTRY AND CONTINENT NAME	BIRTH RATE/1,000	DEATH RATE/1,000
Niger (Africa)	51.45	23.17
Spain (Europe)	9.22	9.03
Hungary (Europe)	9.26	13.34
Ethiopia (Africa)	45.13	17.63
Chad (Africa)	48.81	15.71

m   List three questions that the Birth Rate and Death Rate maps raise in your mind.
Answers will vary.

Step 4   Add the Natural Increase layer

c-1   What is happening to the population in the countries that are red?
Their death rates exceed their birth rates—over time these populations will decline unless migration into the country makes up for the net loss from natural increase.

c-2   Which world region is growing the fastest? Sub-Saharan Africa

c-3   Which world region is growing the slowest? Europe

c-4   Think about what it would mean for a country to have a population that is growing rapidly or one that is growing slowly. Which of these two possibilities (fast growth or slow growth) do you think would cause more problems within the country?
Most problems would occur in countries with fast growth.

On the answer sheet, briefly list some of the problems you would expect to see.
Answers will vary, but students should recognize that a country with rapid growth will have a difficult time keeping up with the constantly increasing need for education, health care, social infrastructure, resources, and jobs.

Step 5   Look at standard of living indicators for Europe and Africa

f   Complete the table below.

INDICATOR	COMPARE SUB-SAHARAN AFRICA AND EUROPE	WHAT DOES THIS "INDICATE" ABOUT THE STANDARD OF LIVING IN THESE REGIONS?
Population > 60 years	Africa: Most countries have 3.15–5.7% in this age group. Europe: Most countries have 17.13–23.91% in this age group.	Africa: Low percent indicates many people die prematurely and do not reach old age. Low standard of living. Europe: High percent indicates more people live into their sixties and beyond because of good health care, sanitation, adequate food supply, and so forth. High standard of living.
GDP per capita	Africa: Most countries have the lowest level of GDP. Europe: Most countries have the highest or second-highest level of GDP.	This is not the same as average income—be sure that students do not make that assumption. A higher GDP per capita does indicate a wealthier country, and that means more money to spend on the infrastructure. High GDP means a high standard of living and enough capital to continue to grow and expand economically.
Infant mortality rate	Africa: All countries have 19.84 infant deaths/1,000 born or higher. Europe: All countries have fewer than 19.84 infant deaths/1,000 born.	High rate of infant mortality indicates a low standard of living. This statistic is typically used to evaluate the health conditions (sanitation, health care, food supply, disease, and so forth) in a country because newborns are much more susceptible to death from such health problems than adults or older children.
Life expectancy	Africa: There is a mix of low to high life expectancy, with most countries having a life expectancy of 37.24–57.42 years. Europe: Most countries have the highest life expectancy of 73.79–83.46 years.	A higher life expectancy reveals a higher standard of living because it reflects prevailing conditions in a country at this time.
Literacy rate	Africa: Literacy rate varies from country to country, with no country at the highest level. Europe: All but one country is at the highest level of literacy.	Information on literacy, while not a perfect measure of education in a country, is probably the most easily available and valid for international comparisons. Low levels of literacy and education in general can impede the economic development of a country in the current rapidly changing, technology-driven world.
Percent of workforce in service sector	Africa: The service sector of the workforce varies from the lowest level to the highest level from country to country. Europe: The service sector of the workforce varies between the top two levels throughout Europe.	A higher percent of the workforce in the service sector indicates a higher standard of living. As a country becomes more developed economically, a larger percent of its workforce is employed in the service sector. The workforce of less developed countries is characterized by higher percent of workers in agriculture and industry.

MODULE 1
ArcMap

MODULE 2
global

MODULE 2
regional

MODULE 3
global

MODULE 3
regional

MODULE 3
advanced

MODULE 4
global

MODULE 4
regional

MODULE 4
advanced

MODULE 5
global

MODULE 5
regional

MODULE 5
advanced

MODULE 6
global

MODULE 6
regional

MODULE 7
global

MODULE 7
regional

MODULE 7
advanced

Step 6   Add the Net Migration layer

a   In step 5 you compared standard of living indicators in Europe and sub-Saharan Africa. Based on your observations of those indicators, which region would you expect to have a negative net migration? A positive net migration?

Negative: Europe (countries with high standards of living)

Positive: Sub-Saharan Africa (countries with low standards of living)

Explain your answer:

Answers will vary, but students should recognize that there will be more out-migration from countries with low standards of living and more in-migration to countries with a higher standard of living.

d   Summarize the overall patterns of net migration in Europe and sub-Saharan Africa in the table below.

NET MIGRATION IN SUB-SAHARAN AFRICA	NET MIGRATION IN EUROPE
Generally, there is a tendency for out-migration from sub-Saharan Africa.	Generally, there is a tendency for in-migration to Europe with the exception of certain Eastern European countries.

e   What are possible political or social conditions or events that could explain any of the migration patterns you see on the map?

Possible answers include Balkan wars, reunified Germany, political unrest in Liberia and Rwanda, and so on.

Step 7   Draw conclusions

h   Based on your map investigations, write a hypothesis about how a country's rate of natural increase affects its standard of living and its net rate of migration.

Answers will vary, but students should note that natural increase has a direct effect on standard of living, and that standard of living creates push–pull factors that influence migration.

i   In the table below, illustrate your hypothesis with data from one European country and one sub-Saharan African country.

Answers will vary depending on the hypothesis that was created in step 7h. However, students should include data for each European and African country that includes natural increase, net migration, and other datasets that support their hypothesis.

EUROPE	DATA	AFRICA
	Country name	
	Natural increase	
	Net migration	

Step 8   Design a layout

n   What are the units of measurement? Kilometers

Module 4   Advanced investigation: Generation Gaps

Step 2   Create new layers

a-1   What world regions have the highest rates of natural increase today? Africa and the Middle East

a-2   What world regions have the lowest rates of natural increase today? Europe and Russia

Step 3   Thematically map world age structure

l-1   What regions of the world have populations with a high percentage below age 15?

Most African countries and some parts of the Middle East

l-2   What regions of the world have populations with a relatively low percentage below age 15?

Europe and North America

l-3   Why do you think there is a much higher percentage of children in some populations than in others?

Student answers will vary. There is a high percentage of children in growing populations because of high birth rates and sustained high fertility rates. As a result, each generation will be larger than the one before it.

l-4   Why are there far more people 60 years of age and older in some populations than in others?

Student answers will vary. Populations with high standards of living and good medical care tend to have large populations over 60 years old.

l-5   Based on the layers in the Generation Gaps data frame, how do you think natural increase is related to age structure?

Where natural increase is high, the population less than 15 years of age is also high. Countries with the greatest percentage of population over age 60 have a low natural increase. A high rate of natural increase leads to an age structure with a large percent of children.

Step 7   Analyze census data for your county

h   Are there identifiable concentrations of children between birth and five years old in your county? How do you explain these concentrations?

Student answers will vary based on their chosen county.

i   Where is the greatest number of this age group found?

Student answers will vary based on their chosen county.

j-1   Do you have any colleges in your county? Any retirement communities? Are these institutions reflected in the census data?

Student answers will vary based on their chosen county.

j-2   Identify patterns of age distribution in your county and suggest explanations for those patterns.

Student answers will vary based on their chosen county.

MODULE 1
ArcMap

MODULE 2
global

MODULE 2
regional

MODULE 3
global

MODULE 3
regional

MODULE 3
advanced

MODULE 4
global

MODULE 4
regional

**MODULE 4
advanced**

MODULE 5
global

MODULE 5
regional

MODULE 5
advanced

MODULE 6
global

MODULE 6
regional

MODULE 7
global

MODULE 7
regional

MODULE 7
advanced

Module 5   Global perspective: Crossing the Line

Step 3   Explore mountain ranges as physiographic boundaries

l   The Pyrenees Mountains are the border between which two countries? Spain and France

n   Complete the table below:

COUNTRIES THAT HAVE MOUNTAIN RANGES AS POLITICAL BOUNDARIES	MOUNTAINS THAT FORM THE BOUNDARY
Italy and Switzerland	Alps
Italy and France	Alps
Poland and Czech Republic	Sudeten Mountains

Step 4   Explore bodies of water as physiographic boundaries

c   In the table below, record the names of three sets of countries that share a boundary that's a river.
Some examples include:

COUNTRIES THAT HAVE RIVERS AS BOUNDARIES	RIVER THAT FORMS THE BOUNDARY
France and Germany	Rhine
Germany and Poland	Oder
Romania and Bulgaria	Danube
Belarus and Ukraine	Dnieper

d   Name three landlocked countries in Western Europe.
Some examples include: Switzerland, Austria, Czech Republic, Slovakia, Hungary, and Serbia and Montenegro.

Step 5   Explore geometric boundaries

e   Record three sets of countries in the table below.
Some examples include:

COUNTRIES THAT ARE SEPARATED BY GEOMETRIC BOUNDARIES		
Egypt and Libya	Niger and Algeria	Namibia and Botswana
Sudan and Chad	Libya and Sudan	Algeria and Mauritania
Libya and Chad	Algeria and Mali	Angola and Zambia

Step 6   Explore anthropographic boundaries based on language and religion

h   Determine the principal language groups in the regions listed below.
South America: Indo-European
Western Europe: Indo-European, Uralic, and Others

j Locate three examples in the world where political boundaries coincide with anthropographic boundaries based on language.
Possible answers include:

ANTHROPOGRAPHIC BOUNDARIES BASED ON LANGUAGE COINCIDE WITH POLITICAL BOUNDARIES BETWEEN		
India and China	Kazakhstan and Russia	Thailand and Laos
Georgia and Russia	Finland and Norway/Sweden	Botswana and Zimbabwe
North Korea and China	Brazil and Paraguay	Venezuela and Guyana

n Determine the principal religions in the following regions:
North America: Protestant, Roman Catholic, Mixed Christian, Mormon, Indigenous
Africa: Sunni Muslim, Indigenous, Roman Catholic, Mixed Christian, Protestant, Eastern, Orthodox

p Locate three examples in the world where political boundaries coincide with anthropographic boundaries based on religion.
Possible answers include:

ANTHROPOGRAPHIC BOUNDARIES BASED ON RELIGION COINCIDE WITH POLITICAL BOUNDARIES BETWEEN		
India and China	Vietnam and Cambodia/Laos	Finland and Russia
India and Myanmar	Kazakhstan and Russia	Ireland and United Kingdom
India and Pakistan	Mongolia and Russia	Bangladesh and India
Thailand and Malaysia	Iran and Pakistan/Iraq/Turkmenistan	Germany and Czech Republic

Step 7 Review physiographic, geometric, and anthropographic boundaries

a Find additional examples of physiographic, geometric, and anthropographic boundaries between countries. Record your findings in the table below.
Answers will vary. See answers above for possible answers to each category.

Step 8 Explore the impact of boundary shape, cultural diversity, and access to natural resources

f Locate another example of each type of country. Record them in the following table in the Example 2 column.
Some examples include:

TYPE OF COUNTRY	EXAMPLE	EXAMPLE 2
Elongated	Chile	Vietnam, Panama
Fragmented	Philippines	Indonesia, Japan
Circular/Hexagonal	France	Uruguay, Zimbabwe
Small/Compact	Bulgaria	Costa Rica, Belgium
Perforated	South Africa	Italy (Vatican City, San Marino), Malaysia (Brunei)
Prorupted	Namibia	Thailand, Afghanistan

g-1 By using language groups as an indicator of cultural uniformity, identify three countries that reflect cultural uniformity.
Answers will vary. Some examples include Japan, France, Argentina, Italy, and Hungary.

g-2   By using language groups as an indicator of cultural diversity, identify three countries that reflect cultural diversity.
Answers will vary. Some examples include Canada, Spain, Nigeria, Burkina Faso, Syria, Turkey, India, Switzerland, Myanmar (Burma), Sudan, Sri Lanka, and Namibia.

i   Use the ArcMap tools and buttons you've learned in this investigation to find an example of a landlocked country on each of the following continents. For a continent that does not have a landlocked country, write "none."
Complete the table:

CONTINENT	LANDLOCKED COUNTRY
North America (including Central America)	None
South America	Bolivia and Paraguay
Africa	Mali, Burkina Faso, Niger, Chad, Central African Republic, Rwanda, Burundi, Uganda, Zambia, Zimbabwe, Botswana, Lesotho
Asia	Jordan, Afghanistan, Nepal, Bhutan, Laos, Mongolia, Kazakhstan, Uzbekistan, Turkmenistan, Kyrgyzstan, Tajikistan, Armenia, Azerbaijan

o-1   Name two Southeast Asian countries that do not have any oil and gas resources within their borders.
Cambodia, Laos, Vietnam

o-2   Name two Southeast Asian countries that have oil and gas resources within their borders.
Indonesia, Brunei, Thailand, Malaysia, Philippines, Myanmar (Burma)

Step 9   Explore boundary changes in the 1990s

c-1   Describe three political boundary changes you see between 1992 and 2004.
Answers will vary, but they should focus on the changes in Eastern Europe and the former USSR.

c-2   Name two countries that existed in 1992, but do not exist in 2004. USSR, Czechoslovakia, Yugoslavia

Step 10   Compare new countries

a   Select three countries from group A and three from group B and complete the table on the answer sheet.
Answers will vary based on countries selected. Refer to Global5.mxd for specific data.

Module 5   Regional case study: A Line in the Sand

Step 3   Identify countries that border the Arabian Peninsula

c   Record the names of the countries on the map that border the Arabian Peninsula to the north.

Jordan, Iraq, Kuwait

Step 4   Investigate the physical characteristics of the Arabian Peninsula

a-1   Is any part of the Arabian Peninsula mountainous? Yes

a-2   If so, where are the mountains located?

The peninsula is mountainous along its west (Red Sea) coast. A second region of mountains can be seen on the northeast (Gulf of Oman) coast.

c-1   Are there any parts of the Arabian Peninsula that do not have any water at all? If so, where are these regions?

Yes. The south-central part of the peninsula has no permanent bodies of water or streams.

c-2   Do you see any relationship between landforms and the availability of water?

Mountains and areas of higher elevation have more surface water.

e   Describe the bodies of water.

The permanent bodies of water look like disconnected fragments of rivers and lakes.

h-1   How many millimeters equal 10 inches? 254 mm

h-2   Based on the amounts of rainfall displayed on the map, do you think there is much farming on the Arabian Peninsula? Explain.

No. Most of the Arabian Peninsula is so dry that agriculture wouldn't be possible without an alternative source of water such as a river. Egypt, for example, is just as dry, but the Nile River provides water for agriculture.

h-3   Approximately what percentage of the Arabian Peninsula is desert?

Approximately 70–90 percent of the Arabian Peninsula is desert.

i   What is the approximate range of temperatures across the Arabian Peninsula during this period?

Sept.–Nov.: 14° C to 30° C; 57° F to 87° F

j-1   Which season is the hottest? Summer (June–August)

j-2   What is the approximate range of temperatures across the Arabian Peninsula during this period?

18° C to 40° C; 66° F to 104° F

k-1   What relationship do you see between the Arabian Peninsula's ecozones as displayed on this map and its patterns of landforms, precipitation, and temperature?

The limited zones of temperate grassland on the Arabian Peninsula are found in the mountains where there is more precipitation and milder temperatures.

MODULE 1
ArcMap

MODULE 2
global

MODULE 2
regional

MODULE 3
global

MODULE 3
regional

MODULE 3
advanced

MODULE 4
global

MODULE 4
regional

MODULE 4
advanced

MODULE 5
global

**MODULE 5
regional**

MODULE 5
advanced

MODULE 6
global

MODULE 6
regional

MODULE 7
global

MODULE 7
regional

MODULE 7
advanced

k-2   Complete the table. List three observations for each physical characteristic.

PHYSICAL CHARACTERISTICS OF THE ARABIAN PENINSULA	
Landforms and bodies of water	The Arabian Peninsula has a narrow coastal plain along the Red Sea that is separated by a low mountain range from the rest of the peninsula. The highest mountains in this range are at the southern point of the peninsula where the Red Sea meets the Gulf of Aden. A second, smaller range of mountains is found on the northeastern point of the peninsula along the Gulf of Oman. There are few permanent bodies of water on the Arabian Peninsula. The region has many intermittent streams.
Climate	Most of the Arabian Peninsula is a desert because it gets less than 10 inches of rainfall per year. Winters are mild in terms of temperature, but in the summer, average temperatures are extremely high.
Ecozones	Most of the Arabian Peninsula has a desert ecosystem. The primary exception to this are the temperate grasslands. They can be found in the mountains and higher elevations of the western part of the peninsula, and in the mountains of the northeast.

k-3   In your opinion, which of the region's physical characteristics would be considered "valuable" in a boundary decision? Explain.

In a boundary decision, the valuable characteristics are grassland ecosystems, areas with greater than 500 mm (19.5 inches) of annual precipitation, and areas with access to permanent bodies of water or springs.

Step 5   Investigate the human characteristics of the Arabian Peninsula

b-1   What is the principal agricultural activity on the peninsula?

The principal agricultural activity of the Arabian Peninsula is nomadic herding.

b-2   Based on what you now know about the physical characteristics of the region, why do you think the agricultural activity is so limited?

There is not sufficient water for farming in most of the region. Livestock can be herded from place to place depending on the seasonal availability of water and pastureland.

c-1   How does Yemen compare to the rest of the Arabian Peninsula in population density?

Southwestern Yemen has the largest area of relatively high population density (greater than 50 people per sq. km.).

c-2   Describe the overall population density of the Arabian Peninsula.

Most of the Arabian Peninsula has fewer than 25 people per sq. km. and at least half of that area has fewer than one person per sq. km.

f-1   Speculate on the most frequent use of the water at these springs and water holes.

Student answers will vary. Because of the large amount of nomadic herding, a logical conclusion is that most of the springs and water holes are used to water livestock.

f-2   Use your answers from previous step 5 questions and analysis of the maps to complete the table. List three observations for each human characteristic.

HUMAN CHARACTERISTICS OF THE ARABIAN PENINSULA	
Agricultural activities	The principal agricultural activity of the Arabian Peninsula is nomadic herding. Farming exists on a very limited basis in desert oases, on irrigated lands along the Red Sea, and in the high elevations of southern Yemen.
Population density and distribution	The Arabian Peninsula is very sparsely populated overall. Most of the population centers are along the coast (with a few exceptions in Saudi Arabia). The largest concentration of people is found in southern Yemen.

f-3   If an international boundary were to be drawn across some part of the Arabian Peninsula, how would these characteristics influence the perception of certain regions as being more "valuable" than others?

Nomadic herders would place a high value on having access to sources of water and grazing land. The population density of a region is a direct reflection of the ability of land to support population. Those areas with very low population density would be least valuable and those with high population density would be most valuable.

Step 6   Locate and describe the Empty Quarter

b-1   Complete the table. List three observations in each column.

THE EMPTY QUARTER	
PHYSICAL CHARACTERISTICS	**HUMAN CHARACTERISTICS**
The Rub' al-Khali or Empty Quarter is a desert region with virtually no permanent bodies of water and less than 2 inches of rainfall per year.	The Rub' al-Khali or Empty Quarter has no agricultural activity at all. The region is virtually uninhabited—most of it has less than one person per sq. km. And about one third has no people at all. The lack of roads in this region indicates minimal human presence.

b-2   What difficulties would an area like this present if an international boundary must cross it?

International boundary difficulties include:

There's no one living in the area to enforce the boundaries.

The lack of permanent landmarks (due to shifting sand dunes) makes the line difficult to mark and see.

Nomadic herders would want easy access throughout the border area.

Step 7   Explore Saudi Arabia's southern boundaries

c-1   Are the boundaries what you expected them to be?   Student answers will vary.

c-2   Which boundary remained unsettled?

The border between Saudi Arabia and Yemen was the only border that remained undefined.

i   What does the area between the green and purple lines represent?

It is claimed by both Saudi Arabia and Yemen—it is the disputed territory between these two countries.

j   What is the principal economic activity of the regions in dispute?

Nomadic herding. The disputed territory is an area of land that borders the Empty Quarter.

k   Describe the population distribution in the disputed territory.

The disputed territory is mostly uninhabited. Most of the disputed territory has fewer than one person per sq. km. The only area with a higher concentration of people is the western part of the territory with 1–25 people per sq. km.

Step 8   Draw the Saudi–Yemeni boundary

b-1   Does the red line go through any cities or towns? (Hint: You may need to zoom in again to answer the question.) If yes, approximately how many does the boundary pass through?

Student answers will vary. There are fewer than 10 villages that the boundary line actually passes through, but if you consider villages within a mile or two of the border (approximately 1.25 km.), there are many more.

b-2   How would you decide which side of the town to put the boundary on? Remember, this decision would determine whether the residents of that village would be citizens of Saudi Arabia or Yemen.

Student answers will vary. One way to decide where to put the boundary line is to survey villagers to find out whether they feel a closer affiliation with Yemen or Saudi Arabia. Such affiliations are based on long-standing tribal traditions.

MODULE 1 ArcMap
MODULE 2 global
MODULE 2 regional
MODULE 3 global
MODULE 3 regional
MODULE 3 advanced
MODULE 4 global
MODULE 4 regional
MODULE 4 advanced
MODULE 5 global
MODULE 5 regional
MODULE 5 advanced
MODULE 6 global
MODULE 6 regional
MODULE 7 global
MODULE 7 regional
MODULE 7 advanced

r   Does the new line seem to favor Yemen or Saudi Arabia? Explain.

The new border seems to favor Yemen. It gained control of all the disputed territory and gained even more territory beyond the previous boundary.

Step 9   Enter the maritime part of the boundary

d   What body of water does the maritime boundary traverse?   The Red Sea

e   How does the actual boundary established by the Treaty of Jeddah compare with the boundary you drew earlier?

Student answers will vary. In most cases, students will find that the Treaty of Jeddah gave more land to Yemen than they predicted.

f   Write three observations about the boundary line created by the Treaty of Jeddah.

Student answers will vary. Here are three possible responses:

The new boundary increased Yemen's territory.

Most of Yemen's new territory is land used by nomadic herders and desert.

The border settlement probably did not have a significant impact on Yemen's overall population as most of the new territory is uninhabited or very sparsely settled.

Step 10   Define the pastoral area

a   How many miles is 20 kilometers? (Hint: 1 kilometer = .6214 miles)   20 km. = 12.428 miles

m-1   In which part of the Saudi–Yemeni border will the pastoral area be most significant? Explain.

The pastoral area will be most significant in the western part of the boundary region (corresponding to Yemen1.shp) because this is the part of the boundary where nomadic herding is the characteristic agricultural activity. The remainder of the new boundary is north of the nomadic herding areas.

m-2   Why do you think the Treaty of Jeddah created a pastoral area?

Student answers will vary. However, students should understand that the establishment of a pastoral area recognizes that nomadic herding is incompatible with fixed and finite boundaries. The pastoral area represents a compromise between the need to clearly define the boundary between Saudi Arabia and Yemen and the reality that the border area is populated by people whose nomadic traditions include territory on both sides of that boundary.

Module 5   Advanced investigation: Starting from Scratch

Step 2   Explore map layers

a   List other important factors that influence boundary decisions.
Historic events (wars and treaties), ethnicity, natural resources are some. Student answers will vary.

c   Use the Religion legend and MapTips to determine three principal religions of South Asia. Record them here.
Hindu, Sunni Muslim, Buddhist

e   The boundary between which two religions corresponds to a physiographic boundary visible in the satellite image?
Hindu and Buddhist

f   Identify the principal language groups in South Asia. Indo-European, Dravidian, Sino-Tibetan

g   The boundary between which two language groups corresponds to a physiographic boundary visible in the satellite image?
Sino-Tibetan and Indo-European

Step 5   Draw the boundaries of new countries

a   Choose the continent you will be working on and record it here. Student answers will vary.

Module 6   Global perspective: The Wealth of Nations

Step 3   Evaluate the legends and patterns of the maps

a-1   What do the darkest colors represent?   A very high percentage in that sector

a-2   What do the lightest colors represent?   A very low percentage in that sector

b-1   What does the description "Very High" refer to in this map?
It refers to the percentage of the workforce that is in the agriculture sector.

b-2   Where are the countries with a high percentage of agricultural workers generally located?   Africa and parts of Asia

b-3   Where are the countries with a low percentage of agricultural workers generally located?
North America and Australia

c-1   Where are the countries with a high percentage of service workers generally located?
North America, eastern South America, Australia, and Europe

c-2   Where are the countries with a low percentage of service workers generally located?   Central Africa and India

c-3   What relationship, if any, do you see between the agriculture and services workforce maps?
They are generally opposite; where one is high, the other is low.

d   What patterns do you see on the map?
The lightest colors are in Africa, the darkest are in Eastern Europe and Argentina.

e   Using the workforce information in all three maps, in what part or parts of the world do you find the greatest number of developing countries?
In Africa and parts of the Middle East and Asia

Step 4   Analyze data on Bolivia

l-1   What percentage of workers in Bolivia are involved in agriculture?   38.9

l-2   What percentage of workers in Bolivia are involved in industry?   20.8

l-3   What percentage of workers in Bolivia are involved in service?   40.3

l-4   Would you classify Bolivia as a developed or developing country? Explain.
Developing. Regardless of their answer, it is important for students to support their answer with the data.

n   In the table below, record the classification (developed or developing) you gave Bolivia in the previous question. Complete the rest of the table by following the instructions in step 5.

COUNTRY	AGRICULTURE	INDUSTRY	SERVICE	DEVELOPING	DEVELOPED
Bolivia	High	Low	High	X	
India	Very High	Low	Low	X	
Australia	Very Low	Moderate	Very High		X
South Korea	Low	Moderate	Very High		X
Ukraine	Moderate	High	High		X
Uruguay	Very Low	Moderate	Very High		X

Step 8   Analyze GDP per capita and energy use data

b-1   What level is the GDP per capita for Bolivia? Low

b-2   Based on this new information and the workforce data, should Bolivia be classified as a developing or developed country? Developing

Why?

Low GDP is consistent with a developing nation. Previous data on workforce statistics also indicate that Bolivia is a developing country.

c-1   What level is the Energy use for Bolivia? Very Low

c-2   Based on this new information and previous data, should Bolivia be classified as a developing or developed country?

Developing

c-3   Why does energy use increase when a country develops?

Energy is used more when countries build infrastructure and establish manufacturing plants than when the primary mode of economic production is agriculture. Industry- and service-oriented production consume more energy than agriculture.

d-1   Complete the table below.

COUNTRY	GDP PER CAPITA	ENERGY USE	DEVELOPED OR DEVELOPING	IS THIS A CHANGE FROM YOUR EARLIER CLASSIFICATION?
Bolivia	Low	Very Low	Developing	No
India	Very Low	Moderate	Developing	No
Australia	Very High	Low	Developed	No
South Korea	High	Low	Developed	No
Ukraine	Very Low	Very Low	Developing	Yes
Uruguay	Moderate	Very Low	Developing	Yes

d-2   Name one country from above that you earlier classified as developed and that has GDP and Energy Use data that indicates it's developing.

Possible answers include Ukraine and Uruguay.

d-3   Based on the data you collected on these six countries, do you feel that the employment criteria are good indicators of a country's economic status? Explain your answer.

Student answers will vary. Employment criteria appear to be good indicators of developing or developed status in some cases, but not all. Students should realize that many factors make up a country's economic status, and as different factors are included as criteria, a country's classification may change. For example, although Ukraine and Uruguay have a high percentage of workers in the service industry, they have a low or average GDP and lower energy use. Inclusion of these additional factors make Ukraine and Uruguay lean toward a "developing" classification.

Module 6   Regional case study: Share and Share Alike

Step 3   Examine the map and attribute table

c-1   Which years does the layer contain data for? 1991 to 2000

c-2   How many attributes are there for each year? Six

c-3   What was the value of goods and services exported from Canada to the United States in 1991? $91,064,000,000

e-1   What is the name of the table? Attributes of NAFTA Countries

e-2   How many rows are there for each country on the map? One

Step 4   Relate another table to the layer table

d-1   How many rows are there for each country? Four

d-2   What information is collected under the field titled "Item" for Canada?
Total Exports to United States
Trade Balance with United States
Total Exports to Mexico
Trade Balance with Mexico

d-3   Describe in general terms the information collected in the field titled "Item."
Total Exports and Trade Balance for each of the NAFTA countries

d-4   How many years of data are represented in the table? 10

g   What happens in the NAFTA Trading Statistics table?
The rows for the selected country (U.S.) become highlighted.

h   What happens in the two tables and the map?
Canada is highlighted (blue) on the map and in the Attributes of NAFTA Countries table. Nothing changes in the NAFTA Trading Statistics table; the rows for the United States remain highlighted.

j   What happens in the two tables and the map?
The Mexico row that was clicked in the NAFTA Trading Statistics table becomes highlighted but Canada remains selected in the attribute table and map.

k   What have you observed about the way the NAFTA Trading Statistics table is tied to the NAFTA Countries attribute table and map layer?
The related table does not automatically reflect changes to a selection in the map or layer attribute table. The relate must be updated in one table or the other to show the related records.

Step 5   Examine export graphs

b-1   Which country exported more goods and services to Canada—Mexico or the United States? United States

b-2   Why is the graph empty in the space for Canada? A country (Canada) doesn't export goods to itself.

d   What happened to the graph? Only the data for Mexico (the selected country) is displayed.

e-1   How many years of data are represented on the graph? 10

e-2   What year does the first bar on the left represent? 1991

e-3   Compare the numbers on the y-axis with those in the two tables. Are the numbers on the graph in thousands, millions, or billions of dollars?

Millions

e-4   Looking at the graph, how would you describe the trend of Mexican exports to Canada over the 10-year period?

They increased dramatically

e-5   What was the approximate value of Mexican exports to Canada in 1991? $2,300,000,000 ($2.3 billion)

In 2000? $8,000,000,000 ($8 billion)

e-6   Approximately how many times greater is the 2000 export figure than the 1991 export figure?

3.5 (or 4, depending on how the student has rounded)

g-1   How would you describe the trend of Mexican exports to the United States over the 10-year period?

They increased steadily and dramatically.

g-2   Approximately how many times greater is the 2000 export figure than the 1991 export figure?

4.6 (or 5, depending on how the student has rounded)

Step 6   Examine a trade balance graph

c-1   Did Mexico have a trade surplus or deficit with the United States for 1992? Deficit

c-2   What was the approximate value of the trade balance for 1992? (Remember, the y-axis is in millions of dollars.)

$–5,000,000,000 ($–5 billion)

c-3   What was the first year that Mexico exported more to the United States than it imported from the United States?

1995

c-4   Describe the trend of Mexico's trade balance with the United States over the 10-year period.

It went from a deficit to a surplus that continued to grow until 2000.

d   Did Canada have a deficit trade balance with the United States anytime during the 10-year period? No

e-1   In 1998, was Canada's trade balance with the United States greater, smaller, or about the same as Mexico's?

About the same

e-2   In 2000, was Canada's trade balance with the United States greater, smaller, or about the same as Mexico's?

Greater

f   Referring to the Attributes of NAFTA_Trading_Statistics table, what was the exact value of Canada's trade balance with the United States in 2000?

$51,897,000,000 surplus

MODULE 1
ArcMap

MODULE 2
global

MODULE 2
regional

MODULE 3
global

MODULE 3
regional

MODULE 3
advanced

MODULE 4
global

MODULE 4
regional

MODULE 4
advanced

MODULE 5
global

MODULE 5
regional

MODULE 5
advanced

MODULE 6
global

MODULE 6
regional

MODULE 7
global

MODULE 7
regional

MODULE 7
advanced

Step 8   Evaluate the effectiveness of NAFTA

b   Use the graphs to find the value of exports between each set of countries for the year 2000. Write the information in the table below.

Student answers may vary slightly because they will be approximating the values from the chart.

DIRECTION OF EXPORT FLOW	VALUE OF EXPORTS (MILLION $)	TOTAL VOLUME BETWEEN PARTNERS (MILLION $)
United States to Mexico	115,000	255,000
Mexico to United States	140,000	
United States to Canada	180,000	410,000
Canada to United States	230,000	
Canada to Mexico	1,000	11,000
Mexico to Canada	10,000	

c   Add the export values together for each pair of countries (for example, United States to Mexico plus Mexico to United States) and write that number in the Total Volume Between Partners column.

Refer to totals in the table above.

d   Rank the trading partners by the overall volume of trade between the two countries. Use 1 for the partners trading the most and 3 for the partners trading the least.

United States–Mexico: 2   United States–Canada: 1   Canada–Mexico: 3

d-1   Do you think that NAFTA had a positive (+), negative (–), or neutral (n) effect on trade volume between each set of partner countries?

United States–Mexico: +   United States–Canada +   Canada–Mexico +

d-2   Do you think that any one of these three countries benefited more than the other two by NAFTA? If so, which country? Explain your answer.

Student answers will vary because they are asked to speculate. Trade volume has increased for all the trading relationships.

g-1   What country has a healthier trade balance with Canada—Mexico or the United States? Mexico

g-2   On what graph do you find a set of bars that looks like a mirror image of those for the U.S. trade balance with Canada?

Canada on the Trade Balance with U.S. graph

h-1   What country had the most dramatic change for the better after NAFTA came into being? (Remember, NAFTA went into effect in 1994.) Mexico

h-2   Estimate the U.S. trade deficit with Canada and Mexico for 1999 and 2000. Express the values in billions of dollars, using round numbers that you estimate from the graphs.

	1999 ($ BILLION)	2000 ($ BILLION)
U.S. trade balance with Mexico	$ –23 billion	$ –25 billion
U.S. trade balance with Canada	$ –30 billion	$ –51 billion
Combined deficit (total)	$ –53 billion	$ –76 billion

h-3   Did the U.S. combined trade balance get better or worse between 1999 and 2000? Worse
By how much? $23 billion

Module 7   Global perspective: Water World

Part 1: A South Pole point of view

Step 3   Look at Antarctica

a   Do you think this map gives you a realistic representation of Antarctica? Explain your answer.

Student answers will vary. They should observe that the map of Antarctica is very skewed in size, shape, and determining distance.

e   Does this projection give you a better view of the region around the South Pole? Why or why not?

No. The region around the South Pole is distorted in size.

f   Do any of these projections work well for viewing Antarctica?

No. None of the projections represent Antarctica in a realistic way.

Part 2: Just Add Water

Step 1   Activate the Water World data frame

f   What significant differences to you see between today's country outlines and the elevation map of 20,000 years ago? List at least three.

Student answers will vary. Possible answers include: Alaska was connected to Russia, Florida was much larger, and Australia was connected to the islands of Indonesia.

Step 2   Analyze global sea levels if Antarctic ice sheets melted

a   Record your general observations of each layer in the table below.

Student answers will vary. Possible answers include:

SEA LEVEL	OBSERVATIONS
Today	Country outline matches up perfectly with the shorelines.
Plus 5 meters	There is no dramatic change. However, some coastal cities in the southern United States (Miami) will be under water.
Plus 50 meters	There is dramatic change. Most of Florida is under water, there is a large gap in the Amazon Basin in South America, and parts of Europe are gone.
Total Thaw (plus 73 meters)	Much of eastern Europe and western Asia are under water. Large portions of Australia, South America, and southeast United States are gone. Africa is the least affected.

Step 3   View changes in water levels

c-1   What kinds of changes do you see in the rivers and lakes? Provide a specific example.

Student answers will vary. In South America, a large lake appears in the north central area. This lake is the result of the increased water level of the Amazon River. The Parana River in Argentina and the Amazon River Basin are significantly shorter in length.

c-2   With a sea level increase of 50 meters, what kinds of consequences do you foresee for the major river ecosystems of South America? Provide a specific example.

Student answers will vary. The Amazon Basin has the possibility of flooding the rain forests of central South America.

c-3   There are several locations around the globe that are on the interior of landmasses and are below sea level. One of them is in South America. Hypothesize how these low-lying areas were formed.

Student answers will vary. One explanation is that plate boundaries are drifting apart at that location. Another is that the land surrounding the Amazon Basin could have risen up due to tectonic activity.

Step 4   View changes in political boundaries

f, g   Record your results in the table below.

REGION	COUNTRIES/AREAS AFFECTED	POSSIBLE CONSEQUENCES
Middle East	Iraq	Boundary disputes over lost land.
Asia	Cambodia	Cambodia is almost completely under water. There will be a large migration of the population moving to neighboring countries.
Europe	Netherlands	Netherlands would be completely submerged. As a result, a large migration of the population would move to neighboring countries.
Africa	Guinea-Bissau	Guinea-Bissau would be completely submerged. As a result, a large migration of the population would move to neighboring countries. There might be a climate change to the drier areas of the continent.
Oceania	Australia	There would be a loss of major cities and economic areas. These people would move inland.
North America	United States	There would be a loss of most major southern ports: Houston, New Orleans, Miami—and most of Florida. These people would move inland.
Latin America	Brazil	The loss of much of the Brazilian rain forest could have negative repercussions on the global environment.

Step 5   Look for additional data to explore

Based on your previous observations, list other possible layers of data you would like to analyze to study the impact of this phenomenon. In your final assessment, you will have the opportunity to explore many other datasets; this list will help to guide you in further explorations.

Student answers will vary. For a complete list of available data, refer to the assessment data sheet.

Module 7   Regional case study: In the Eye of the Storm

Part 1: The calm before the storm

Step 3   Focus on the capital cities of Central America

g   Record the capital cities for each of the countries in Central America in the Populated Places column in the table on the next two pages.

The table on the following page presents the correct answers.

Step 4   Focus on Central America prior to Hurricane Mitch

b   Record this data in the Populated Places and Transportation Network columns in the answer sheet on the next two pages.

Answers will vary. The table on the following page presents a sampling of possible answers.

m   Analyze the precipitation for each country and record the precipitation data in the Average Precipitation column in the table on the next two pages.

Answers will vary. The table on the following page presents a sampling of possible answers.

p-1   Which country has the most area devoted to agriculture? El Salvador

p-2   Which country has the most area covered by mountains? Honduras

p-3   Which country has the most extensive transportation network? El Salvador

MODULE 1
ArcMap

MODULE 2
global

MODULE 2
regional

MODULE 3
global

MODULE 3
regional

MODULE 3
advanced

MODULE 4
global

MODULE 4
regional

MODULE 4
advanced

MODULE 5
global

MODULE 5
regional

MODULE 5
advanced

MODULE 6
global

MODULE 6
regional

MODULE 7
global

MODULE 7
regional

MODULE 7
advanced

Centra America Prior to Hurricane Mitch

COUNTRY	POPULATED PLACES	TRANSPORTATION NETWORK	MAJOR EXPORTS	AGRICULTURAL LAND USE	AVERAGE PRECIPITATION	PHYSICAL LANDMARKS
Belize	Capital: Belmopan Distribution: Throughout the country, but concentrated around the capital	Roads: Sparse road network Railways: None Airports: 1 civilian	Bananas	Primarily forest with some irrigated land and little cropland	Primarily 1,401–2,800 mm	Maya Mountains
Guatemala	Capital: Guatemala Distribution: Concentrated near southern coast	Roads: Well devl. network, esp. along southern coast Railways: Primarily in south Airports: 1 civilian, 7 total	Coffee	About 1/2 forest mixed with irrigated land, cropland, and forested wetlands along the south coast	Mixed from less than 1,000 to 5,600 mm	Sierra De Santa Cruz Sierra De Los Cuchumatanes Sierra Madre
Honduras	Capital: Tegucigalpa Distribution: Concentrated near the capital; bare in the east	Roads: Well devl. along eastern half of country Railways: Along northern coast Airports: 12 total, 1 civilian	Coffee	1/4 forest, some cropland, some grazing land, 1/2 nonirrigated land	Mixed from less than 1,000 mm in center of country to 4,000 mm along the coast	11 mountains Montana De Botaderos Montana
El Salvador	Capital: San Salvador Distribution: Concentrated around the capital	Roads: Well developed Railways: Well devl. network Airports: 4 total	Coffee	Primarily cropland with some grazing, forested wetlands along the coast, little forest	Primarily 1,401–2,800 mm	Volcan De Santa Ana Cordillera De Celague Volcan De San Miguel
Nicaragua	Capital: Managua Distribution: Coast communities and concentration near capital; bare patches in central	Roads: Well devl. in most of country, esp. west coast Railways: Sparse; near capital Airports: 7 total	Coffee	One-third forest, some cropland, grazing land, some forested wetland along the east coast	Mixed from less than 1,000 mm in the northwest to 5,600 mm along the east coast	3 mountains, 2 volcanoes Volcan Cosiguina Cordillera Dariense
Costa Rica	Capital: San Jose Distribution: Concentrated near Pacific coast and capital	Roads: Along west coast Railways: Through center of country and capital Airports: 13 total, 1 civilian	Garments	Little agriculture: mostly nonirrigated land, small amounts of cropland and forest	2,800–5,600 mm along most of coastline with the exception of less than 1,400 mm in the northwest	4 mountains, 2 volcanoes Volcan Barva Cerro Chirripo
Panama	Capital: Panama Distribution: Concentrated in southern coast and capital; bare in southeast region	Roads: Southwest coast, west of capital Railways: Sparse Airports: 16 total, 1 civilian	Bananas	No agricultural places identified	Primarily a range from 2,000–4,000 mm	6 mountains, 1 volcano Cerro Santiago Volcan De Chiriqui

MODULE 1
ArcMap

MODULE 2
global

MODULE 2
regional

MODULE 3
global

MODULE 3
regional

MODULE 3
advanced

MODULE 4
global

MODULE 4
regional

MODULE 4
advanced

MODULE 5
global

MODULE 5
regional

MODULE 5
advanced

MODULE 6
global

MODULE 6
regional

MODULE 7
global

MODULE 7
regional

MODULE 7
advanced

Part 2: The storm

Step 2   Track Hurricane Mitch

d-1   At what time was Tropical Storm Mitch at this location? 10/22/21Z

d-2   What does the "z" mean in the time?
Zulu time. (The time at 0° longtitude. It is used as a standard reference for anywhere on the globe.)

d-3   What was Mitch's wind speed at this location? 40 mph

f-1   What are the latitude and longitude coordinates for Hurricane Mitch at this location?
14.3 latitude, –77.7 longitude

f-2   At what time was Hurricane Mitch at this location? 10/24/09Z

f-3   What was Mitch's wind speed at this location? 80 mph

h-1   At what time was Hurricane Mitch at this location? 10/27/21Z

h-2   What was Mitch's wind speed at this location? 135 mph

p   Write down the times for each event and determine the time difference. The time is written in this format: Month/Day/Hour (of 24).
Hurricane - 5: 10/26/12Z
Tropical_Storm: 10/24/03Z
Time Difference: 00/02/09 (it took 2 days and 9 hours)

r   Examine the attribute table further and identify the maximum wind speed. 155 mph

Step 3   Measure the size of the storm

g   What is the diameter of the eye of Hurricane Mitch? 25 miles

h, q   Record your measurements and observations in the table below.
Note: Answers in this table are approximate values. Student values will differ depending on the precise location of each measurement taken.

	DIAMETER OF EYE	DIAMETER OF STORM	DISTANCE OF EYE TO COASTLINE OF HONDURAS	HOW HAS THE STORM CHANGED FROM THE PREVIOUS IMAGE?
Mitch2sat.tif	25 miles	830 miles	110 miles	Not applicable.
Mitch3sat.tif	13 miles	995 miles	50 miles	Storm appears more intense and enlarged. It's closer to the coastline.
Mitch4sat.tif	0 (not visible)	825 miles	0 (on shore)	The eye is not visible, clouds are much thicker, but the spiral shape is still visible.
Mitch5sat.tif	0 (not visible0	875 miles	0 (on shore)	There is still a large amount of clouds, but the spiral shape is gone.

MODULE 1
ArcMap

MODULE 2
global

MODULE 2
regional

MODULE 3
global

MODULE 3
regional

MODULE 3
advanced

MODULE 4
global

MODULE 4
regional

MODULE 4
advanced

MODULE 5
global

MODULE 5
regional

MODULE 5
advanced

MODULE 6
global

MODULE 6
regional

MODULE 7
global

**MODULE 7
regional**

MODULE 7
advanced

Step 4   Analyze rainfall from Hurricane Mitch

c-1   What pattern do you notice in the amount of rainfall within the storm?

The greatest amount of precipitation is on the southwest arm of the storm.

c-2   Is this a pattern you expected to find? Why or why not?

Student answers will vary depending on their familiarity with hurricanes.

e-1   At the Mitch4 location, what was the highest range of rainfall measured? 24–29 inches

e-2   Which country received the majority of this heavy rain? Nicaragua

e-3   Describe the difference between the rainfall patterns on October 30 and October 31, 1998.

The rainfall pattern on October 30 was centered heavily over the western coast of Nicaragua and southern El Salvador, with other bands extending due north and one off the eastern coast of Nicaragua and Honduras. On October 31, the main rain center was much larger in area, but less intense in rainfall. The outside bands appear to have merged with the main rain band from October 30.

e-4   What kind of damage do you expect to find with this type of storm? What aspects of the region will be most affected? Elaborate on your answer using your table, Central America Prior to Hurricane Mitch, as a resource.

Student answers will vary. They should mention the possibility of flooding, potential problems with landslides, damaged utility lines, and power outages due to high winds. Nicaragua, Honduras, and El Salvador were severely affected by the rainfall and wind.

Module 7   Advanced investigation: Data Disaster

Step 2   Add the mystery data

b-1   What is the name of the feature class that the layer is based on? WWF_Eco

b-2   What is the name of the geodatabase that contains the feature class? World7.mdb

d   What do you think this map is showing?

Student answers will vary. They should observe that the data addresses the environment in some way. The legend displays the names of different biomes.

Step 3   Evaluate the attribute table and map two layers

b-1   Write each field name and the type of data it contains (numeric or text) in the table below.

FIELD NUMBER	FIELD NAME	FIELD TYPE
1	SHAPE	Text
2	ECOREGION	Text
3	MHT_NAME	Text
4	BDI	Numeric
5	THREAT	Numeric
6	FINAL	Numeric

b-2   Write a hypothesis on what you think the data represents.

Student answers will vary, but they should include some of the following information:

The numeric fields appear to be some sort of scale. It's unclear which value is considered most critical—1 or 5.

"0" is more than likely a "no data" value.

There is a relationship between the numeric fields and the string fields.

The numeric fields represent the status of the various ecoregions.

c   Which field does this layer map? MHT_NAME

p   Record the similarities and differences between these two layers.

The WWF_Eco (ECOREGION) layer contains much more detail than the WWF_Eco1 (MHT_NAME) layer.

Step 4   Map the data and analyze it

e-1   After mapping out all the data fields, what additional conclusions can you make about the data?

Possible answers include:

The numeric fields appear to be some sort of scale.

"0" is likely to be a "no data" value.

There is a relationship between the numeric and string fields.

e-2   Use this new information to revise your hypothesis about what the data represents.

Student answers will vary and depend on their response to the previous question.

f Read the table below and use the mapped data to complete the missing pieces. The definitions in the table belong to the six fields you identified in step 3b-1. Fill in the table with the appropriate field name from the attribute table.

FIELD NAME	DEFINITION
FINAL	The final assessment of the ecoregion as the estimated threat to the ecoregion over the next 20 years.
ECOREGION	Descriptive name for the _____ that are relatively large areas of land or water that share a large majority of their species, dynamics, and environmental conditions.
BDI	The biological distinctiveness index. It is based on the species richness, endemism, rareness, and so forth.
SHAPE	Represent global terrestrial and freshwater areas defined as ecoregions.
MHT_NAME	The major habitat for the area.
THREAT	Degree of threat to the ecoregion. Some examples include logging, conversion to agriculture/ urbanization, and so on.

g Complete the following tables by filling in the numeric values for each description. Keep in mind that the definitions may not be listed in numeric order and don't forget about zero. The first table has been filled in for you.

THREAT	
5	Relatively stable in degree of threat
4	Relatively intact in degree of threat
3	Vulnerable in degree of threat
0	Not assessed or unknown
2	Endangered in degree of threat
1	Critical in degree of threat

FINAL	
0	Not assessed or unknown
1	Critical in estimated threat over the next 20 years
2	Endangered in estimated threat over the next 20 years
4	Relatively intact in estimated threat over the next 20 years
3	Vulnerable in estimated threat over the next 20 years
5	Relatively stable in estimated threat over the next 20 years

Step 5   View metadata documentation

j-1   How would you describe this dataset? It represents the boundaries for the continents of the world.

j-2   Who made this data? ESRI (Environmental Systems Research Institute, Inc.)

j-3   When was this data published? 2004

j-4   Which section explains whether or not you can reuse the data for a project or publish a work that uses the data?
       The question "Are there legal restrictions on access or use of the data?" displays the use constraints of the data.
       It includes information on redistribution rights and the data license agreement for continents.

k   How would you describe the data in the WWF_Eco layer that you investigated?
       Student answers will vary and depend on their response in previous questions 3b-2 and 4e-2.

Step 6   Create metadata documentation

e–s   Type information into the metadata document.
       Student answers will vary. You can view the actual metadata document for WWF_Eco from within the module 7
       data folder (C:\MapWorld9\Mod7\Data). The file is wwf_eco.htm.

Bibliography

Advanced National Seismic System. 2000. *ANSS worldwide earthquake catalog, 2000* [data file]. Available from Northern California Earthquake Data Center Web site, quake.geo.berkeley.edu/anss/catalog-search.html.

Bureau of Economic Analysis. 2003. BEA's Regional Accounts. Retrieved from www.bea.doc.gov/bea/regional/articles.cfm.

Chandler, T. 1989. *Four thousand years of urban growth: An historical census.* Lewiston, NY: Edwin Mellen Press.

CountryWatch. 2000. Various demographic datasets [data file]. Available from Country Watch Web site, www.countrywatch.com.

DeBlij, H. J., and P. O. Muller. 1992. *Geography: Regions and concepts.* New York: John Wiley & Sons, Inc.

ESRI. 1996. ArcAtlas: Our earth [computer software]. Redlands, CA: ESRI.

———. ESRI Data & Maps 1999 [data CD]. 1999. Redlands, CA: ESRI.

———. ESRI Data & Maps Media Kit (Version 8) [data CD]. 2000. Redlands, CA: ESRI.

———. ESRI Data & Maps 2004 [data CD]. 2004. Redlands, CA: ESRI.

———. ESRI Data & Maps Media Kit (Version 9) [data DVD]. 2004. Redlands, CA: ESRI.

———. ESRI Schools & Libraries Program. 2000. Mapping GLOBE visualizations. Retrieved October 29, 2001, from gis.esri.com/industries/education/arclessons/arclessons.cfm.

———. *Getting to know ArcView GIS.* 3rd ed. 1999. Redlands, CA: ESRI Press.

———. *Getting to know ArcGIS Desktop.* 2nd ed. 2004. Redlands, CA: ESRI Press.

Fellmann, J., A. Getis, and J. Getis. 1992. *Human geography: Landscapes of human activities.* 3rd ed. Dubuque, IA: Wm. C. Brown Publishers.

Ferrigno, J. G., and R. S. Williams. 1999. *Satellite image atlas of glaciers of the world.* U.S. Geological Survey Fact Sheet 130-02. Retrieved December 16, 2004, from pubs.usgs.gov/fs/fs 130-02/fs 130-02.html.

Food and Agricultural Organization of the United Nations. n.d. *FAO home page.* Retrieved July 2001, from www.fao.org.

Guiney, J. L., and M. B. Lawrence. 1999, January 28. *Preliminary report: Hurricane Mitch 22 October–05 November 1998.* National Hurricane Center. Retrieved September 17, 2001, from www.nhc.noaa.gov/1998mitch.html.

Hardwick, S. W., and D. G. Holtgrieve. 1996. *Geography for educators: Standards, themes, and concepts.* Upper Saddle River, NJ: Prentice Hall. 82–112.

Instituto Nacional de Estadistica, Geographica e Informatica, Mexico. 2001. Export and import data [data file]. Available from Instituto Nacional de Estadistica, Geographica e Informatica, Mexico, Web site, www.inegi.gob.mx.

International Society for Technology in Education. 1998. *National educational technology standards for all students.* Eugene, OR: Author.

Kennedy, H., ed. 2001. *The ESRI Press dictionary of GIS terminology.* Redlands, CA: ESRI Press.

Kious, J., and R. I. Tilling. 1996. *This dynamic earth: The story of plate tectonics.* Retrieved June 7, 2001, from pubs.usgs.gov/publications/text/dynamic.html.

McFalls, J. A., Jr. 1998. Population: A lively introduction. *Population Reference Bureau Population Bulletin* 53 (3).

National Aeronautics and Space Administration. n.d. *NASA visible earth: Cryosphere.* Retrieved December 16, 2004, from visibleearth.nasa.gov/Cryosphere/Sea_Ice/Icebergs.html.

National Council for Teachers of Mathematics. 2000. *Principles and standards for school mathematics.* Reston, VA: Author.

National Geographic Society. 1994. *The National Geography Standards, 1994.* Washington, DC: The Geography Education Standards Project.

National Oceanic and Atmospheric Administration. 2001. *The south geographic pole* (Image ID: Corp1566, NOAA Corps Collection) [data file]. Available from NOAA Photo Library Web site, www.photolib.noaa.gov/corps/corp1566.htm.

National Research Council—National Academy of Sciences. 1995. *National science education standards.* Washington, DC: National Academy Press.

National Snow and Ice Data Center. n.d. *Antarctic ice shelves and icebergs.* Retrieved December 16, 2004, from nsidc.org/iceshelves.

NOVA. 1998, April. *Warnings from the ice.* Retrieved December 15, 2004, from www.pbs.org/wgbh/nova/warnings. WGBH Educational Foundation.

Population Reference Bureau. n.d. *PRB home page.* Retrieved December 2004 from www.prb.org.

Schofield, Richard. Abridged version of speech given March 31, 1999, published in *British–Yemeni Society Journal* (July 2000) and retrieved from www.al-bab.com/bys/articles/schofield00.htm.

Sheets, B., and J. Williams. 2001. *Hurricane watch: Forecasting the deadliest storms on earth.* New York: Vintage Books.

The GLOBE Program. n.d. *GLOBE sites visualizations.* Retrieved from viz.globe.gov/viz-bin/home.cgi?l=en&b=g&rg=n.

United Nations. 1985. Estimates and projections of urban, rural and city populations (ST/ESA/Ser.R/50).

U.S. Census Bureau. 2001. Export and import data [data file]. Available from U.S. Census Bureau Foreign Trade Web site, www.cenus.gov/foreign-trade/www.

U.S. Geological Survey. 2001. Global GIS database: Digital atlas of Central and South America (Digital Data Series DDS-62-A) [computer software and data]. U.S.: Author.

U.S. Geological Survey. 2001. Global GIS database: Digital atlas of the Middle East [computer software and data]. Unpublished, Author.

U.S. Geological Survey. n.d. *USGS earthquake hazards program.* Retrieved June 7, 2001, from earthquake.usgs.gov.

U.S. Geological Survey. n.d. *USGS Hurricane Mitch program.* Retrieved September 10, 2001, from mitchnts1.cr.usgs.gov.

U.S. Geological Survey. 2000, February. *USGS TerraWeb: Antarctica.* Retrieved December 16, 2004, from terraweb.wr.usgs.gov/projects/Antarctica.

U.S. Geological Survey. n.d. *USGS volcano hazards program.* Retrieved December 15, 2004, from volcanoes.usgs.gov.

Whitaker, B. 2000, June 12. Translation of the Treaty of Jeddah. Retrieved from www.al-bab.com/yemen/pol/int5.htm.

Whitaker, B. 2000, July 1. Commentary on the border treaty. *YEMEN Gateway.* Retrieved from www.al-bab.com/yemen/pol/border000629.htm.

White, Frank. 1998, September. *Overview effect: Space exploration and human evolution.* 2nd ed. Reston, VA: American Institute of Aeronautics and Astronautics.

Williams, J. 2001. Ice shelves float on the sea. *USA Today.com.* Retrieved September 6, 2001, from www.usatoday.com/weather/antarc/aiceshlf.htm.

Williams, J. 2001. Warming effect on sea level unsure. Edited summary of Intergovernmental Panel on Climate Change's book *Climate Change 2001.* USA Today.com. Retrieved September 6, 2001, from www.usatoday.com/weather/antarc/iceipcc.htm.

World Climate.com. n.d. Various world climate datasets [data file]. Available from World Climate.com Web site, www.worldclimate.com.

World geography: Building a global perspective. 1998. Upper Saddle River, NJ: Prentice Hall.

World Wildlife Fund. n.d. *Conservation science: Global 200 ecoregions.* Retrieved December 15, 2004, from www.worldwildlife.org/science/ecoregions/g200.cfm.

License Agreement

Permitted Uses:

- Licensee may install the Software, Data, and Related Materials or portions of the Data, **to be used solely in conjunction with the exercises and context of this book,** onto permanent storage device(s) and thereafter use, copy, reproduce, and distribute the Software, Data, and Related Materials, including Documentation, in quantities sufficient to meet Licensee's own internal needs.

- Licensee may make only one (1) copy of the original Software, Data, and Related Materials for archival purposes during the term of this Agreement.

- Licensee may modify the Data and merge other data sets with the Data for Licensee's own internal use. The portions of the Data merged with other data sets will continue to be subject to the terms and conditions of this Agreement.

- Licensee may use, copy, alter, modify, merge, reproduce, and/or create derivative works of the online Documentation for Licensee's own internal use. The portions of the online Documentation merged with other software, data, hard copy, and/or digital materials shall continue to be subject to the terms and conditions of this Agreement and shall provide the copyright attribution notice acknowledging ESRI and its Licensor(s) proprietary rights in the online Documentation as instructed in the accompanying metadata file.

- Licensee may use, copy, reproduce, and/or redistribute the Data or any derived portion(s) of the Data in published hard-copy and/or in static, electronic (i.e., .gif, .jpeg, etc.) formats, provided that Licensee affixes a legend statement acknowledging the appropriate Data Publisher as the source of the portion(s) of the Data displayed, printed, or plotted.

- Licensee may customize the Software using any (i) macro or scripting language, (ii) open application programming interface (API), or (iii) source or object code libraries, but only to the extent that such customization is described in the Documentation.

- Licensee may use the Data only as described in the Distribution Rights section of the help files or metadata files delivered with the Software, Data, and Related Materials.

Uses Not Permitted:

- Licensee shall not sell; rent; lease; sublicense; lend; assign; time-share; or act as a service bureau or Application Service Provider (ASP) that allows third party access to the Software, Data, and Related Materials except as provided herein; or transfer, in whole or in part, access to prior or present versions of the Software, Data, or Related Materials, any updates, or Licensee's rights under this Agreement.

- Licensee shall not redistribute the Software, in whole or in part, including, but not limited to, extensions, components, or DLLs without the prior written approval of ESRI as set forth in an appropriate redistribution license agreement.

- Licensee shall not reverse engineer, decompile, or disassemble the Software, Data, or Related Materials, except to the extent that such activity is expressly permitted by applicable law notwithstanding this restriction in order to protect ESRI and its Licensor(s) trade secrets and proprietary information contained in the Software, Data, or Related Materials.

- Licensee shall not make any attempt to circumvent the technological measure(s) (e.g., License Manager, etc.) that controls access to or use of the Software, Data, and Related Materials, except to the extent that such activity is expressly permitted by applicable law notwithstanding this restriction.

- Licensee shall not use the Software to transfer or exchange any material where such transfer or exchange is prohibited by copyright or any other law.
- Licensee shall not remove or obscure any ESRI or its Licensor(s) patent, copyright, trademark, or proprietary rights notices contained in or affixed to the Software, Data, or Related Materials.

Term: The license granted by this Agreement is for a period of one (1) year and shall commence upon installation of the Software. In the event Licensee licenses a copy of ESRI Software or the K-12 Bundle, Licensee may continue to use the Data and Related Materials after termination of this Agreement. The Agreement shall automatically terminate without notice if Licensee fails to comply with any provision of this Agreement. Licensee shall then return the Software, Data, and Related Materials to ESRI. The parties hereby agree that all provisions that operate to protect the rights of ESRI and its Licensor(s) shall remain in force should breach occur.

Limited Warranty: ESRI warrants that the media upon which the Software, Data, and Related Materials are provided will be free from defects in materials and workmanship under normal use and service for a period of ninety (90) days from the date of receipt. The Software, Data, and Related Materials are excluded from the warranty, and Licensee expressly acknowledges that the Data contain some nonconformities, defects, errors, or omissions. ESRI and its Licensor(s) do not warrant that the Data will meet Licensee's needs or expectations, that the use of the Data will be uninterrupted, or that all nonconformities, defects, or errors can or will be corrected. ESRI and its Licensor(s) are not inviting reliance on these data, and Licensee should always verify actual data.

EXCEPT FOR THE LIMITED WARRANTIES SET FORTH ABOVE, THE SOFTWARE, DATA, AND RELATED MATERIALS CONTAINED THEREIN ARE PROVIDED "AS IS," WITHOUT WARRANTY OF ANY KIND, EITHER EXPRESS OR IMPLIED, INCLUDING, BUT NOT LIMITED TO, THE IMPLIED WARRANTIES OF MERCHANTABILITY AND FITNESS FOR A PARTICULAR PURPOSE.

Exclusive Remedy and Limitation of Liability: ESRI and/or its Licensor(s) entire liability and Licensee's exclusive remedy shall be to terminate the Agreement upon Licensee returning the Software, Data, and Related Materials to ESRI with a copy of Licensee's invoice/receipt and ESRI returning the license fees paid to Licensee.

IN NO EVENT SHALL DATA PUBLISHERS AND/OR THEIR LICENSOR(S) BE LIABLE FOR COSTS OF PROCUREMENT OF SUBSTITUTE GOODS OR SERVICES, LOST PROFITS, LOST SALES OR BUSINESS EXPENDITURES, INVESTMENTS, OR COMMITMENTS IN CONNECTION WITH ANY BUSINESS, LOSS OF ANY GOODWILL, OR FOR ANY INDIRECT, SPECIAL, INCIDENTAL, EXEMPLARY, OR CONSEQUENTIAL DAMAGES ARISING OUT OF THIS AGREEMENT OR USE OF THE SOFTWARE, DATA, AND RELATED MATERIALS, HOWEVER CAUSED, ON ANY THEORY OF LIABILITY, AND WHETHER OR NOT DATA PUBLISHERS AND/OR THEIR LICENSOR(S) HAVE BEEN ADVISED OF THE POSSIBILITY OF SUCH DAMAGE. THESE LIMITATIONS SHALL APPLY NOTWITHSTANDING ANY FAILURE OF ESSENTIAL PURPOSE OF ANY EXCLUSIVE REMEDY.

No Implied Waivers: No failure or delay by ESRI and/or its Licensor(s) in enforcing any right or remedy under this Agreement shall be construed as a waiver of any future or other exercise of such right or remedy by ESRI and/or its Licensor(s).

U.S. Government Restricted/Limited Rights: Any software, Documentation, and/or data delivered hereunder is subject to the terms of the License Agreement. In no event shall the U.S. Government acquire greater than RESTRICTED/ LIMITED RIGHTS. At a minimum, use, duplication, or disclosure by the U.S. Government is subject to restrictions as set forth in FAR §52.227-14 Alternates I, II, and III (JUN 1987); FAR §52.227-19 (JUN 1987) and/or FAR §12.211/12.212 (Commercial Technical Data/Computer Software); and DFARS §252.227-7015 (NOV 1995) (Technical Data) and/or DFARS §227.7202 (Computer Software), as applicable. Contractor/Manufacturer is ESRI, 380 New York Street, Redlands, CA 92373-8100 USA.

Export Control Regulations: Licensee expressly acknowledges and agrees that Licensee shall not export, reexport, or provide the Software, Data, or Related Materials, in whole or in part, to (i) any country to which the United States has embargoed goods; (ii) any person on the U.S. Treasury Department's list of Specially Designated Nationals; (iii) any person or entity on the U.S. Commerce Department's Table of Denial Orders; or (iv) any person or entity where such export, reexport, or provision violates any U.S. export control law or regulation. Licensee shall not export the Software, Data, and/or Related Materials or any underlying information or technology to any facility in violation of these or other applicable laws and regulations. Licensee represents and warrants that it is not a national, resident, located in or under the control of, or acting on behalf of any person, entity, or country subject to such U.S. export controls.

Severability: If any provision(s) of this Agreement shall be held to be invalid, illegal, or unenforceable by a court or other tribunal of competent jurisdiction, the validity, legality, and enforceability of the remaining provisions shall not in any way be affected or impaired thereby.

Governing Law: This Agreement shall be construed and enforced in accordance with and be governed by the laws of the United States of America and the State of California without reference to conflict of laws principles.

Entire Agreement: The parties agree that this constitutes the sole and entire agreement of the parties as to the matter set forth herein and supersedes any previous agreements, understandings, and arrangements between the parties relating hereto.

Installing the Exercise Data

**Part 1.
About the exercise
data installation**

Be sure to follow the instructions in part 2, "How to install the exercise data," and do not copy the files directly from the CD to your hard drive. A direct file copy does not remove write protection from the files. In addition, a direct file copy will not enable the automatic uninstall feature.

Allow about ten minutes to install the exercise data. Actual time will vary depending on processor, hard drive, and CD drive speeds. The data uses about 340 megabytes of disk space on your computer.

**Part 2.
How to install the
exercise data**

The screen graphics in this part reflect the standard appearance of the Windows 2000 operating system. If you have a different operating system, such as Windows XP, your screen images will look slightly different. The differences do not affect the installation steps.

a Put the MOWGLE 9 Exercise Data CD in your computer's CD drive.

b Right-click the Start button in the Windows taskbar. On the context menu, click Explore to open Windows Explorer.

c In Windows Explorer, navigate to your CD drive, called ESRI (D:). (The drive letter may be different.) Click the drive icon to display the contents of the CD.

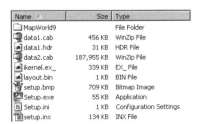

d Double-click the Setup.exe file.

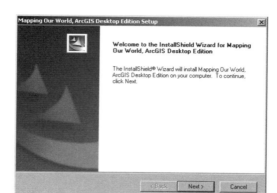

e Click Next.

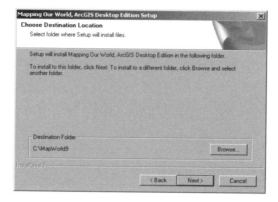

f Accept the default destination folder (C:\MapWorld9), unless you have a good reason to choose another folder.

If you want to choose a different folder, click the Browse button and navigate to the location you want (for example, C:\TEMP). Click OK on the Choose Folder dialog box. No matter which location you choose, a MapWorld9 folder will be created inside it and "MapWorld9" will become the final part of the path name. Be advised, however, that exercise instructions for opening map documents and adding data refer to C:\MapWorld9 as the data location.

g Click Next and let the installation process run.

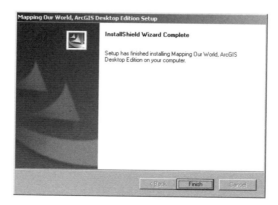

h Click Finish.

If ArcGIS 9 Desktop software (either ArcView, ArcEditor, or ArcInfo) is already installed on your computer, you are ready to begin the book. If you are installing the ArcView 9 Demo Edition software that comes with this book, go to "Installing and Registering the ArcView 9 Demo Edition Software."

Part 3.
How to uninstall
the exercise data

a Click the Start button on your Windows taskbar, point to Settings, and click Control Panel.

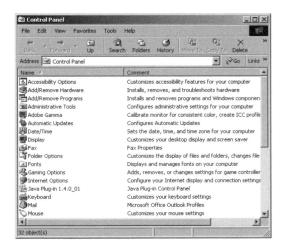

b In the Control Panel, double-click Add/Remove Programs.

c In the Add/Remove Programs window, click on Mapping Our World, ArcGIS Desktop Edition.

d Click Change/Remove.

e On the wizard panel, click the option to "Remove," as shown.

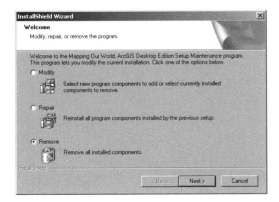

f Click Next. Click OK when you are prompted to confirm the uninstall. Let the uninstallation process run.

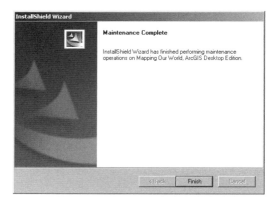

g Click Finish.

Installing and Registering the ArcView 9 Demo Edition Software

Part 1. About the software installation

The ArcView 9.0 Demo Edition software included with this book is intended for use only with *Mapping Our World: GIS Lessons for Educators.* Once installed and registered, ArcView 9 will run for 365 days. The time limit cannot be extended, nor can the software be reinstalled on the same machine once the time limit has expired.

The ArcView 9 Demo Edition software may be installed on more than one computer in accordance with the provisions contained in the "License Agreement" section (page 521) of this book.

> Subsequent installations of the ArcView 9 Demo Edition software are subject to the expiration date of the first installation. For example, if you make a second installation ten days after the first, the evaluation period for the second installation will be 355 days. If you intend to install the Demo Edition software on more than one computer, plan accordingly.

If ArcGIS 9 Desktop software (either ArcView, ArcEditor, or ArcInfo) is already installed on your computer, you do not need to—and, in fact, cannot—install the Demo Edition software. Use the book with the software that is already installed.

If ArcGIS 8 Desktop software (any 8.x version of ArcView, ArcEditor, or ArcInfo) is already installed on your computer, you must uninstall it before you can install the Demo Edition software. (You can subsequently reinstall it.) The exercises require ArcGIS version 9 software.

If ArcView 3.x software is already installed on your computer, you can install the Demo Edition software. ArcView 3 and ArcGIS Desktop 9 can coexist on your computer.

If you have problems with the software installation or registration process, go to the following Web site for frequently asked questions and troubleshooting tips:

www.esri.com / mappingourworld

or request help by sending an e-mail to this address:

workbook-support@esri.com

**Part 2.
System
requirements**

Before you install the Demo Edition software, you should make sure that your computer system satisfies the following minimum requirements.

Processor	Intel® Pentium® 800 MHz minimum
Operating system	Microsoft Windows Server 2003 Microsoft Windows XP Microsoft Windows 2000 Microsoft Windows NT—Intel 4.0 Service Pack 6a (collectively referred to as Microsoft Windows)
Internet Explorer	Microsoft Internet Explorer 6.0 (If your computer does not have Microsoft Internet Explorer 6.0 or a later version, the ArcView 9 Demo Edition setup program will notify you of the missing requirements and exit. After installing the appropriate version, you can run the setup program again.)
RAM	256 MB minimum
Disk space	Approximately 600 MB

**Part 3.
How to install
the Demo Edition
software**

The screen graphics in this part reflect the standard appearance of the Windows 2000 operating system. If you have a different operating system, such as Windows XP, your screen images will look slightly different. The differences do not affect the installation steps.

a Close all open applications.

b Put the Demo Edition software CD in your computer's CD drive.

c Right-click the Start button in the Windows taskbar. On the context menu, click Explore to open Windows Explorer.

d In Windows Explorer, navigate to your CD drive, called ESRI (D:). (The drive letter may be different.) Click the drive icon 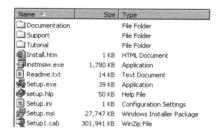 to display the contents of the CD.

Name	Size	Type
Documentation		File Folder
Support		File Folder
Tutorial		File Folder
Install.htm	1 KB	HTML Document
instmsiw.exe	1,780 KB	Application
Readme.txt	14 KB	Text Document
Setup.exe	39 KB	Application
setup.hlp	50 KB	Help File
Setup.ini	1 KB	Configuration Settings
Setup.msi	27,747 KB	Windows Installer Package
Setup1.cab	301,941 KB	WinZip File

e Double-click the Setup.exe file and wait for the ArcView Setup program to load.

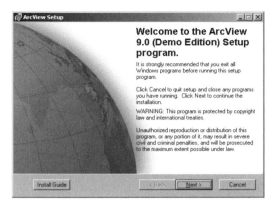

If the program doesn't load and you are instead prompted to install **Microsoft Windows Installer,** click Yes and restart your computer when prompted. If ArcView 9 Demo Edition does not then resume installation on its own, navigate to the Setup.exe file and double-click on it again.

f Click Next. Read the license agreement. If you agree to the terms, click "I accept the license agreement," as shown. (If you do not agree to the terms, click Cancel to end the installation.)

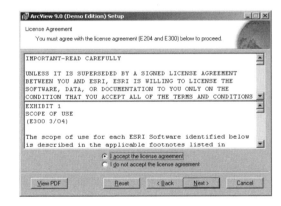

g Click Next.

h Make sure the installation type is set to Typical. Click Next.

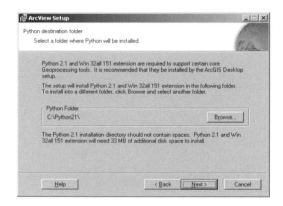

i Accept the default installation folder (C:\Program Files\ArcGIS), unless you have a good reason to choose another folder.

> If you want to choose a different folder, click the Browse button and navigate to the location you want. Click OK on the ArcView 9.0 (Demo Edition) Setup dialog box.

j Click Next.

k Accept the default installation folder for the Python program (C:\Python21), unless you have a good reason to choose another folder.

> Python® is a scripting language used by some ArcGIS geoprocessing functions. If the Python application is already installed on your computer, the setup program bypasses this step.

l Click Next.

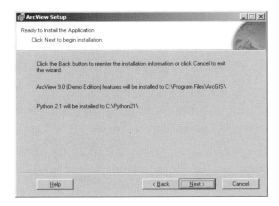

m Click Next and let the installation process run. It takes several minutes.

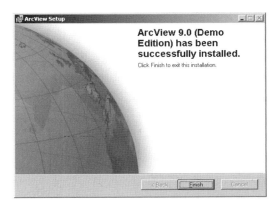

n Click Finish. Uncheck the ArcGIS Tutorial Data box, as shown.

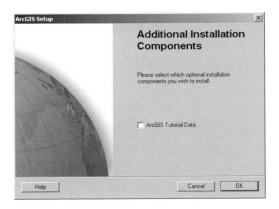

o Click OK.

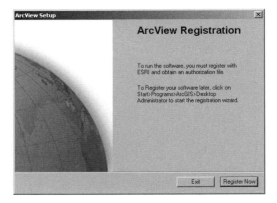

You are ready to continue with part 4.

Part 4.
About the
software
registration

ArcView 9 Demo Edition software cannot be used until it has been registered.

If you are installing the software on more than one computer, each installed copy must be registered separately.

The recommended way to register the software is over the Internet. This is the simplest and fastest way. The instructions for Internet registration are given below in part 5, "How to register ArcView 9 Demo Edition software over the Internet."

The software can also be registered in other ways, as listed in part 7, "Other ways to register the software."

The 365-day software evaluation period begins when your registration is processed by ESRI, not when you first use the software. If you do not intend to use the software soon, you should exit the registration process and return to it later. For instructions on resuming the registration process at a later date, see part 8, "How to resume the registration process."

Part 5.
How to register
ArcView 9 Demo
Edition software
over the Internet

a Make sure you are connected to the Internet.

b Click Register Now.

c Make sure the following option is selected: "I have installed ArcView and need to register the software." Click Next.

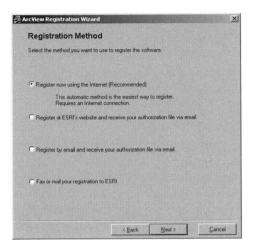

d Make sure the following option is selected: "Register now using the Internet." Click Next.

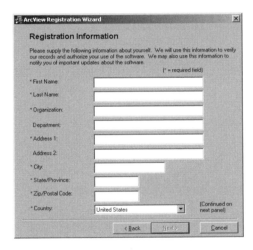

e Fill out the first part of the registration form with your information. Only English-language characters are accepted. Click Next.

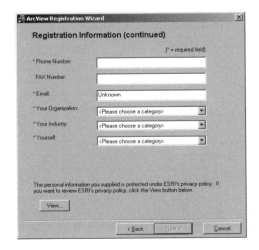

f Finish the registration form. Click Next.

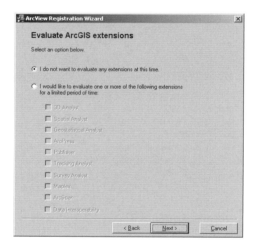

g Enter the registration code printed on the software CD jacket on the inside back cover of your copy of *Mapping Our World.* It will be a text string such as *EVA123456789.* Click Next.

h Make sure the following option is selected: "I do not want to evaluate any extensions at this time." Click Next.

i The software is registered and ready to use. Click Finish.

If the registration fails, make a note of the problem and use the help resources listed at the end of part 1, "About the software installation." You may also want to try a different registration method (see part 7, "Other ways to register the software").

j Continue with part 6, "How to test the installation."

Part 6.
How to test the
installation

ArcView 9 contains the ArcMap and ArcCatalog applications. To confirm that the installation was successful, you should test both applications.

a Click the Start button, point to Programs, point to ArcGIS, and click ArcMap.

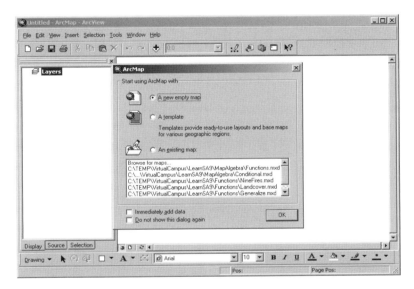

b Click OK to start using ArcMap with a new, empty map.

c Click the Close button (marked with a small ✕) in the upper-right corner of the ArcMap window.

d Click the Start button, point to Programs, point to ArcGIS, and click ArcCatalog.

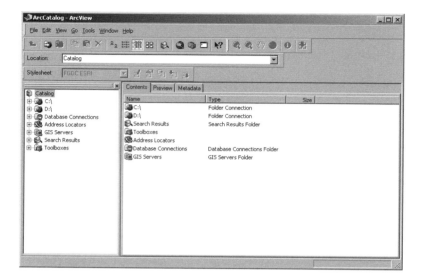

e Click the Close button (marked with a small ✗) in the upper-right corner of the ArcCatalog window.

If either application does not open as expected, contact *workbook-support@esri.com.*

f Your ArcView 9 Demo Edition software installation and registration are complete. You do not need to read the remaining instructions until you are ready to uninstall the software.

**Part 7.
Other ways to
register the
software**

The ArcView 9 Demo Edition software can be registered in four ways:
- Over the Internet (described in part 5, above)
- By e-mail
- At the ESRI Web site
- By fax or regular mail

Instructions for each of these methods are found in the file software_reg.pdf, located on the ArcView 9 Demo Edition CD (and also at *www.esri.com/mappingourworld*). To read this file, you need Adobe Reader® 6.0 or higher software. Adobe Reader can be downloaded free of charge from www.adobe.com.

a Make sure the Demo Edition software CD is in your computer's CD drive.

b In Windows Explorer, navigate to your CD drive: ESRI (D:). Click the drive icon to display the contents of the CD.

c Double-click the file software_reg.pdf to open it. Print the document.

d Follow the instructions in the software_reg.pdf document for the registration method you want to use. (After Internet registration, the option to register by e-mail is the most common method.)

**Part 8.
How to resume
the registration
process**

Follow the instructions below if you have installed the ArcView 9 Demo Edition software but exited the registration process before completing it.

a Click the Start button in the Windows taskbar. Point to Programs, point to ArcGIS, and click Desktop Administrator.

b In the left-hand panel of the Desktop Administrator, click the "Register Single Use and Extensions" folder.

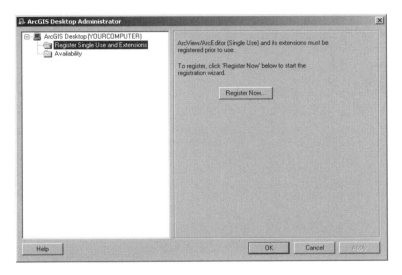

c In the right-hand panel, click Register Now.

d Go to part 5, "How to register ArcView 9 Demo Edition software over the Internet," or to part 7, "Other ways to register the software."

**Part 9.
How to uninstall
the ArcView 9
Demo Edition
software**

It is recommended that you uninstall the software when the 365-day time limit expires. You must uninstall it before you can install a standard licensed version of ArcGIS 9 Desktop software.

a Click the Start button on your Windows taskbar, point to Settings, and click Control Panel.

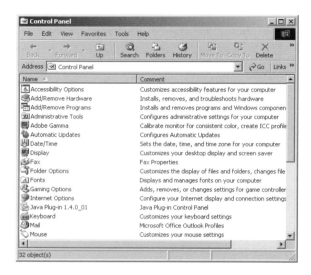

b In the Control Panel, double-click Add/Remove Programs.

c In the Add/Remove Programs window, click on ArcView 9.0 (Demo Edition).

d Click Remove. Click Yes at the prompt to confirm the uninstall. Let the uninstallation process run.

e Uninstall the "Python 2.1" and "Python 2.1 combined Win32 extensions" programs in the same way.

f When the software has been removed from your computer, click the Close button (marked with a small ✕) in the upper-right corner of the Add/Remove Programs window.

GIS Terms QUICK REFERENCE

ArcGIS Computer software for implementing a geographic information system (GIS).

ArcView Desktop GIS software that includes ArcMap for displaying and interacting with maps and layouts, and ArcCatalog for previewing data and metadata.

attribute A piece of information that describes a geographic feature on a GIS map. The attributes of an earthquake might include the date it occurred, its latitude and longitude, depth, and magnitude.

attribute table A table that contains all of the attributes for like features on a GIS map, arranged so that each row represents one feature and each column represents one feature attribute. In a GIS, attribute values in an attribute table can be used to find, query, and symbolize features. The attribute table for the Top 10 Cities, 1950 layer includes attributes for each of the ten cities listed.

Attributes of Top 10 Cities, 1950

OBJECTID*	Shape*	CITY NAME	COUNTRY NAME	HISTORIC NAME	RANK_1950	POP_1950
1	Point	Moscow	Russia	Mocow	6	5,100,000
2	Point	London	United Kingdom	London	2	8,860,000
3	Point	Paris	France	Paris	4	5,900,000
4	Point	Chicago	United States	Chicago	8	4,906,000
5	Point	New York	United States	New York	1	12,463,000
6	Point	Tokyo	Japan	Edo	3	7,000,000
7	Point	Shanghai	China	Shanghai	5	5,406,000
8	Point	Calcutta	India	Calcutta	10	4,800,000
9	Point	Buenos Aires	Argentina	Buenos Aires	7	5,000,000
10	Point	Essen	Germany	Ruhr	9	4,900,000

axis The vertical (y-axis) or horizontal (x-axis) lines in a graph on which measurements can be illustrated and coordinated with each other. Each axis in a GIS graph can be made visible or invisible and labeled.

bookmark In ArcMap, a shortcut you can create to save a particular geographic extent on a map so you can return to it later. Also known as a spatial bookmark.

color selector The window that allows you to change the color of geographic features and text on your GIS map.

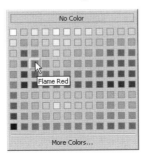

comma-delimited values file (.csv)
A data table in text form where the values are separated by commas. This is a popular format for transferring data from one program to another, for example between spreadsheet programs and ArcMap. These programs use the commas to determine where a new piece of data stops and starts.

coordinate system
A system of intersecting lines that is used to locate features on surfaces such as the earth's surface or a map. In ArcMap, each feature class (layer) of data has a coordinate system that tells ArcMap where on the map to draw the features. A feature class may also have a map projection. (*See also* feature class; map projection.)

data
Any collection of related facts, from raw numbers and measurements to analyzed and organized sets of information.

data folder
A folder on the hard drive of your computer or your network's computer that is available for storage of GIS data and map documents that you create.

data frame
A map element that defines a geographic extent, a page extent, a coordinate system, and other display properties for one or more layers in ArcMap. In data view, only one data frame is displayed at a time; in layout view, all a map's data frames are displayed at the same time.

data frame, active
In ArcMap, the active data frame is the target for many tools and commands. In data view, the active data frame is bold in the table of contents and visible in the display area. In layout view, the active data frame has a dashed line around it to show it is the active one.

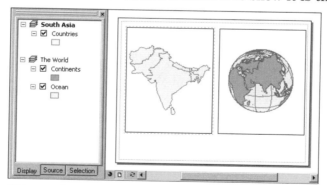

This map document, shown in layout view, has two data frames: South Asia and The World. South Asia is the active data frame.

data source	The data referenced by a layer or a layer file in ArcMap or ArcCatalog. Examples of data sources are a geodatabase feature class, a shapefile, and an image.

The data source for this World Phone Lines layer is the geodatabase feature class "phones" found in the World1 geodatabase. The geodatabase is located in the C:\MapWorld\Mod1\Data folder.

data view	A view in ArcMap for exploring, displaying, and querying geographic data. This view hides all map elements, such as titles, north arrows, and scale bars. *Compare* layout view.

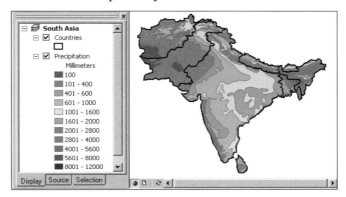

decimal degrees	Degrees of latitude and longitude expressed in decimals instead of in degrees, minutes, and seconds. Decimal degrees converts the degrees, minutes, and seconds into a decimal number using the mathematical formula below. In a GIS, decimal degrees is more efficient than degrees, minutes, and seconds because they make digital storage of coordinates easier and computations faster.

decimal degrees = degrees + (minutes/60) + (seconds/3,600)

73° 59′ 15″ longitude is equal to 73.9875 decimal degrees.

feature	A geographic object on a map represented by a point, a line, or a polygon.

- A point feature is a point on a map that represents a geographic object too small to show as a line or polygon. A point feature might represent a tree, or a phone line, or even a city viewed from a satellite.

- A line feature is a line on a map that represents a geographic object too narrow to show as a polygon at a particular scale. A line feature might represent a river on a world map or a street on a city map. In ArcGIS, another name for a line feature is a polyline feature.

feature (continued)

- A polygon feature is an area on a map that represents a geographic object too large to show as a point or a line. A polygon feature might represent a lake, or a city viewed from an airplane, or a whole continent viewed from a satellite.

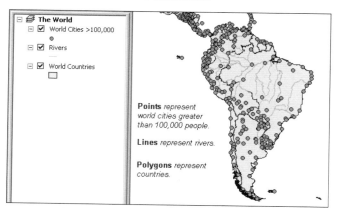

feature class

A collection of geographic features with the same geometry type (point, line, or polygon), the same attributes, and the same spatial reference (coordinate system and map projection).

field

The column in a table that contains the values (information) for a single attribute of each geographic feature in a GIS layer.

ObjectID*	Shape*	City Name	Country Name	Population
1	Point	Guatemala	Guatemala	1,400,000
2	Point	Tegucigalpa	Honduras	551,606
3	Point	San Salvador	El Salvador	920,000
4	Point	Managua	Nicaragua	682,000
5	Point	San Jose	Costa Rica	670,000
6	Point	Belmopan	Belize	4,500
7	Point	Panama	Panama	625,000

In this table, the City Name field contains the name for each city in this layer. The Population field contains the population value for each city.

field name

The column heading in an attribute table. Because field names are often abbreviated, ArcGIS allows you to create a an alternative name, or alias, that can be more descriptive. In the graphic above, City Name and Country Name are aliases for fields named "NAME" and "COUNTRY."

Find button

An ArcMap button used for locating one or more map features that have a particular attribute value.

folder connection

A shortcut that allows you to navigate to a folder without having to enter the entire path.

geodatabase

A database used to organize and store geographic data in ArcGIS.

georeference

To assign coordinates from a reference system, such as latitude/longitude, to the page coordinates of an image or map.

graduated color map

A map that uses a range of colors to show a sequence of numeric values. For example, on a population density map the more people per square kilometer the darker the color.

graph A graphic representation of tabular data.

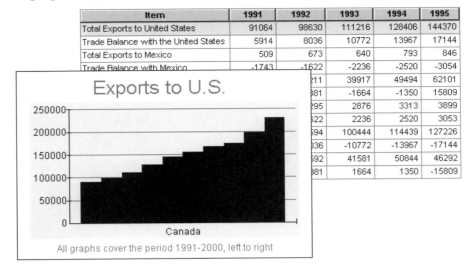

Item	1991	1992	1993	1994	1995
Total Exports to United States	91064	98630	111216	128406	144370
Trade Balance with the United States	5914	8036	10772	13967	17144
Total Exports to Mexico	509	673	640	793	846
Trade Balance with Mexico	-1743	-1622	-2236	-2520	-3054
		211	39917	49494	62101
		881	-1664	-1350	15809
		295	2876	3313	3899
		322	2236	2520	3053
		594	100444	114439	127226
		036	-10772	-13967	-17144
		692	41581	50844	46292
		881	1664	1350	-15809

Exports to U.S.

Canada

All graphs cover the period 1991-2000, left to right

Identify tool An ArcMap tool used to display the attributes of features in the map.

image A graphic representation of data such as a photograph, scanned picture, or a satellite photograph.

join An operation that appends the fields of one table to those of another through an attribute field common to both tables. A join is usually used to attach more attributes to the attribute table of a map layer so that these attributes can be mapped. For example, you could join a country table with population data to a country layer attribute table. *Compare* relate.

label Text placed next to a geographic feature on a map to describe or identify it. Feature labels usually come from an attribute field in the attribute table.

layer A layer is a set of geographic features of the same type along with its associated attribute table, or an image. Example layers are "Major Cities," "Countries," and "Satellite Image." A layer references a specific data source such as a geodatabase feature class or image. Layers have properties, such as a layer name, symbology, and label placement. They can be stored in map documents (.mxd) or saved individually as layer files (.lyr). *See also* data source.

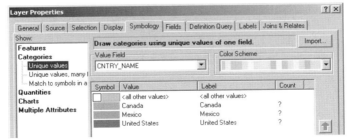

A layer has many properties, including Symbology properties. Some of the properties for the NAFTA Countries layer are pictured here.

layer, turn on   Turning on a layer allows the layer to display in the map. In ArcMap, a layer is turned on by placing a check mark in the box next to the layer name in the table of contents.

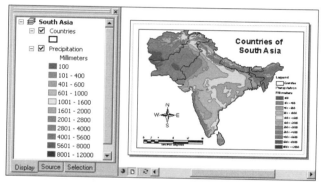

layer file   In ArcGIS, a file with a .lyr extension that stores the path to a data source and other layer properties, including symbology.

layout   In ArcMap, an on-screen presentation document that can include maps, graphs, tables, text, and images.

layout view   A view in ArcMap in which geographic data and map elements, such as titles, legends, and scale bars are placed and arranged for printing.

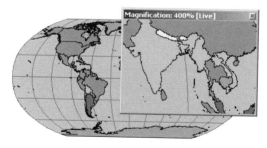

legend   A list of symbols on a map that contains a sample of each symbol as well as text that identifies what the symbol represents.

line   *See* feature.

Magnifier window   A window in ArcMap data view that shows a zoomed-in view of a small area of the main map. Moving the Magnifier window around does not change the extent of the map underneath.

map document   In ArcMap, the file that contains one or more date frames and the associated layers, tables, graphs, and reports. Map document files have a .mxd extension.

map projection   A method by which the curved surface of the earth is portrayed on a flat map. Every map projection distorts distance, area, shape, direction, or some combination thereof. Map projections are made using complex mathematical formulas that are part of ArcGIS software's automatic functions.

MapTip   In ArcMap, a pop-up label for a map feature that displays when the mouse is paused over that feature. The label comes from a field in the layer attribute table.

Measure tool   An ArcMap tool used to measure distance on a map.

metadata   Information about the content, quality, condition, and other characteristics of data. Metadata may include a brief description of the data and its purpose, the names of the authors or compilers of the data, the date it was collected or created, the meaning of attribute fields, its scale and its spatial reference (coordinate system and map projection).

pan   To move your map up, down, or sideways without changing the viewing scale.

point   *See* feature.

polygon   *See* feature.

polyline   *See* feature.

projection   *See* map projection.

record   A row in an attribute table that contains all of the attributes values for a single feature.

OBJECTID*	Shape*	City Name	Country Name	Population
1	Point	Guatemala	Guatemala	1,400,000
2	Point	Tegucigalpa	Honduras	551,606
3	Point	San Salvador	El Salvador	920,000
4	Point	Managua	Nicaragua	682,000
5	Point	San Jose	Costa Rica	670,000
6	Point	Belmopan	Belize	4,500
7	Point	Panama	Panama	625,000

This table has seven records. The fourth record is highlighted. It contains the all of the attributes for the point feature representing the city of Managua, Nicaragua.

relate   An operation that establishes a temporary connection between records in two tables using a field common to both. Unlike a join operation, a relate does not append the fields of one table to the other. A relate is usually used to associate more records and their attributes to the attribute table of a map layer. For example, you could relate a table listing large cities to a layer attribute table of countries. For example, you could join a world cities table to a country layer attribute table. *Compare* join.

scale   The relationship between a distance or area on a map and the corresponding distance or area on the ground, commonly expressed as a fraction or ratio. A map scale of 1/100,000 or 1:100,000 means that one unit of measure (e.g., one inch) equals 100,000 of the same unit on the earth.

The map on the left has a map scale of 1:80,000,000. The map on the right, which is zoomed in, has a map scale of 1:6,000,000.

selected feature   A geographic feature that is chosen and put into a subset so that various functions can be performed on the feature. In ArcMap, a feature can be selected in a number of ways, such as by clicking it on the map with the Select Features tool or based on one or more of its attributes. When a geographic feature is selected it is outlined in blue on the map. Its corresponding record in the attribute table is highlighted in blue.

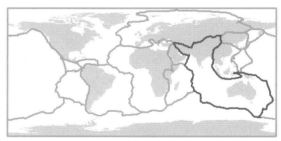

The Indo-Australian plate is selected in this map of the earth's tectonic plates.

shapefile (.shp)   A data storage format for storing the location, shape, and attribute information of geographic features. A shapefile is stored in a set of related files and contains one feature class.

source data   *See* data source.

sort ascending   To arrange an attribute table's rows in order from the lowest values to the highest values in a field. For example, number values would be ordered from 1 to 100, and alphabetical values would be ordered from A to Z.

sort descending   To arrange an attribute table's rows in order from the highest to the lowest values in a field For example, number values would be ordered from 100 to 1, and alphabetical values would be ordered from Z to A.

Symbol Selector The dialog in ArcMap for selecting symbols and changing their color, size, outline, or other properties.

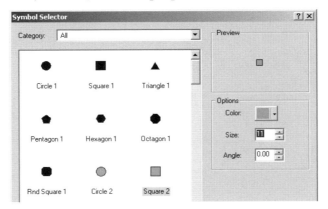

table of contents A list of data frames and layers on a map that may also show how the data is symbolized.

toolbar A set of commands that allow you to carry out related tasks. The Main Menu toolbar in ArcMap has a set of menu commands; other toolbars typically have buttons. Toolbars can float on the desktop in their own window or may be docked at the top, bottom, or sides of the main window.

vertex One of the points that defines a line or polygon feature.

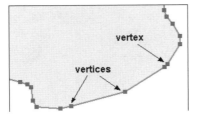

zoom To display a larger or smaller extent of a GIS map or image.

ArcMap Toolbar QUICK REFERENCE

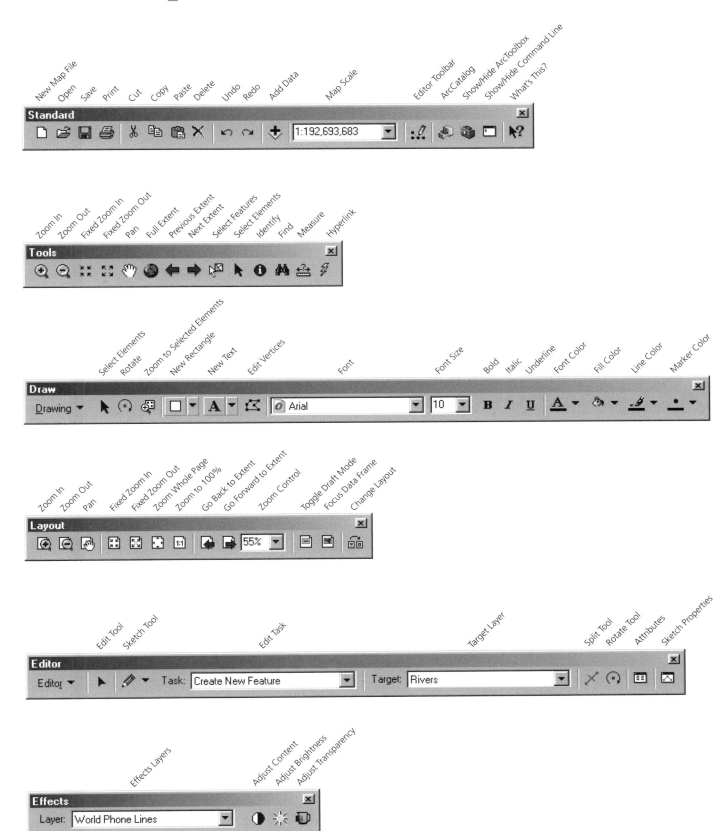

New Map File · Open · Save · Print · Cut · Copy · Paste · Delete · Undo · Redo · Add Data · Map Scale · Editor Toolbar · ArcCatalog · Show/Hide ArcToolbox · Show/Hide Command Line · What's This?

Standard
1:192,693,683

Zoom In · Zoom Out · Fixed Zoom In · Fixed Zoom Out · Pan · Full Extent · Previous Extent · Next Extent · Select Features · Select Elements · Identify · Find · Measure · Hyperlink

Tools

Select Elements · Rotate · Zoom to Selected Elements · New Rectangle · New Text · Edit Vertices · Font · Font Size · Bold · Italic · Underline · Font Color · Fill Color · Line Color · Marker Color

Draw
Drawing · Arial · 10 · B I U

Zoom In · Zoom Out · Pan · Fixed Zoom In · Fixed Zoom Out · Zoom Whole Page · Zoom to 100% · Go Back to Extent · Go Forward to Extent · Zoom Control · Toggle Draft Mode · Focus Data Frame · Change Layout

Layout
55%

Edit Tool · Sketch Tool · Edit Task · Target Layer · Split Tool · Rotate Tool · Attributes · Sketch Properties

Editor
Editor · Task: Create New Feature · Target: Rivers

Effects Layers · Adjust Content · Adjust Brightness · Adjust Transparency

Effects
Layer: World Phone Lines

ArcMap
Zoom and Pan Tools

As you work with a map, you might want to zoom and pan around the data to investigate different areas and features. The tools for navigating your map are found on the Tools toolbar.

Use the Zoom In tool to get a closer look at the data. Click the Zoom In tool, then click a spot on the map or drag a box around an area to zoom in on it. When the map redraws, the point or area you selected will appear in the center of your display.

Use the Zoom Out tool to see a larger area of the data. Click the Zoom Out tool, then click a spot on the map or drag a box around an area to zoom out on it. When the map redraws, the point or area you selected will appear in the center of your display.

Use the Fixed Zoom In button to zoom in a fixed amount from the center of the map. Click the Fixed Zoom In button one or more times depending on how much you want to zoom in.

Use the Fixed Zoom Out button to zoom out a fixed amount from the center of the map. Click the Fixed Zoom Out button one or more times depending on how much you want to zoom out.

Use the Go Back To Previous Extent button to return to a previous view of the data. Click the Go Back To Previous Extent button one or more times to return to previous displays, all the way back to the display when you first opened the map.

Use the Pan tool to reposition the map. Click the Pan tool and hold your cursor over the map display (it turns into a hand). Click a spot on the map, hold down the mouse button, and drag it to a new location. For example, you might want to click on a spot along the edge of the map and drag it to the center.

Use the Full Extent button to view the entire map. Click the Full Extent button once to zoom your map to the extent of all the layers.

Books from

ESRI Press

Advanced Spatial Analysis: The CASA Book of GIS *1-58948-073-2*
ArcGIS and the Digital City: A Hands-on Approach for Local Government *1-58948-074-0*
ArcView GIS Means Business *1-879102-51-X*
A System for Survival: GIS and Sustainable Development *1-58948-052-X*
Beyond Maps: GIS and Decision Making in Local Government *1-879102-79-X*
Cartographica Extraordinaire: The Historical Map Transformed *1-58948-044-9*
Cartographies of Disease: Maps, Mapping, and Medicine *1-58948-120-8*
Children Map the World: Selections from the Barbara Petchenik Children's World Map Competition *1-58948-125-9*
Community Geography: GIS in Action *1-58948-023-6*
Community Geography: GIS in Action Teacher's Guide *1-58948-051-1*
Confronting Catastrophe: A GIS Handbook *1-58948-040-6*
Connecting Our World: GIS Web Services *1-58948-075-9*
Conservation Geography: Case Studies in GIS, Computer Mapping, and Activism *1-58948-024-4*
Designing Better Maps: A Guide for GIS Users *1-58948-089-9*
Designing Geodatabases: Case Studies in GIS Data Modeling *1-58948-021-X*
Disaster Response: GIS for Public Safety *1-879102-88-9*
Enterprise GIS for Energy Companies *1-879102-48-X*
Extending ArcView GIS (version 3.x edition) *1-879102-05-6*
Fun with GPS *1-58948-087-2*
Getting to Know ArcGIS Desktop, Second Edition Updated for ArcGIS 9 *1-58948-083-X*
Getting to Know ArcObjects: Programming ArcGIS with VBA *1-58948-018-X*
Getting to Know ArcView GIS (version 3.x edition) *1-879102-46-3*
GIS and Land Records: The ArcGIS Parcel Data Model *1-58948-077-5*
GIS for Everyone, Third Edition *1-58948-056-2*
GIS for Health Organizations *1-879102-65-X*
GIS for Landscape Architects *1-879102-64-1*
GIS for the Urban Environment *1-58948-082-1*
GIS for Water Management in Europe *1-58948-076-7*
GIS in Public Policy: Using Geographic Information for More Effective Government *1-879102-66-8*
GIS in Schools *1-879102-85-4*
GIS in Telecommunications *1-879102-86-2*
GIS Means Business, Volume II *1-58948-033-3*
GIS Tutorial: Workbook for ArcView 9 *1-58948-127-5*
GIS, Spatial Analysis, and Modeling *1-58948-130-5*
GIS Worlds: Creating Spatial Data Infrastructures *1-58948-122-4*
Hydrologic and Hydraulic Modeling Support with Geographic Information Systems *1-879102-80-3*
Integrating GIS and the Global Positioning System *1-879102-81-1*
Making Community Connections: The Orton Family Foundation Community Mapping Program *1-58948-071-6*
Managing Natural Resources with GIS *1-879102-53-6*
Mapping Census 2000: The Geography of U.S. Diversity *1-58948-014-7*
Mapping Our World: GIS Lessons for Educators, ArcView GIS 3.x Edition *1-58948-022-8*
Mapping Our World: GIS Lessons for Educators, ArcGIS Desktop Edition *1-58948-121-6*
Mapping the Future of America's National Parks: Stewardship through Geographic Information Systems *1-58948-080-5*
Mapping the News: Case Studies in GIS and Journalism *1-58948-072-4*
Marine Geography: GIS for the Oceans and Seas *1-58948-045-7*
Measuring Up: The Business Case for GIS *1-58948-088-0*
Modeling Our World: The ESRI Guide to Geodatabase Design *1-879102-62-5*
Past Time, Past Place: GIS for History *1-58948-032-5*

Continued on next page

When ordering, please mention book title and ISBN (number that follows each title)

Books from ESRI Press (continued)

Planning Support Systems: Integrating Geographic Information Systems, Models, and Visualization Tools *1-58948-011-2*
Remote Sensing for GIS Managers *1-58948-081-3*
Salton Sea Atlas *1-58948-043-0*
Spatial Portals: Gateways to Geographic Information *1-58948-131-3*
The ESRI Guide to GIS Analysis, Volume 1: Geographic Patterns and Relationships *1-879102-06-4*
The ESRI Guide to GIS Analysis, Volume 2: Spatial Measurements and Statistics *1-58948-116-X*
Think Globally, Act Regionally: GIS and Data Visualization for Social Science and Public Policy Research *1-58948-124-0*
Thinking About GIS: Geographic Information System Planning for Managers (paperback edition) *1-58948-119-4*
Transportation GIS *1-879102-47-1*
Undersea with GIS *1-58948-016-3*
Unlocking the Census with GIS *1-58948-113-5*
Zeroing In: Geographic Information Systems at Work in the Community *1-879102-50-1*

Forthcoming titles from ESRI Press

Arc Hydro: GIS for Water Resources, Second Edition *1-58948-126-7*
A to Z GIS: An Illustrated Dictionary of Geographic Information Systems *1-58948-140-2*
Charting the Unknown: How Computer Mapping at Harvard Became GIS *1-58948-118-6*
Finding Your Customers: GIS for Retail Management *1-58948-123-2*
GIS for Environmental Management *1-58948-142-9*
GIS for the Urban Environment *1-58948-082-1*
GIS Methods for Urban Analysis *1-58948-143-7*
The GIS Guide for Local Government Officials *1-58948-141-0*

Ask for ESRI Press titles at your local bookstore or order by calling 1-800-447-9778. You can also shop online at www.esri.com/esripress. Outside the United States, contact your local ESRI distributor.

ESRI Press titles are distributed to the trade by the following:

In North America, South America, Asia, and Australia:
Independent Publishers Group (IPG)
Telephone (United States): 1-800-888-4741 • Telephone (international): 312-337-0747
E-mail: frontdesk@ipgbook.com

In the United Kingdom, Europe, and the Middle East:
Transatlantic Publishers Group Ltd.
Telephone: 44 20 8849 8013 • Fax: 44 20 8849 5556 • E-mail: transatlantic.publishers@regusnet.com

ESRI Press • 380 New York Street • Redlands, California 92373-8100 • www.esri.com/esripress

The ArcView 9 software accompanying this book is provided to get schools started. For use beyond the one-year time limit, users must obtain fully licensed ArcView 9.x software. ArcView is licensed to K–12 schools as a building site license without time restrictions. Other license options are available such as for individuals or for higher-education institutions.

For more information about licensing ArcView for your school or classroom, visit *www.esri.com/k-12* or send e-mail to *k12-lib@esri.com*.

For help with installing this book's software and data, send e-mail to *workbook-support@esri.com*.